T0190194

Lecture Notes in Electrical Engineering

Volume 380

About this Series

"Lecture Notes in Electrical Engineering (LNEE)" is a book series which reports the latest research and developments in Electrical Engineering, namely:

- Communication, Networks, and Information Theory
- Computer Engineering
- Signal, Image, Speech and Information Processing
- Circuits and Systems
- Bioengineering

LNEE publishes authored monographs and contributed volumes which present cutting edge research information as well as new perspectives on classical fields, while maintaining Springer's high standards of academic excellence. Also considered for publication are lecture materials, proceedings, and other related materials of exceptionally high quality and interest. The subject matter should be original and timely, reporting the latest research and developments in all areas of electrical engineering.

The audience for the books in LNEE consists of advanced level students, researchers, and industry professionals working at the forefront of their fields. Much like Springer's other Lecture Notes series, LNEE will be distributed through Springer's print and electronic publishing channels.

More information about this series at http://www.springer.com/series/7818

Ahmed El Oualkadi · Fethi Choubani
Ali El Moussati
Editors

Proceedings of the Mediterranean Conference on Information & Communication Technologies 2015

MedCT 2015 Volume 1

 Springer

Editors
Ahmed El Oualkadi
National School of Applied Sciences
   of Tangier
Abdelmalek Essaadi University
Tangier
Morocco

Ali El Moussati
National School of Applied Sciences, Oujda
Mohammed Premier University
Oujda
Morocco

Fethi Choubani
Technopark el Ghazala
Sup'Com
Ghazala
Tunisia

ISSN 1876-1100   ISSN 1876-1119   (electronic)
Lecture Notes in Electrical Engineering
ISBN 978-3-319-80774-4   ISBN 978-3-319-30301-7   (eBook)
DOI 10.1007/978-3-319-30301-7

Preface

The Mediterranean Conference on Information & Communication Technologies (MedICT 2015) was held at the wonderful Moroccan Blue Pearl city of Saidia, Morocco during 7–9 May 2015. MedICT is a Mediterranean premier networking forum for leading researchers in the highly active fields of Information and Communication Technologies. The general theme of this edition is Information and Communication Technologies for sustainable development of Mediterranean countries.

MedICT provides an excellent international forum to the researchers and practitioners from both academia as well as industry to meet and share cutting-edge development. The conference has also a special focus on enabling technologies for societal challenges, and seeks to address multidisciplinary challenges in Information and Communication Technologies such as health, demographic change, well-being, security and sustainability issues.

The proceedings publish high-quality papers which are closely related to the various theories, as well as emerging and practical applications of particular interest to the ICT community.

This volume provides a compact yet broad view of recent developments in the Devices, technologies and Processing, and covers exciting research areas in the field including Microwave Devices and Printed Antennas, Advances in Optical and RF Devices and Applications, Signal Processing and Information Theory, Wireless and Optical Technologies and Techniques, Computer Vision, Optimization and Modeling in Wireless Communication Systems, Modeling, Identification and Biomedical Signal Processing, Photovoltaic Cell and Systems, RF Devices and Antennas for Wireless Applications, RFID, Ad Hoc and Networks Issues.

Ahmed El Oualkadi
Fethi Choubani
Ali El Moussati

Contents

Part VII   Modeling, Identification and Biomedical Signal Processing

Part VIII   Photovoltaic Cell & Systems

Contents

Part I
Microwave Devices and Printed Antennas

A Novel Design of a Low Cost CPW Fed Multiband Printed Antenna

Issam Zahraoui, Jamal Zbitou, Ahmed Errkik, Elhassane Abdelmounim, Lotfi Mounir, Mandry Rachid and Mohamed Latrach

Abstract This work comes with a novel study on the design of a CPW-Fed multi band planar antenna. This structure can be integrated easily with passive and active elements. The antenna validated is suitable for GPS, UMTS and WiMAX bands. Its entire area is 70.4×45 mm^2 and is printed on an FR-4 substrate. Simulation results show that the antenna has a good input impedance bandwidths for S11 ≤ -10 dB, covering the GPS, UMTS and WiMAX bands. This antenna is optimized, miniaturized and simulated by using ADS "Advanced Design System", with a comparison with another software CST Microwave Studio.

Keywords Multi-band antenna · CPW-fed · GPS band · UMTS band · Wimax band

I. Zahraoui (✉) · A. Errkik · L. Mounir · M. Rachid
LMEET Laboratory, FST of Settat Hassan 1st University, Settat, Morocco
e-mail: zahraoui.issam84@gmail.com

A. Errkik
e-mail: ahmed.errkik@uhp.ac.ma

L. Mounir
e-mail: lotfi.mounirr@hotmail.com

M. Rachid
e-mail: mandryr@yahoo.fr

J. Zbitou
LMEET Laboratory, FST/FPK of Settat Hassan 1st University, Settat, Morocco
e-mail: jazbitou@gmail.com

E. Abdelmounim
ASIT Laboratory, FST of Settat Hassan 1st University, Settat, Morocco
e-mail: abdelmou@hotmail.com

M. Latrach
Microwave Group, ESEO, Angers, France
e-mail: mohamed.latrach@eseo.fr

© Springer International Publishing Switzerland 2016
A. El Oualkadi et al. (eds.), *Proceedings of the Mediterranean Conference on Information & Communication Technologies 2015*, Lecture Notes in Electrical Engineering 380, DOI 10.1007/978-3-319-30301-7_1

1 Introduction

In recent years, we find a growing interest in the use of lightweight and compact dual band or multi band printed planar antennas for wireless communication systems. Among these bands we have the global system for mobile communication GSM [1], Digital Communication System DCS [2], Global Positioning System GPS [3], Universal Mobile Telecommunications System UMTS [4] and the Worldwide Interoperability for Microwave Access WiMAX [5]. For such applications, coplanar waveguide CPW-fed planar structures have many attractive features, such as having a low radiation loss, less dispersion, single metallic layer for the feed network and the antenna element has an easy integration with passive and active elements [6]. To achieve multiband antennas we can find the techniques using L-shaped shorted strip which is connected on both sides of the ground plane with CPW-fed [7–9], and tuning stub technique [10, 11]. The antenna validated in this study has a geometry structure composed from different arms of different lengths and an L-shaped shorted strip on the both sides of the ground plane. This antenna is designed and optimized for GPS, UMTS and WiMAX bands.

The following parts of this paper will present the antenna design and optimization with a comparison study between ADS and CST simulation results including radiation pattern and current distribution.

2 Antenna Design

The geometry of the proposed multi-band antenna is shown in Fig. 1. This antenna is printed on a low cost FR-4 substrate with a total area of 70.4 × 45 mm^2 (Ls × Ws), and a dielectric constant ε_r = 4.4, a thickness h = 1.6 mm, a loss tangent tan

Fig. 1 Geometry of the proposed antenna

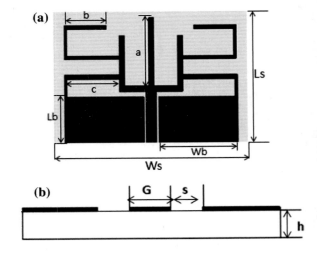

Table 1 Dimensions of the proposed antenna (unit: in mm)

Parameter	Value
Ls	45
Ws	70.4
Lb	15.5
Wb	30
a	23
b	16
c	19.7

(δ) = 0.025 and a metallization thickness t = 0.035 mm. The antenna is fed by a 50 Ω CPW line with a fixed strip G = 4 mm and a gap S = 0.7 mm. The antenna structure and feeding CPW line are implemented on the same plane. Therefore, only one single sided layer metallization is used. The optimization of this antenna is done by using Momentum electromagnetic software integrated into ADS, which provides different techniques and calculation methods. After many series of optimization and miniaturizations by using ADS, the final dimensions of the proposed antenna are shown in Table 1. To compare the obtained results, we have conducted another study by using CST Microwave Studio where the numerical analysis is based on the Finite Integration Time Domain FITD.

To design the multiband antenna, we have studied some important parameters which can influence the input impedance bandwidth and the return loss of the antenna. Therefore we have found some important parameters such as the antenna shape, the feed line width and the dimensions of the ground plane that can give good return loss and good matching input impedance. As shown in Fig. 2, we can see that the length 'a' is a critical parameter which permits to adjust the desired frequency band. As depicted in Fig. 2, we have obtained an optimized antenna with the first resonant frequency of 1.274 GHz and a bandwidth (1.245–1.308 GHz), that can match to the GPS band, the second resonant mode occurs at the frequency of 2.097 GHz with a bandwidth (1.984–2.193 GHz), which is the UMTS band, and the third resonant occurs at 2.52 GHz with a bandwidth (2.464–2.591 GHz), that tends to the WiMAX band.

Fig. 2 Return loss (S11)/dB for different values of "a" parameter on ADS

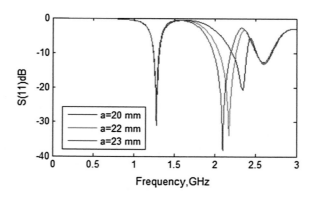

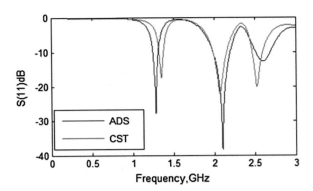

Fig. 3 Comparison of return loss (S11)/dB between ADS and CST-MW

To compare these results, we have used another electromagnetic software, CST-MW. As presented in Fig. 3, we can conclude that we have a good agreement between the simulation results obtained in ADS and CST-MW.

The simulated Far-field radiation patterns in CST-MW for the three resonant frequencies are shown in Fig. 4.

The current distributions of the proposed antenna for three resonant frequencies are shown in Fig. 5.

As we can see from Fig. 5. The current distribution affects different parts of the patch depending on the frequency level. When it reaches 1.274 GHz, the current distribution is concentrated in the L shape. When the frequency reaches 2.097 GHz,

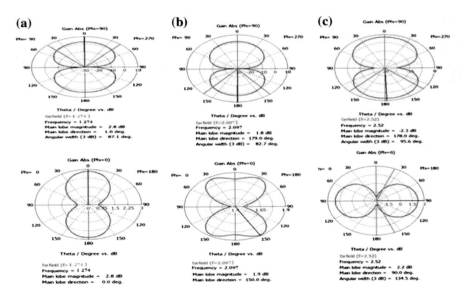

Fig. 4 Radiation pattern of the proposed antenna in CST: **a** @ 1.274 GHz, **b** @ 2.097 GHz, **c** @ 2.520 GHz

(a) **(b)**

(c)

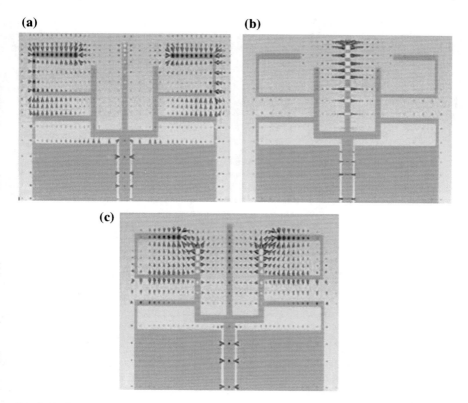

Fig. 5 Surface current distributions of the designed antenna: **a** @ 1.274 GHz, **b** @ 2.097 GHz, **c** @ 2.52 GHz

the density of the current distribution is in the middle of the patch along the feed line. At the frequency of 2.520 GHz the distribution of the current is concentrated between the feed line and on both L shapes.

3 Conclusion

In this study, a novel compact CPW-fed coplanar antenna has been successfully designed and optimized by using ADS "Advanced Design System" and CST Studio Suite Microwave Studio. This antenna has a Low cost and simple fabricated structure and a compact size of 70.4 × 45 mm². The simulated results obtained show that the antenna structure is validated in three frequency bands (1.245–1.308 GHz), (1.984–2.193 GHz) and (2.464–2.591 GHz), which cover the GPS, the UMTS and the WiMAX bands. The use of a CPW Fed for this antenna structure permits it to be associated easily with passive and active microwave integrated circuits.

Acknowledgments We thank Mr. Mohamed Latrach Professor in ESEO, engineering institute in Angers, France, for allowing us to use all the equipments and electromagnetic solvers available in his laboratory.

References

1. Zhong, J., Edwards, R.M., Ma, L., Sun, X.: Multiband slot antenna for metal back cover mobile handsets. Progress Electromagn. Lett. **39**, 115–126 (2013)
2. Mahatthanajatuphat, C., Saleekaw, S., Akkaraekthalin, P.: Arhombic patch monopole antenna with modified Minkowski fractal geometry for UMTS, WLAN and mobile WiMAX application. Progress Electromagn. Lett. **89**, 57–74 (2009)
3. Zhang, Y.Q., Li, X., Yang, L., Gong, S.X.: Dual-band circularly polarized antenna with low wide angle axial-ratio for tri-band GPS application. Progress Electromagn. Lett. **32**, 169–179 (2012)
4. Zheng, X., Zhengwei, D.: A novel wideband antenna for UMTS and WlAN bands: employing a Z-shaped slot and folded T-shaped ground branch to widen bandwidth. IEEE Antennas Wirel. Propag. Lett. 1–4 (2008)
5. Huang, S.S., Li, J., Zhao, J.Z.: A novel compact planar triple-band monopole antenna for WLAN/WiMAX applications. Progress Electromagn. Lett. **50**, 117–123 (2014)
6. Liu, W.C., Wu, C.M., Chu, N.C.: A compact CPW-fed slotted patch antenna for dual-band operation. IEEE Antennas Wirel. Propag. Lett. **9**, 110–113 (2010)
7. Lee, C.H., Chang, Y.H., Chiou, C.E.: Design of multi-band CPW-fed antenna for triple frequency operation. IEEE Antennas Wirel. Propag. Lett. **48**, 543–545 (2012)
8. Zhao, Q., Gong, S.X., Jiang, W., Yang, B., Xie, J.: Compact wide-slot tri-band antenna for WLAN/WiMAX application. Progress Electromagn. Lett. **18**, 9–18 (2010)
9. Chen, W.S., Lee, B.Y.: A meander pda antenna for GSM/DCS/PCS/UMTS/WLAN applications. Progress Electromagn. Lett. **14**, 101–109 (2010)
10. Purahong, B., Jearapradikul, P., Archevapanich, T., Anantrasirichai, N., Sangaroon, O.: CPW-fed slot antenna with inset U-strip tuning stub for wideband. IEEE Antennas Wirel. Propag. Lett. 1781–1784 (2008)
11. Jearapraditkul, P., Kueathaweekun, W., Anantrasirichai, N., Sangaroon, O., Wakabayashi, T.: Bandwidth enhancement of CPW-fed slot antenna with inset tuning stub. IEEE Antennas Wirel. Propag. Lett. 14–17 (2008)

Probes Correction for Antennas Near Field Measurement

Chakib Taybi, Mohammed Anisse Moutaouekkil, Rodrigues Kwate Kwate, Bachir Elmagroud and Abdelhak Ziyyat

Abstract The probe is one of the major components building an indoor system for antenna near field measurement. This device allows to collect the distribution of the field radiated by the antenna under test (AUT), and to transform it into available voltage at its output. Thus, because of the response of the probe to other components of the field radiated by the AUT, to the influence of the geometric parameters of this one and to the problems related to the measuring surface. An unwanted error appear between the voltage and the field, and a step of post-processing is therefore necessary to correct this one. In this paper we address the problem for the correction by the technique of the deconvolution, a probe based on an Open Ended Rectangular WaveGuide (OERWG) is used to validate this approach. Thus, the spatial response of the probe is calculated in module and validated by three AUT, an open ridges rectangular waveguide, an open circular waveguide and a dipole antenna. The results obtained show that the calculated reference field and the reconstructed one are confused for the three antennas under test.

Keywords Near field measurement · Near field probe · Waveguide probe · Probes correction · Deconvolution technique spatial response

1 Introduction

The antennas have their fields of application extend to all industries, whether military or civilian [1]. Thanks to the various technological advances, their developments grow up quickly and safely. To ensure both cohabitation without the

C. Taybi (✉) · R. Kwate Kwate · B. Elmagroud · A. Ziyyat
Electronic and Systems (LES) Laboratory, Faculty of Sciences,
Mohammed First University, 60000 Oujda, Morocco
e-mail: c.taybi@ump.ma

M.A. Moutaouekkil
Information Technology Laboratory, National School of Applied Sciences,
Chouab Doukkali University, 24002 El Jadida, Morocco

© Springer International Publishing Switzerland 2016
A. El Oualkadi et al. (eds.), *Proceedings of the Mediterranean Conference on Information & Communication Technologies 2015*, Lecture Notes in Electrical Engineering 380, DOI 10.1007/978-3-319-30301-7_2

9

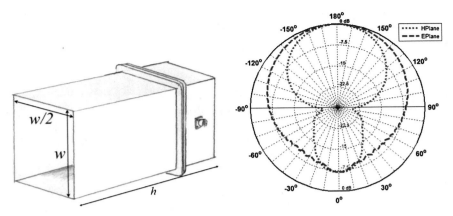

Fig. 1 Architecture and the radiation pattern calculated at a frequency $f_o = 10$ GHz of an open ended rectangular waveguide

antennas risk, reliable operation and the respect of the international electromagnetic compatibility norms (EMC), it is crucial to know precisely their electromagnetic field. Therefore, various techniques have been developed to implement a setup for measuring the electromagnetic field near the antenna under test (AUT) [2, 3], in order to derive the antenna characteristics such as radiation pattern and gain [4]. Or, for dosimetric investigations such as the calculation of the specific absorption rate (SAR) or security perimeter [5].

Measurements in these systems are mainly based on the use of probe devices, which collects the electric field, the magnetic field, or a combination of both in a fixed point of space [6].

The Open Ended Rectangular Wave Guide (OERWG) Fig. 1, is one of the most probes used in near field antennas measurements, particularly in the planar, cylindrical or spherical systems [7–9], and, in the systems designed to measure SAR [10, 11]. Thanks to their simplicity, robustness and low manufacturing cost.

If a perfect probe should normally measure the local field in a single point and only responds to one of its component, in practice, the OERWG measure the distribution of the field on a finite region of space and also responds to more than one of its components as shown in Fig. 2. Thus, the geometric dimensions of the sensor influence on the measurement [12]. These points are reflected by a difference between, the electromagnetic field radiated by the AUT and the voltage v retrieved by the probe. This difference is calculated in module and represented on Fig. 3 for the OERWG probe at a frequency $f_o = 10$ GHz and for two distances $d_1 = 0.85\lambda_o$ and $d_2 = \lambda_o$.

To correct this problem, a detailed studies of the influence of various geometrical parameters of the probe on the measure field is presented in [12]. This last one allows, to minimize this perturbation and to approach at best real profiles of fields. However, this geometric study is often followed by a step of post-treatment, generally called, as correction of a probe is necessary to extract the exact distribution of the radiated field by AUT. One of these methods is the deconvolution technique [13].

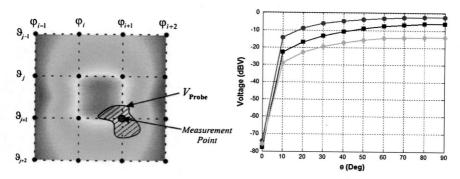

Fig. 2 Measuring surface problem and the selectivity of the probe calculated for a frequency $f_o = 10$ GHz and a variation of the polarization angle of $0°$ to $90°$

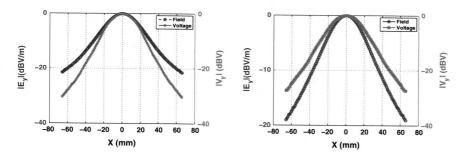

Fig. 3 Error in module between the field e and the voltage v at the output of the probe for distance $d_1 = 0.85\lambda_o$ and $d_2 = \lambda_o$ and a frequency $f_o = 10$ GHz

In this work we report on the problem for the correction by the technique of the deconvolution, a probe based on an open ended rectangular waveguide (OERWG) is used to validate this approach. Thus, the spatial response of the probe is calculated in module and validated by three AUT, an open ridges rectangular waveguide, an open circular waveguide and dipole antenna. Our results show that the calculated reference field and the reconstructed field are confused for our test three antennas.

2   Technique of Deconvolution

In this technique, the voltage recovered at the output of the probe $v(x, y, d, f)$, for a distance d and a given frequency f, can be considered as a convolution of the exact distribution of the $e(x, y, d, f)$ with the spatial response of the probe $h(x, y, d, f)$ possibly tainted by the noise $n(x, y, d, f)$ introduced during the measurement.

$$v(x,y,d,f) = e(x,y,d,f) * h(x,y,d,f) + n(x,y,d,f) \tag{1}$$

$$v(x,y,d,f) = \int\limits_{-\infty}^{+\infty} \int\limits_{-\infty}^{+\infty} e(x,y,d,f)h(x-x',y-y',d,f)dx'dy' + n(x,y,d,f) \tag{2}$$

The convolution integral is easier to evaluate in the frequency domain, since it is a simple multiplication of two functions Fourier Transforms.

$$V(k_x,k_y,d,f) = \int\limits_{-\infty}^{+\infty} \int\limits_{-\infty}^{+\infty} e(x,y,d,f)h(x,y,d,f)e^{-j(k_x x + k_y y)}dk_x dk_y + N(k_x,k_y,d,f) \tag{3}$$

$$V(k_x,k_y,d,f) = E(k_x,k_y,d,f) \times H(k_x,k_y,d,f) + N(k_x,k_y,d,f) \tag{4}$$

The relation of convolution between the exact e field and the voltage v recovered at the output of the probe presented in (2) is similar to the classical formula for the near field probes compensation. Our model replaces the terms vector by those scalars by adding a term of noise mainly present in the practical measures. As a result, the scalar model of near-field measurement can be used for any components of the electric field distribution e_x, e_y, e_z.

The capital letters in (4) denote the spatial Fourier transformed of the field e, voltage v, noise n and the spatial response of the probe h. In practice, the data that represent these functions are available with a limited number of samples, the passage from the spatial domain to the spectral range is provided by the Discrete Fourier Transform (DFT) defined by the (5):

$$D(k_{xp},k_{yq},d) = \sum_{m=0}^{M-1} \sum_{n=0}^{N-1} d(x_m,y_n,d)e^{-j(k_{xp}x_m + k_{yq}y_n)} \tag{5}$$

In (5) k_{xp} and k_{yq} are defined by:

$$[k_{xp},k_{yq}] = \left[\frac{2\pi p}{M\Delta x},\frac{2\pi q}{M\Delta y}\right]_{M \times N} \tag{6}$$

with: $p = 0\ldots, M-1$ and $q = 0\ldots, N-1$.

The function d could be replaced by any set of discrete data obtained from the measurement, the simulation or the post-processing.

The spatial response of the probe is determined from Eq. (4) by an inverse Discrete Fourier Transformation (iDFT) and by changing the AUT by the probe itself as show in Fig. 4, while assuming that the measurement is made without noise. In the fact, the ripples in the quiet zone due to the parasite reflections on domain walls of the anechoic chamber, is minimised by the Least squares filtering

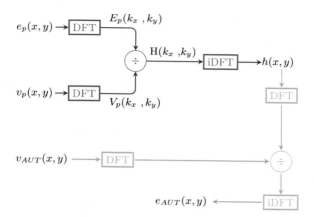

Fig. 4 Exaction (*blue*) and validation (*green*) process of the spatial response

algorithm [13]. The lowercase letter in Eq. (8) which represent the spatial response
are obtained by an iDFT of H defined by (7).

$$H(k_x, k_y, d, f) = \frac{V(k_x, k_y, d, f)}{E(k_x, k_y, d, f)} \tag{7}$$

$$h(x, y, d, f) = \text{iDFT}\left[\frac{V(k_x, k_y, d, f)}{E(k_x, k_y, d, f)}\right] \tag{8}$$

Thus, the amplitude of spatial response of the probe OEWG for two distances
$d_1 = 0.85\lambda_o$ and $d_2 = \lambda_o$ and for a frequency $f_o = 10$ GHz is shown in Fig. 5.

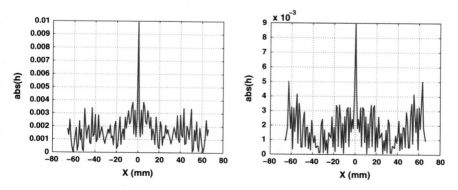

Fig. 5 Module of the spatial response of the probe at a frequency $f_o = 10$ GHz and a distance
$d_1 = 4\lambda_o$ and $d_2 = 5\lambda_o$

3 Results and Discussion

To validate the spatial response represented on Fig. 5, we calculate in this section, the voltage v recovered by the OERWG probe from two AUTs (Fig. 4), The first one is an OERWG loaded by two ridges and the second is a circular waveguide. The radiations patterns of these antennas under test are calculated for a frequency $f = 10$ GHz and they are represented in Fig. 6.

In Figs. 7 and 8, for d_1, d_2 and for f_o, we represent the field reconstructed (e_{rec}) from the voltage v recovered by the probe and the reference field e_{ref} calculated without the AUT. These results show that the reconstructed field from the voltage confused with the reference field for the AUT that have a pattern similar to that of the probe, and good cohabitation with the second type antenna proposed in this study.

To validate the case of AUT that present a wide variation in the diagram relative to the probe, we represent in Fig. 9, for a half wave dipole antenna the reconstructed field from the voltage and the reference field for a frequency $f = 10$ GHz and for both distance $d_1 = 0.85\lambda_o$ and $d_2 = \lambda_o$.

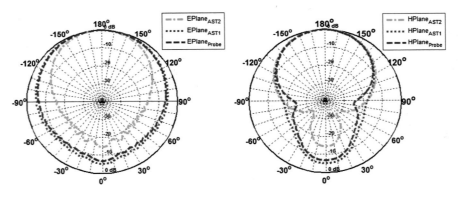

Fig. 6 Radiation pattern of the antenna under test and probe for a frequency $f_o = 10$ GHz

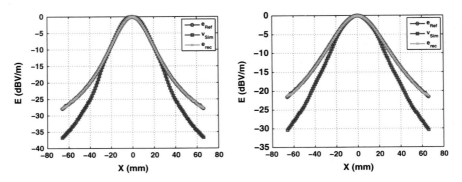

Fig. 7 Comparison between the reconstructed field, the reference field and the voltage at the output of the probe for the ridges OERWG antenna

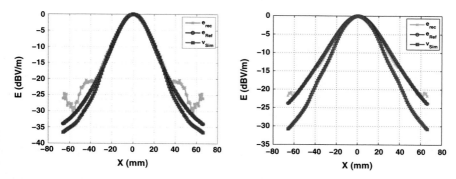

Fig. 8 Comparison between the reconstructed field, the reference field and the voltage at the output of the probe for circular wave guide

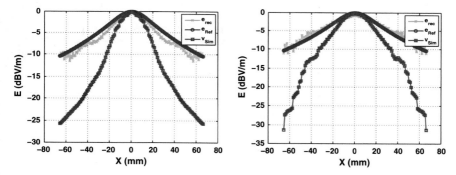

Fig. 9 Comparison between the reconstructed field, the reference field and the voltage at the output of the probe for dipole antenna

4 Conclusion

In this work we studies the correction of the OERWG probe used to measure the tangential component of the electric field. Because of the response of the probe to the other components of the field and the influence of the geometric parameters of this one and to the problems related to the measuring surface, an error appear between the real profile of the field radiated by the AUT and the voltage at the output of the probe. Therefore, a post-processing step is needed to determine the exact distribution of the field. The deconvolution is one of these methods. In our work we reported the fact that the amplitude of the response of the OERWG probe is to determine for two different distances. A test procedure is launched in the latter part of the studies to validate this response. This latter is tested before three antennas under test, a circular guide, a rectangular guide loaded by ridges and a dipole antenna, the results showed a coincidence between the reconstructed eld and calculated without the probe.

References

1. Balanis, C.A.: Antenna Theory: Analysis and Design. Wiley, Canada (2005)
2. Slater, D.: Near-field Antenna Measurement Techniques. Artech House, Boston (1990)
3. Yaghjian, A.D.: An overview of near-field antenna measurements. IEEE Trans. Antennas Propagat. **34**, 30–45 (1986)
4. Johnson, R.C., Ecker, H.A., Hollis, J.S.: Determination of far-field antenna patterns from near-field measurements Proc. IEEE **61**(12), 1668–1694 (1973)
5. Gai-Mai, W.: A new near-field Sar imaging algorithm. In: IEEE 3rd International Conference on Communication Software and Networks (ICCSN) (2011)
6. Moutaouekkil, M.A., Taybi, C., Ziyyat, A.: Caracterisation dune sonde Monople utilise dans la mesure de champ proche. Congers Mediterraneen des telecommunications, Fes (2012)
7. Yaghjian, A.D.: Approximate formulas for the far field and gain of open ended rectangular waveguide. IEEE Trans. Antennas Propagat. **32**, 378–384 (1984)
8. Newell, A.: The effect of the absorber collar on open-ended waveguide probes. In: Antenna Measurement Techniques Association (AMTA), Salt Lake City, Utah, USA, Nov 2009
9. Gregson, S., McCormick, J., Parini, C.: Principales of Planar Near Field Antenna Measurement. IET, London (2007)
10. Qingxiang, Li, Gandhi, O.P., Kang, G.: An open-ended waveguide system for SAR system validation or probe calibration for frequencies above 3 GHz. Phys. Med. Biol. **49**, 4173 (2004)
11. Porter, S.J., Manning, M.I.: A method validates SAR measurement systems. Microwaves RF **44**(3), 68 (2005)
12. Taybi, C., Moutaouekkil, M.A., Kwate, K.R., Elmagroud, B., Ziyyat, A.: Probes characterization for antennas near field measurements. In: 14th Mediterranean Microwave Symposium, IEEE Conference, pp. 12–14. Marrakech, Morocco (2014)
13. Moutaouekkil, M.A.: Least square filtering algorithm for reactive near field probe correction. Progress Electromagn. Res. B, **42**, 225–243 (2012)

Directive Beam-Steering Patch Array Antenna Using Simple Phase Shifter

Hayat Errifi, Abdennaceur Baghdad, Abdelmajid Badri and Aicha Sahel

Abstract Beam-steering antennas are the ideal solution for a variety of system applications, it is most commonly achieved through using phased arrays, where phase shifters are used to control the relative of the main-beam. In this paper, a low-cost directive beam-steering phased array (DBS-PA) antenna using switched line phase shifters is demonstrated. The proposed DBS-PA antenna has four micro-strip patch antennas, three power dividers and four phase shifters printed on the same Rogers RT-Durroid substrate with a dielectric constant of 2.2 with dimensions of 8×3.5 cm. The phased array antenna has a directivity of 11.92 dBi and the main beam direction can be switched between the angles of $\pm25°$ with a 3 dB beam-width of 23°. All design and simulations have been carried out using Ansoft HFSS software tool. The frequency considered for the operation is 10 GHz.

Keywords Phase shifters · Switched line · Beam steering · Patch antenna · Phased arrays · Return loss · Directivity

1 Introduction

The phase of an electromagnetic wave of a given frequency can be shifted when propagating through a transmission line by the use of Phase Shifters [1]. They have many applications in various equipment such as beam forming networks, power dividers and phased array antennas or reconfigurable antennas [2].

Reconfigurable antenna has gain a lot of attentions in wireless communication system recently. It can be classified into three major fields which are reconfigurable frequency, polarization and radiation pattern [1]. The reconfigurable antenna can be realized via RF switches such as PIN diodes, MEMs and GaAs FETs. By changing

H. Errifi (✉) · A. Baghdad · A. Badri · A. Sahel
EEA & TI Laboratory, Faculty of Sciences and Technologies, Hassan II University, Casablanca, Morocco
e-mail: errifi.hayat@live.fr

© Springer International Publishing Switzerland 2016
A. El Oualkadi et al. (eds.), *Proceedings of the Mediterranean Conference on Information & Communication Technologies 2015*, Lecture Notes in Electrical Engineering 380, DOI 10.1007/978-3-319-30301-7_3

17

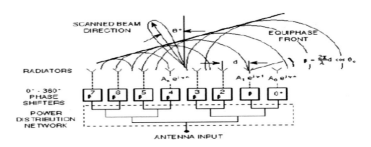

Fig. 1 Phased array antenna system [1]

the switch state to either 'ON' or 'OFF' mode, this determined either feed line would receive or not the radio frequency (RF) signal [3].

Many antenna system applications require that the direction of the beam's main lobe be changed with time, or scanned. This is usually done by mechanically rotating a single antenna or an array with fixed phase to the element. However, mechanical scanning requires a positioning system that can be costly and scan too slowly. For this reason, electronic scanning antennas which are known as phased array antennas are used. It can sweep the direction of the beam by varying electronically the phase of the radiating element, thereby producing a moving pattern with no moving parts, as shown in Fig. 1.

In this paper, the investigation on amount of beam-steering of main beam by using microstrip switched line phase shifters is presented. The switched line phase shifter is dependent only on the length of line used. An important advantage of this circuit is that the phase shift will be approximately a linear function of frequency. This enables the circuit to operate at a broader frequency range [4].

This research is focuses on the beam-steering patch antenna radiation pattern which suitable for point-to-point communication system that requires high gain and directivity characteristics. All designs and simulations have been carried out using High Frequency Structure Simulator tool (HFSS). The rest of the paper is organized as follow: Sect. 2 presents the design specifications of microstrip phased array antenna. The simulated results are discussed in Sect. 3 and finally Sect. 4 provides the conclusion and future works.

2 Design of Microstrip Phased Array Antenna

Microstrip patch antennas are important as single radiating elements but their major advantages are realized in application requiring moderate size arrays. The primary radiator microstrip antenna is designed at frequency of 10 GHz which gives single patch antenna (SPA) as shown in Fig. 2. The dimensions of the patch are calculated using formulae found in [5, 6] and their values are stored in Table 1.

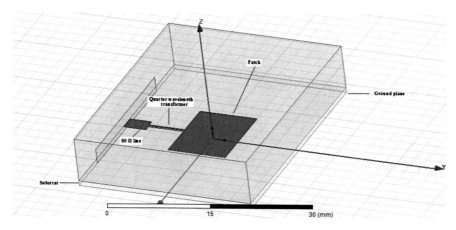

Fig. 2 Geometry of SPA

The proposed directive beam-steering phased array (DBS-PA) antenna is developed from four SPA as shown in Fig. 3. All elements are homogenous which has similar dimensions and characteristics as well [7]. The dimension of the DBS-PA antenna is 3.5 × 8 cm. The inter-element spacing (IES) of radiating elements must be determined precisely in order to get the better antenna performance. The IES is realized to half of wavelength ($\lambda/2$) in most array cases. The feed network is composed of three T-junction power dividers and four microstrip switched line phase shifters printed on the same Rogers RT-Durroid substrate with a dielectric constant of 2.2 and thickness of 0.79 mm [8]. The phase shifters are outlined by red circles and there are labelled as S1, S2, S3 and S4.

With in-phase feeding to the patch antennas, the beam direction of the array is perpendicular to the substrate (0°). To achieve beam steering, the microstrip switched line phase shifters are used to reconfigure the delay feeding line between the patches.

The basic schematic of the proposed microstrip switched-line phase shifter is shown in Fig. 4, it is comprised of two microtrip line segments of different length (L_1 and L_2) selectively connected to the transmission line, RF input, RF output, four

Table 1 Dimensions of SPA

Parameter	Value (mm)
Substrate type	Roger RT-Durroid
Substrate length	27.93
Substrate width	35.58
Substrate thickness	0.79
Patch width	11.86
Patch length	9.31
Feed-line width	2.408
Quarter wavelength transformer width	0.5

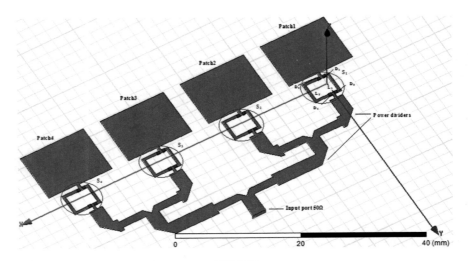

Fig. 3 Simulated geometry of the proposed DBS-PA antenna

PIN diodes D_1, D_2, D_3, and D_4. Only one arm should be ON at a time. When the PIN diodes D_1 and D_3 are ON while PIN diodes D_2 and D_4 are OFF, the reference delay line L_1 is in the circuit. When the PIN diodes D_2 and D_4 are ON while PIN diodes D_1 and D_3 are OFF, the delay line L_2 is in the circuit [9]. By switching the signal between two lines of different lengths $L_1 = 5.7$ mm and $L_2 = 15$ mm, it is possible to realize a specific phase shift given by the formula (1).

$$\Delta\varphi = \frac{2\pi(L2 - L1)}{\lambda} \tag{1}$$

Fig. 4 Basic schematic of the proposed microstrip switched line phase shifter

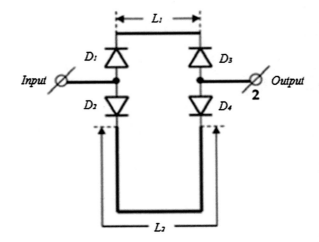

However, since this design is implemented using microstrip technology [10], the physical length, ΔL, is determined by:

$$\Delta L = L2 - L1 = \frac{\beta c}{2\pi f \sqrt{\varepsilon \, \text{eff}}}.$$ (2)

where c is the speed of light, β is the propagation constant of the line, f is the operating frequency and ε_{eff} is the effective dielectric constant.

3 Simulation Results and Discussion

In antenna design, firstly, the researcher needs to determine the application of the antenna development in order to fix the antenna's operation frequency. This research focused on the 10 GHz operating frequency of telecommunication satellite and radar systems. The simulation results of the proposed phased-array system are obtained using Ansoft high frequency structure simulator (HFSS).

The presented DBS-PA antenna is competent to perform beam shape ability with sustain frequency operating. This can be done by the RF switches configuration as summarized in Table 2. We can choose from any one of L_1 or L_2 as a reference line and the other is delay line. There are five parametric analysis to be considered which are operating frequency, directivity, beam-width at −3 dB, lobe direction and radiation efficiency.

Table 2 PIN diode switches configuration

Type of phase shifter	Number of phase shifter	Phase shifter status		
		Case 1	Case 2	Case 3
Microstrip switched line phase shifter	S_1	D_1 and D_3 ON (L_1)	D_1 and D_3 ON (L_1)	D_1 and D_3 ON (L_1)
	S_2	D_1 and D_3 ON (L_1)	D_2 and D_4 ON (L_2)	D_1 and D_3 ON (L_1)
	S_3	D_1 and D_3 ON (L_1)	D_1 and D_3 ON (L_1)	D_2 and D_4 ON (L_2)
	S_4	D_1 and D_3 ON (L_1)	D_2 and D_4 ON (L_2)	D_2 and D_4 ON (L_2)
Operating frequency (GHz)		10.00	9.84	9.90
Lobe direction (°)		0°	13°	25°
Directivity (dBi)		11.41	11.51	11.32
HPBW (°)		25°	23°	23°
Radiation efficiency (%)		90	93	89

H. Errifi et al.

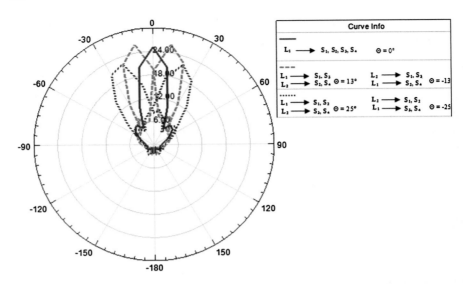

Fig. 5 Comparison of radiation pattern by certain phase shifters configuration in polar plot

Directional beam steering from −25° to 25° has been observed. To illustrate this beam steering, five different beam steering angles at −25°, −13°, 0°, 13°, and 25° are simulated and presented. The E-plane radiation patterns of the DBS-PA antenna are shown in Figs. 5 and 6. The scanning ability is 50° with two delay paths (L_1 and L_2).

There is an improvement in terms of directivity between case 1 and case 2. This has been clearly compared in Fig. 6 which case 2 has high directivity of up to 11.51 dBi while case 1 has directivity of 11.41 dBi. Single patch provide directivity up to 7 dBi, here directivity is increased significantly by 4.5 dBi using array structure. Besides, the half power beam width (HPBW) of case 2 is smaller than case 1, the radiation beam becomes narrow (at 23°) and more directive.

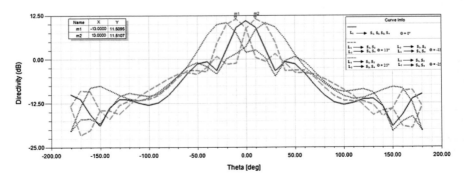

Fig. 6 Comparison of radiation pattern by certain phase shifters configuration in rectangular plot

Fig. 7 **a** 3D radiation plot at 10 GHz for 0° beam steering. **b** 3D radiation plot 10 GHz for 13° beam steering

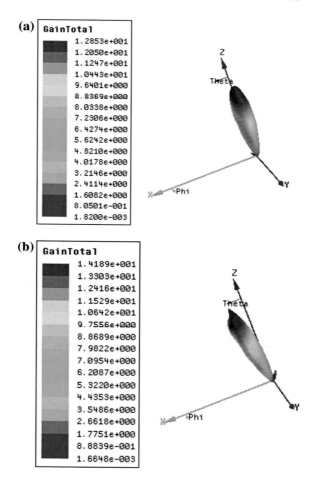

Our goal is to maximize the antenna gain at the specified angle. As an example, Fig. 7 shows the 3D radiation pattern obtained at 0° direction and when steering the radiation beam at 13°, we obtained a maximum gain at the desired angle, it is about 14 dB.

It is important for the phased array to have good impedance matching at all beam steering angles. To verify this impedance matching, return losses for the proposed DBS-PA antenna at beam steering angles of 0°, 13° and 25° are plotted in Fig. 8.

This figure illustrates the reflection coefficient result under the tolerable S_{11} of less than −10 dB for certain phase shifters configuration. It is observed that very good impedance matching is maintained across all the beam steering angles with small discrepancy in frequency due to the shift line lengths.

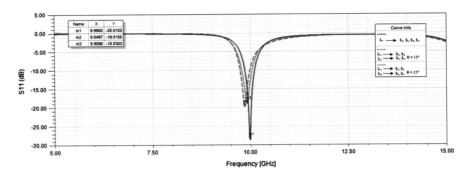

Fig. 8 Comparison of reflection coefficient by certain phase shifters configuration

4 Conclusion

The paper presented the design and simulation of the DBS-PA antenna using microstrip switched line phase shifters for the X band frequency. A patch antenna is selected due to the easiness to feed each element via feed network. The proposed antenna which is rectangular in shape and has a dimension of 8 × 3.5 cm can be considered small in size. Besides, the simulation results show that the proposed antenna has successfully achieved the scanning ability of 50° including beam steering angles of 0°, 13° and 25° in the right and left as well as improved directivity and good impedance matching are maintained across all the beam steering angles. With all capability demonstrated, the presented DBS-PA antenna has a potential to be implemented in smart antenna system and also for military applications. Future research on implementing this antenna and achieving larger beam steering angles is being conducted by the authors.

Acknowledgments Our sincere thanks to the Faculty of Science and Technology, Hassan II University, Casablanca-Mohammedia, Morocco, for providing us an opportunity to carry out our said work in a well-equipped laboratory (EEA & TI). We are also thankful to all our colleagues who helped us while we were working on this project.

References

1. Skolnik, Merrill.I.: Radar Handbook, 3rd edition, p. 13.51, Copyright by The McGraw-Hill Companies, The United States of America (2008)
2. www.qsl.net/va3iul/Phase_Shifters/Phase_Shifters.pdf
3. Chen, Q., Kurahashi, M., Sawaya, K.: Dual-mode patch antenna switched by PIN diode. In: IEEE Topical Conference on Wireless Communication Technology, pp. 148–149 (2003)
4. Koul, K.S., Bhat, B.: Microwave and Millimeter Wave Phase Shifters, vol. 1: Semiconductor and Delay Line Phase Shifters. Artech House, Boston (1991)
5. Constantine, A.: Balanis, Antenna Theory—Analysis and Design. Wiley, Singapore (2002)

6. Errifi, H., Baghdad, A., Badri, A.: Effect of change in feedpoint on the antenna performance in edge, probe and inset-feed microstrip patch antenna for 10 GHz. Int. J. Emerg. Trends Eng. Dev. Jan 2014. ISSN 2249–6149.9-6149
7. Errifi, H., Baghdad, A., Badri, A., Sahel, A.: Design and simulation of microstrip patch array antenna with high directivity for 10 GHz applications. In: International Symposium on Signal Image Video and Communications, ISIVC-2014, Marrakech, Morocco, 19–21 Nov 2014
8. Errifi, H., Baghdad, A., Badri, A., Sahel, A.: Design and analysis of directive microstrip patch array antennas with series, corporate and series-corporate feed network. Int. J. Electron. Electr. Eng. (in press)
9. Clenet, M., Morin, G.: Visualization of radiation-pattern characteristics of phased arrays using digital phase shifters. IEEE Antennas Propag. Mag. **45**(2), 20–35 (2003)
10. Pozar, D.M.: Microwave Engineering, 2nd edn. Wiley, Newyork (1998)

Part II
Advances in Optical and RF Devices and Applications

Design of a V-Band Buffer Amplifier for WPAN Applications

Maryam Abata, Moulhime El Bekkali, Said Mazer, Catherine Algani and Mahmoud Mehdi

Abstract This paper presents a V-band buffer amplifier based on 0.15 µm pHEMT from UMS foundry for WPAN applications. This amplifier will be used as a part of a frequency quadrupler in the millimeter wave band at 60 GHz. The amplifier circuit can deliver a gain up to 13 dB at 60 GHz with a low noise figure of 2.4 dB. An input return loss of 18 dB and an output return loss of 10 dB are also achieved by this amplifier. The input P1 dB and IP3 are 0.3 and 27 dBm respectively.

Keywords Buffer amplifier · Mm-wave band · Local oscillator · WPAN applications · Frequency quadrupler

1 Introduction

During the last few years, the wireless communications have made out a great development due to the flexibility of their interface. The strong demand from users and the growing need for high speed broadband, the communication systems are gradually shifted to the mm-waves band [1]. The millimeter band is characterized by significant signal attenuation due to absorption by oxygen that reaches 15 dB/km [2]. This allows the re-use of frequencies in very small sized cells and reduces the co-channel interferences. These advantages of the mm-waves band make them suitable for WPAN applications. A major challenge for electronic systems operating in the mm-wave band is the signal generation with a stable carrier frequency and a low level of phase noise [3]. To overcome this problem, the signal can be generated by combining a low frequency local oscillator (LO) and several stages of frequency

M. Abata (✉) · M.E. Bekkali · S. Mazer
LTTI Laboratory, Sidi Mohamed Ben Abdellah University, Fez, Morocco
e-mail: maryam.abata@usmba.ac.ma

C. Algani
ESYCOM Laboratory-CNAM, National Conservatory of Arts and Crafts, Paris, France

M. Mehdi
Microwaves Laboratory, Lebanese University, Beirut, Lebanon

© Springer International Publishing Switzerland 2016
A. El Oualkadi et al. (eds.), *Proceedings of the Mediterranean Conference on Information & Communication Technologies 2015*, Lecture Notes in Electrical Engineering 380, DOI 10.1007/978-3-319-30301-7_4

multipliers [4, 5], but the isolation problem of the local oscillator from the other stages is always present, which leads to have various mechanisms that can disturb the oscillator performance [6], so a buffer stage is often included in oscillator outputs. In this work, we present the study and design of a buffer amplifier in the mm-wave band based on 0.15 μm pHEMT from UMS foundry. This amplifier will play a fundamental role in the design of a frequency quadrupler around 60 GHz.

2 Local Signal Generation Around 60 GHz

The theory of frequency multiplication based on field-effect transistors has been detailed in several publications [7, 8]. The transistor should be operating in the non-linear region: close to the pinch-off zone where the gm transconductance takes a minimum value. In this case, the transistor promotes the generation of even harmonics. In the second zone, close to the saturation region, the gm transconductance reaches a maximum value, which leads to generate the odd harmonics.

In the design of a frequency quadrupler around 60 GHz based on PHEMT transistor, the principle of harmonics generation is the operation of the transistor in the non-linear zone: close to the pinch-off area where Vgs = −0.7 V to generate the even harmonics, in the saturation region where the Gm transconductance reaches its maximum value which corresponds to Vgs = 0 V to obtain the odd harmonics. After determining the optimal bias point that maximizes the generation of even harmonics, the spectral study of frequency quadrupler shows that the power level of the first three harmonics is higher than that of the fourth harmonic. In order to have a good frequency quadrupler, the fourth harmonic must be greater than the other harmonics; therefore we propose the structure of two poles high-pass filter [9].

After inserting the filter we find that the power levels of the first three harmonics have dropped significantly. However, the power level of the fourth harmonic is improved. We notice that the power level of the fourth harmonic is still low, which implies the need to add an amplifier to the circuit (Fig. 1). The inserted amplifier is a buffer type; its role is to increase the power level of the fourth harmonic in the range of (56–64 GHz), reject the unwanted harmonics and insure the isolation between stages.

This amplifier will also improve the frequency quadrupler parameters namely the conversion gain and the rejection ratio.

Fig. 1 Location of buffer amplifier in the multiplication chain

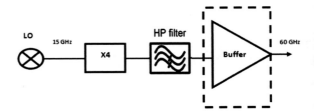

3   The V-Band Buffer Amplifier at 60 GHz

3.1   Description of the Circuit

A three-stage buffer amplifier is used for amplification of the output signal from the frequency quadrupler. The electrical scheme of this amplifier is shown in Fig. 2.

This amplifier is composed of three RC feedback stabilized amplifier stages. The amplifying circuit consists of three similar transistors with their own bias circuits (Vds = 3 V and Vgs = −0.1 V). The used transistors are PHEMT type with a gate width of 2 * 70 μm and a length of 0.15 μm.

The input and the output matching circuits are realized by distributed elements in order to achieve a good performance.

3.2   Simulated Results

The amplifying circuit was designed using ADS from Agilent. Figure 3 shows the Noise figure; we can see that around 60 GHz, it achieves a value of 2.4 dB.

From Fig. 4 that shows the stability factor, we can observe that at 60 GHz the amplifying circuit is unconditionally stable (Stability Factor over 1).

Figure 5 shows the S-parameters simulation results of the buffer amplifier. We can see from the figure that S12 which is the isolation factor is lower than −30 dB, this factor is very important for the LO buffer amplifier.

We can also see an input return loss (S11) of −18 dB and an output return loss (S22) of −10 dB at 60 GHz. Concerning the frequency response of gain (S21), it reaches a value up to 13 dB at the operating frequency.

Fig. 2 Buffer amplifier circuit

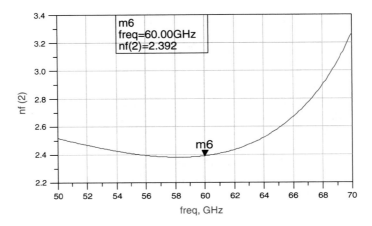

Fig. 3 Noise figure

The output power and the gain of the buffer amplifier as a function of the input power are depicted in Fig. 6.

The linear zone can be defined using the concept of 1 dB compression point, which is the point of the characteristic that is 1 dB away from the linear zone. To obtain the curve of 1 dB compression point, the input power is varied and the power behavior of the fundamental frequency at the output of the amplifier is observed. From Fig. 6, the input and output P1 dB are 0.3 and 13.16 dBm respectively.

The third Interception point (IP3) is used to characterize the linearity of an amplifier. Figure 7 shows the output power versus the input signal power for the

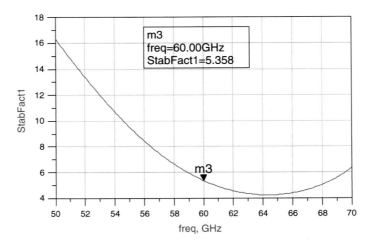

Fig. 4 Stability factor of the designed V-band buffer amplifier

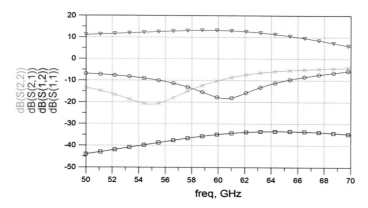

Fig. 5 S-parameters of the designed V-band buffer amplifier. S11 (*open circle*), S22 (*multiplication sign*), S21 (*open triangle*), S12 (*open square*)

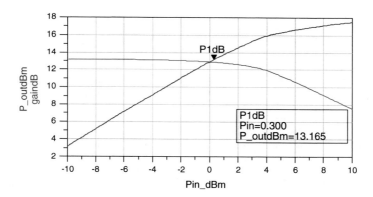

Fig. 6 Output power (*Blue*) and gain (*Red*) versus input power

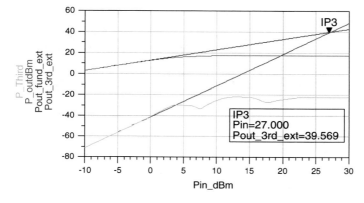

Fig. 7 The third intercept point (IP3)

fundamental frequency and the third harmonic. The IP3 is defined by the intersection of the asymptotes of the two curves. This IP3 is shown in Fig. 7:

From Fig. 7, the third Interception point (IP3) is (Pin = 27 dBm, Pout = 39.5 dBm).

4 Conclusion

In this paper, we have presented and designed a V-band buffer amplifier at 60 GHz for WPAN applications. The designed buffer amplifier shows excellent performance including input and output return loss of −18 and −10 dB respectively, a gain up to 13 dB, and a noise figure of 2.4 dB. The circuit has a high linearity ($P1dB_{in}$ = 0.3 dBm, IP3 = 27 dBm) which make it suitable as a buffer amplifier between LO and mixer for WPAN applications.

References

1. Abata, M., El Bekkali, M., Mazer, S., El Hamdani, W., Algani, C.: Design of a MMIC frequency quadrupler in millimeter-wave band. In: 14th International Conference on Multimedia Computing and Systems. Marrakesh, Morocco, 14–16 Apr 2014
2. Richardson, A.J., Watson, P.A.: Use of the 55–65 GHz oxygen absorption band fort short-range broadband radio networks with minimal regulatory control. IEEE Proc. **137** (1990)
3. Zirath, H., Kozhuharov, R., Ferndahl, M.: A balanced InGaP-GaAs colpitt-VCO MMIC with ultra low phase noise. In: Proceedings of 12th Gallium Arsenide and Other Compound Semiconductors Application Symposium, pp. 37–40. Amsterdam, The Netherlands, Oct 2004
4. Piernas, B., Hayashi, H., Nishikawa, K., Kamogawa, K., Nakagawa, T.: A broadband and miniaturized V-Band PHEMT frequency doubler. IEEE Microwave Guided Wave Lett. **10**(7), 276 (2000)
5. Kangaslahti, P., Alinikula, P., Porra, V.: Miniaturized artificial-transmission-line monolithic millimeter—wave frequency doubler. IEEE Microwave Theory Tech. **48**(4), 510–518 (2000)
6. Crandall, M.: Buffer Amplifiers Solve VCO Problems. www.maxim-ic.com
7. Maas, S.A.: Nonlinear Circuit, pp. 397–416. Artech House, Boston, MA (1988)
8. Camargo, E.: Design of FET Frequency Multipliers and Harmonic Oscillators. Artech House, Boston (1998)
9. Abata, M., El Bekkali, M., Mazer, S., Algani, C.: Structure of a frequency multiplier by 4 for WPAN networks applications. In: 14th Mediterranean Microwave Symposium. Marrakesh, Morocco, 12–14 Dec 2014

Simulation and Experimentation of an RFID System in the UHF Band for the Reliability of Passive Tags

Sanae Taoufik, Ahmed El Oualkadi, Farid Temcamani,
Bruno Delacressonniere and Pascal Dherbécourt

Abstract This paper presents the simulation of a UHF RFID (Radio Frequency Identification) system for passive tags, using Advanced Design System software from Agilent Company. Temporal measurements based on a commercial RFID reader are achieved which permit to analyze and to validate the principles of communication between the reader and the tag. Also a test bench is developed to evaluate the effects of high temperature on the reliability of passive UHF RFID tags. The obtained test results show that the thermal storage has a marked effect on the performances of the RFID tags.

Keywords RFID · Passive tag · UHF band · Simulation ADS · Reliability

1 Introduction

Radio Frequency Identification (RFID) is a generic term used to describe a system which transmits an object or a person identity by using a radiofrequency link. Nowadays, the RFID technologies are used in many domains such as transport, medical treatment, smart cards...

Generally, RFID systems consist of two basic elements, a reader and a tag. The reader is composed of a high-frequency module, a control unit and an antenna [2].

S. Taoufik (✉) · A.E. Oualkadi
Laboratory of Information Technology and Communication,
National School of Applied Sciences of Tangier,
Abdelmalek Essaadi University, Tétouan, Morocco
e-mail: sanae.taou@gmail.com

S. Taoufik · P. Dherbécourt
Materials Physics Group, UMR CNRS 6634 University of Rouen,
Avenue de l'université B.P 12, 76801 Saint Etienne du Rouvray, France

F. Temcamani · B. Delacressonniere
ECS-Lab EA3649, National School of Electronics and Its Applications,
6 Avenue du Ponceau, 95014 Cergy Pontoise, France

© Springer International Publishing Switzerland 2016
A. El Oualkadi et al. (eds.), *Proceedings of the Mediterranean Conference on Information & Communication Technologies 2015*, Lecture Notes in Electrical Engineering 380, DOI 10.1007/978-3-319-30301-7_5

The tag attached to an object, consists an antenna and a silicon chip, containing an identification code. According to the working frequency, RFID systems are classified in four frequency bands: low frequencies (LF) around 125 kHz, high frequencies (HF) at 13.56 MHz, ultra-high frequencies (UHF) at 860–960 MHz, and the microwaves at 2.4 GHz [3]. For the LF and HF systems, the communication distance is shorter than one meter. For UHF RFID systems, it reaches around five meters. The RFID systems are also classified according to the tag technology. Active RFID tags have a transmitter and their own power source (typically a battery). The power source is used to run the microchip's circuitry and to broadcast a signal to a reader. Passive tags have no battery. Instead, they draw power from the reader, which sends out electromagnetic waves that induce a current in the tag's antenna [4]. This paper shows the design of a passive UHF RFID system. The system respects the EPC class1 Generation2 which is an air interface protocol first published by EPC global in 2004, it defines the physical and logical requirements for an RFID system of interrogators and passive tags, operating in the 860–960 MHz UHF range. Over the past decade, EPC class1 Gen2 has been considered as the standard for UHF implementations across multiple sectors, and is at the heart of more and more RFID implementations [5].

2 Description of the RFID System Architecture

The reader consists of a transceiver, with an antenna that communicates with a passive tag. The tag returns its identification code by retro-modulation. The RFID system, shown in Fig. 1, is simulated using the co-simulation between the numeric front end and the radio front end.

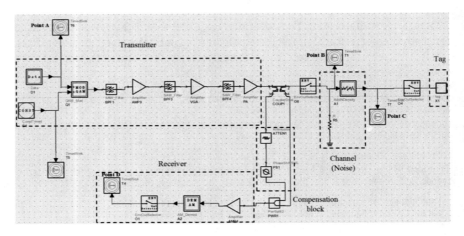

Fig. 1 Block diagram of the RFID communication system

Fig. 2 EPC Class1 protocol
signer encoding

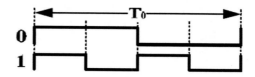

The transmitter part in Fig. 1, is mainly made up of a digital data source and an ASK (Amplitude Shift Keying) modulator with internal oscillator. Filtering and amplification stages are also included, with a bandwidth in the (860–960) MHz range. Filters are used to reject the out-of-band spectrum.

The receiver part is also presented in Fig. 1. The received signal is first amplified by a low-noise amplifier (LNA), and then filtered and demodulated by an ASK demodulator. The emitted and received signals are separated by a directional coupler. A compensation block is implemented to attenuate and phase shift the signal in order to reduce the isolations defects at reception [6].

The communication channel defect is modelled by an additive Gaussian filter to introduce sources of noise representing the environment.

Considering EPC Class 1 protocol, tags respond to the reader by retro-modulation sending the encode form signal shown in Fig. 2. Two transitions are observed for a binary zero and four transitions are observed for a binary one during one elementary cell. The data rate of the returned signal is 140.35 Kbps (in North America) and 30 Kbps (in Europe) [7].

3 Simulation Results

In this section, significant simulation results for each part of the system are presented. Figure 3 shows the simulation of PRBS (Pseudo Random Binary Sequence) data generated by the data-source from the reader transmitter (point A on Fig. 1). The data rate is chosen equal to 100 Kbit/s corresponding to a bit time equal to 10 μs. Figure 4 shows the modulated signal (point B on Fig. 1) using DSB-ASK (Double-side band amplitude shift keying) modulation sent by the reader transmitter.

Considering the operational working, the modulated signal is transmitted to the tag, which returns its coded identification. In order to simulate this process, the tag is considered as a data autonomous source. Figure 5 shows the encoded data sent by the tag (point C on Fig. 1). Figure 6 (point D on Fig. 1) shows the demodulated signal using the encoding according to Class 1 protocol. The data represent the binary sequence "111111100100001...". The modulated signal is transmitted through the compensation block. The data are thus available at the output of the receiver circuit. Figure 6 shows that the received data correspond to the data sent by the tag, as we expected.

Fig. 3 Digital data to be
transmitted

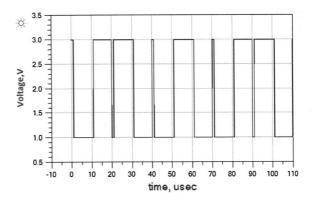

Fig. 4 Carrier modulated by
the data

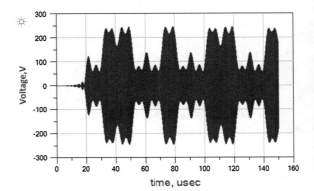

Fig. 5 Signal sent by the tag

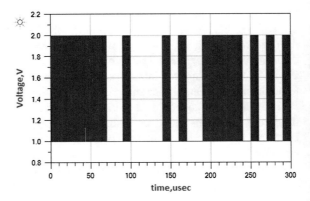

Fig. 6 Signal received by the reader

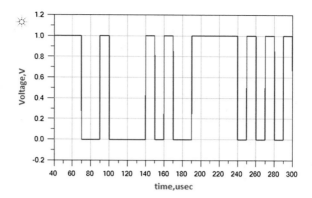

4 Description of Measurement Bench

In order to study the robustness and the reliability of passive tags submitted to severe environmental constraints, we have developed the measurement bench shown in Fig. 7. The commercial reader SPEEDWAY [8] is controlled by the PC

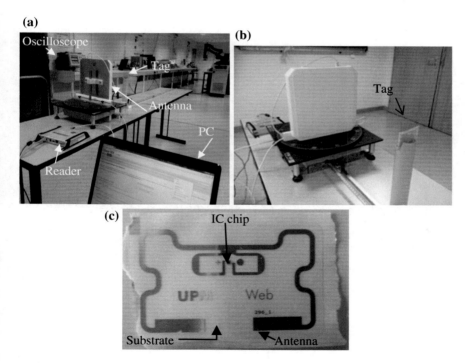

Fig. 7 **a** View of the test bench. **b** Tag mounted on the sliding support. **c** View of the UPM Web tag

(hot), it is connected to an antenna placed on a turntable 360°. The tag to be tested is placed on a stand disposed on a slide rail for a distance of up to 2 m. Placed in an open environment of obstacles, this bench allows readings of the power reflected by the tag depending on the distance.

The bench can operate with four possible frequencies in the UHF band, corresponding to the European standard: 866.30–866.90–867.5 MHz or 865.7 MHz, the emitted power of the reader can reach 30 dBm, the gain of the antenna is 6 dBi.

The passive tag chosen under test is shown in Fig. 7c. Its size is 40 mm × 54 mm and it can be easily attached to the object to be stored. It meets the requirements of EPC Class 1 Generation 2 (Gen 2) ISO 18 000-6C, and has an antenna attached to an integrated memory chip containing its unique identifier whose memory is 96 bits, or 12 bytes.

5 Time Measurements for the Communication Reader/Tag

Figure 8 shows chronograms recorded using the oscilloscope "Agilent Infinium" with a 6 GHz bandwidth, covering the UHF range. Figure 8 shows the signal sent by the reader with the presence of a tag placed in the reading field. As expected, a DSB-ASK modulation signal measured.

As shown in chronograms, part A of the signal represents the continuous wave (CW) signal suitable to feed the tag. Part B represents the instruction sent to the tag.

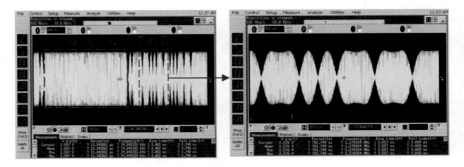

Fig. 8 Time measurements for communication between reader/tag

6 Experimental Results for Tag Reliability Study

6.1 Experimental Process Conditions

In this section some tests on tags are performed and presented, using the bench describe in paragraph 4. The operating frequency band depends mainly on the RFID reader used. The working frequency is equal to 865.7 MHz. The reader generates a radio frequency signal transmitted by a coaxial cable to the antenna. The tag under test is positioned on a rail. The distance can be adjusted in a 200 cm range. The tag receives the signal and returns its identification code, the reflected power measured and displayed on the screen of a PC connected to the RFID reader.

To get representative values, fourteen fresh tags for first series were tested. The results of these tests are shown in Fig. 9. The power emitted by the reader is fixed to 27 dBm. The measurements reveal no significant variations of the reflected power (standard deviation 2.07 %) for the different tags, showing a similar behavior. These results show that the bench is fully operational to meet the needs of this study.

6.2 Aging Reliability Tests

Passive UHF RFID tags are used for object identification in various environmental conditions which may affect their reliability [9, 10]. In order to evaluate the degradation of the performances, the effect of different temperatures is studied [11].

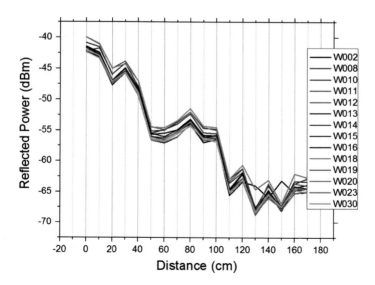

Fig. 9 Reflected power by the tags versus distance

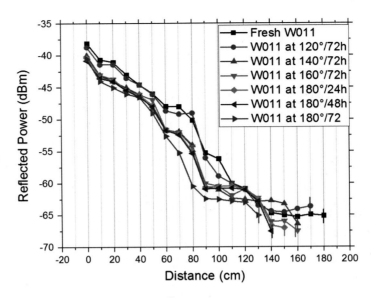

Fig. 10 Reflected power by W011 versus distance

According to the manufacturer [12] the tags under study can operate efficiently in the temperature range between −40 and 85 °C. To study the impact of accelerated life tests, storages under high temperatures are performed on three tags. The starting temperature was set at 100 °C corresponding to an overtaking close to 20 % of the maximum value recommended by the manufacturer. The selected step is 20 ° C; storage time is 72 h by step. The results for storage temperatures are shown in Fig. 10.

The performance parameters of the tag are significantly changed during the tests, the reflected power decreases after each test which strongly influence the range of the tag. Harder accelerated life tests are underway and the stress levels are in progress to accelerate the failures mechanisms.

7 Conclusion and Perspectives

The purpose of this work is to study the robustness and reliability of RFID systems in the UHF band under stern environmental conditions. A failure analysis will be implemented in order to determine accurately the type of defect. Moreover, the metallization of the antenna should be examined with an optical microscope to identify any cracks which will affect the performance of the tags. In addition, with a series of thirty devices, a statistical analysis using the predicted reliability calculation tools should be implemented to determine the mechanisms failures acceleration. Based on the simulations developed under ADS software, our objective is developing a reader based on discrete components. The system will be able to communicate with the tags that will be tested by varying the transmit power and the frequency throughout the whole UHF band.

References

1. Agilent EEsof EDA, Advanced Design System, The Industry's Leading RF, Microwave and High-Speed Design Platform
2. Dobkin, D.: The RF in RFID Passive UHF RFID in Practice. Elsevier (2008). ISBN 978-07506-8209-1
3. Lim, M.K., Bahr, W., Leung, S.C.H.: RFID in the warehouse: a literature analysis (1995–2010) of its applications, benefits, challenges and future trends. Int. J. Prod. Econ. **145**(1), 409–430 (2013)
4. Hauet, P.: L'identification par radiofréquence (RFID) techniques et perspectives, article invité, REE No. 10, Nov 2006
5. EPC™ Radio-Frequency Identity Protocols, Generation-2 UHF RFID, Specification for RFID Air Interface, Protocol for Communications at 860 MHz–960 MHz. Version 2.0.0 Ratified
6. You, B., Yang, B., Wen, X., Qu, L.: Implementation of low-cost UHF RFID reader front-ends with carrier leakage suppression circuit. Int. J. Antennas Propag. **135203**, 8 (2013)
7. Jin, L., Cheng, T.: Analysis and simulation of UHF RFID system. In: IEEE ICSP2006 Proceedings (2006)
8. UHF GEN 2 RFID, SpeedwayR installation and operations guideversion 5.2, Copyright © 2012–2014 Impinj, Inc. All rights reserved. http://www.impinj.com
9. Viswanadham, P., Singh, P.: Failure Modes and Mechanism in Electronic Packages, p. 370. Chapman & Hall, London (1998)
10. Suhir, E.: Accelerated life testing (ALT) in microelectronics and photonics: its role, attributes, challenges, pitfalls, and interaction with qualification tests. ASME J. Electron. Packag. **124**(3), 281–291 (2002)
11. Lahokallio, S., Saarinen-Pulli, K., Frisk, L.: Effects of different test profiles of temperature cycling tests on the reliability of RFID tags. Microelectron. Reliab. **55**(1), 93–100 (2015). ISSN 0026-2714
12. UPM RFID, Datasheet web-RFID-Tag, EPCglobal. www.upmrfid.com

Cryptographic Algorithm Approach for Low Cost RFID Systems

Zouheir Labbi, Ahmed Maarof, Mohamed Senhadji
and Mostafa Belkasmi

Abstract Today the security of RFID systems is a very important issue especially for low cost RFID tags. Then, it is difficult for several researchers to adopt a strong encryption algorithm due to the extremely limited computational, storage and communication abilities of the tag. So, the priority in RFID systems is to ensure the integrity, confidentiality and authentication of messages using different algorithms (such as encryption and hashing algorithm). In this paper, we propose a lightweight symmetric key encryption approach that includes integrity as part of the encryption process in order to meet the resource limitations of low cost RFID tags. Furthermore we do not need to use separate hash functions to verify message integrity, thus leading to increase the computational efficiency of the system.

Keywords RFID · Lightweight encryption algorithm · Low cost RFID tags

1 Introduction

RFID (Radio Frequency Identification) systems are very useful and convenient to identify objects automatically through wireless communication channel. It has been used mainly to replace the legacy bar codes. RFID systems are generally consisted of tags, readers, and the back-end server connected to the reader. In these days, RFID system is increasing [1] and applied to various applications such as access control, supply chain management, inventory control, smart labels, etc. However,

Z. Labbi (✉) · A. Maarof · M. Senhadji · M. Belkasmi
TSE Laboratory, ENSIAS, Mohammed V University, Rabat, Morocco
e-mail: zouhir.labbi@gmail.com

A. Maarof
e-mail: ahmed.maarof@gmail.com

M. Senhadji
e-mail: m.senhadji@um5s.net.ma

M. Belkasmi
e-mail: m.belkasmi@um5s.net.ma

© Springer International Publishing Switzerland 2016
A. El Oualkadi et al. (eds.), *Proceedings of the Mediterranean Conference on Information & Communication Technologies 2015*, Lecture Notes in Electrical Engineering 380, DOI 10.1007/978-3-319-30301-7_6

since the wireless communication channel is not secure from the various security attacks, a security mechanism is needed which necessitating the use of data encryption algorithms [2]. However, the less computational power and storage capabilities of RFID systems restrict and limit to perform sophisticated crypto-graphic operations [3] like RSA, AES and DES which provide a strong confiden-tiality. Although they are very heavy algorithms in terms of computation costs and required space and they do not support authentication and integrity alone. As a result, a several lightweight encryption algorithms were proposed by researchers such as Hummingbird [4], RBS (Redundant Bit Security algorithm) [5], etc.

Basically, the security has different goals (such as authenticity, confidentiality, integrity...), most of which can be achieved by encrypting the message using cryptographic algorithms [6]. Encryption algorithms are categorized into different types, like Symmetric Key Encryption (SKE), Asymmetric Key Encryption (AKE) and hashing algorithms. A message is first encrypted by symmetric or asymmetric key algorithm to provide confidentiality. Then, the message will be hashed using a one-way hashing function to provide authentication and integrity. Therefore, every message is subjected to multiple processes as encryption and hashing that make conventional cryptosystems computationally less efficient [7]. This motivated us to propose an approach that can offer both integrity and confi-dentiality with less computation. In this work, we do not consider authentication.

In this paper, we not combine the encryption and hashing process, but introduces an additional process in the encryption to offer an implicit integrity check, without need for a separate hashing operation. As a result, computation is expected to be less.

This paper is organized as follows. In Sect. 2, we present a literature review of existing algorithms. In Sect. 3, we explain our proposed algorithm. In Sect. 4, we provide a security analysis of our approach and compare our approach with existing algorithms. Section 5 concludes the paper.

2 Background Information and Related Work

AKE is considered as more secure compared to SKE. But AKE is a very heavy algorithm in terms of computation costs and required space which cannot be accomplished by RFID systems. Otherwise, the Implementation of SKE requires less computational complexity according to [7, 8]. Batina et al. [3] discussed the feasibility of the AKE for RFID systems. They identified that AKE works slower than SKE. Furthermore conclude that public key Schnorr's identification protocol is less expensive than Okamoto's technique for RFID systems.

On 2010, Engels et al. [4] proposed an SKE algorithm which designed for RFID systems called Hummingbird. It encrypts the message block-by-block and has internal registers for the key stream update. It has 64-bit block size, and 256-bit key size that is split into four 64-bit keys. This algorithm works on a principle of permutation and substitution operation. It is resistant to linear as well as differential attacks. Kutuboddin and Krupa [9] have extended on 2013 the Hummingbird algorithm to increase computational efficiency, for which they included an additional Exclusive-OR (XOR) operation. Finally, their proposed approach had a better computational efficiency compared to the original version.

Hashing is a mathematical one-way function. A variable sized message is converted to a fixed size output (say n = 196, 256, 512 bits depending on the hashing algorithm) called the Message Digest (MD). The receiver will compute the MD using the received message to verify the integrity. However, encryption (or decryption) and hashing are disjoint operations. Therefore, use of the hashing algorithm along with an encryption will increase the computational overhead on the system [7, 10].

In 2013, Jeddi et al. [5] proposed a SKE for RFID systems using low-cost tags called redundant bit security algorithm (RBS) to offer confidentiality with integrity. The redundant bits for RBS are generated by using the MAC algorithm, to provide integrity to the message. Redundant bits generated are inserted into the original message while encrypting. The encryption is performed with a smaller number of cycles as compared to traditional encryption schemes and that leads to less computational complexity.

In summary, the proposed algorithms offer any one security goals independently or more security goal when integrated with other algorithms (e.g. RBS uses MAC [5]). Our proposed approach provides confidentiality with integrity to the message independently, without support from any other algorithms.

3   Proposed Approach

We propose a SKE for RFID systems, which is composed of two parts: an encryption part and a key generation part. The encryption part uses a basic XOR operation, an expansion function and a bit flipping function. Further, the bit flipping function is also used for key generation. Our approach provides confidentiality with integrity in a single encryption cycle, rather than requiring separate MAC computation following encryption like RBS, which reduces the logic gate usage.

3.1   Communication Architecture

The communication among tag, reader and Server is proceeds as follows. The reader first queries the tag, and tag responds with an encrypted response. This encryption is based on a key that is pre-shared with the server. The reader then forwards this message to the server, which decrypts it and retrieves information about the object represented by the tag. It then sends this information with acknowledgment to the reader, which forwards the acknowledgment to the tag. The tag will update the new key only if it receives the acknowledgment. Figure 1 shows the architecture of our approach. We assume that an initial key is securely shared

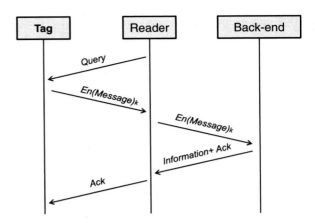

Fig. 1 Approach architecture

between a server and an RFID tag, while successive keys for every message transmission are generated at both ends from the initial key without direct communication.

3.2 Encryption

Encrypt a message to generate a ciphertext using symmetric or asymmetric between legitimate users, offers confidentiality. But to avoid and detect the attacks (such as, man-in-the-middle attack) caused by an unauthorized entity, data integrity verification and entity authentication are employed. In our approach, we concentrate on the integrity aspect to ensure that the receiver efficiently detects data modification on the message (ciphertext).

In our work, we used expansion function and a new operation called bit flipping. The expansion function uses a logical exclusive-OR gate to offer integrity to the message that doubles the original message size. Bit flipping is introduced only as a supplement to the expansion function that offers integrity. These two functions act as core components to our approach are applicable to encryption first, followed by the same for decryption. We will discuss them in the following sub-sections.

Our protocol for RFID systems first employs the XOR operation, the expansion function to generate integrity parameters, followed by a bit flipping operation and finally an additional XOR operation. The message size is 64-bit and the key size is $(64 + n)$ bit including n-bit nonce. A last n bit nonce is used to generate the key for the next communication but first 64-bit is used to encrypt the message for the current request from the reader. Initially, the 64-bit message is XORed with the first 64-bits of the key, which is then followed by expansion.

Expansion function, $Exp(m)$: The goal of the expansion function in our approach is to introduce additional bits to the original message that will generate a pattern to offer message integrity. A logical XOR operation is used to form these patterns. To control the expansion rate for the message m, we restrict it to produce an 8-bit output for every 4-bit input (i.e. every 4-bit in the message starting from most significant bit (MSB) is converted to 8-bit output after an expansion function) that converts a 64-bit input message into a 128-bit output.

Our algorithm computes the integrity parameter for bit pairs beginning with the MSB bit, m_0, until the least significant bit (LSB), m_{l-1} where l represents the length of the message in bits. The expansion function, $Exp(m)$, is defined as follows:

$$Exp(m_i, m_{i+1}, m_{i+2}, m_{i+3}) = m_i||m_1'||m_{i+3}||m_{i+1}||m_{i+2}||m_2'||m_{i+1}||m_{i+3}. \quad (1)$$

Where $m_1' = XOR(m_i, m_{i+1})$ and $m_2' = XOR(m_{i+2}, m_{i+3})$;
m_i is the ith bit of the message m;
$XOR(m_i, mi+1)$ Represents the XOR operation on the m_ith, m_{i+1}th bit positions;
$Exp()$ represents expansion function for encryption; $||$ represents concatenation operation.

After the expansion function, the message block will be twice the length of the input block. This process is followed by bit flipping, discussed in the next section.

Bit flipping function for encryption, $BF_{enc}(m)$: The bit flipping function in our approach provides an extra layer of assurance that data modification is detected efficiently. Our protocol applies the bit flipping function on 128-bit output of the expansion function beginning with the MSB, m_0, until the least significant bit, m_{2l-1} (note that the length of the message increases to $2l$ after expansion). The bit flipping function, $BF_{enc}(m)$ generates the same number of bits as output from $Exp(m_{2l-1})$.

$$m_x = \neg m_x. \quad (2)$$

Where, $\neg$ represents bit inversion or bit flipping;
$x = (i + j + 2) \% 128; \forall m_i \in m, i = [0...127], j = [0...3]$;
i and $i + 1$ were the indication of bit positions;
j indicates flipping bit with respect to indication bit i. for each i j = decimal (m_i, m_{i+1});
m_x represents the bit to flip in a message m.

We use the value of i and j to find the flipped bit that is represented by m_x. A variable i indicates the bit position in a message and j would be calculated by bit combination of i and $i + 1$. Any bit that is flipped in a block is reflected only within that block (i.e. the bit operations performed in a block are cyclic).

In summary, 128-bit output from the expansion function is considered for bit flipping to provide a means for integrity check. After the bit flipping function, the 128-bit is XORed with a 128-bit key to generate a ciphertext. The key generation process, which performs the second round of XOR is discussed in the next section.

3.3   Key Generation

In our algorithm, we use two keys. A present key K_p (Key for first round of current message) and a future key K_f. Both keys K_p and K_f are $(64 + n)$-bit long (that includes an n-bit nonce), where K_f is directly derived from K_p through bit flipping. The first 64-bits from the key K_p are XORed with the message for the first round. For the last round of encryption, the first 64-bits from the keys K_p and K_f are combined to generate a 128-bit, which is XORed with the 128-bit output from bit flipping function. The final XORed output is the ciphertext. For the next communication, the key K_f will be used as K_p and vice versa. Figure 2 illustrates the encryption process.

3.4   Decryption

The decryption process is the inverse of encryption. The 128-bit ciphertext is XORed with a 128-bit key that is extracted from the keys K_p and K_f. Initially, we apply a bit flipping function followed by a compression operation. After the compression process, 128-bit is restored to 64-bit. Finally, the 64- bit output is XORed with the key K_p. The key generation process for decryption is the same as the encryption process.

Fig. 2 Overview of encryption process

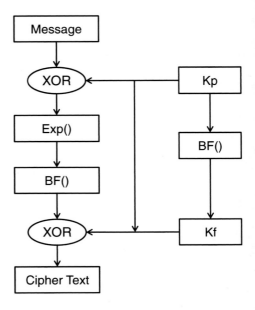

Bit flipping function for decryption, $BF_{dec}(m)$: Our protocol performs the following bit flipping operation in a manner reverse to the encryption, i.e. beginning with the LSB, m_{2l-1}, until the MSB, m_0. $BF_{dec}(m)$ consumes 128-bits and generates the same number of bits as output from the XOR operation, which is similar to encryption.

$$m_x = \neg m_x. \tag{3}$$

Where, $\neg$ represents bit inversion or bit flipping;
$x = (i + j + 2) \% 128; \forall m_i \in m, i = [127...0], j = [0...3];$
i is the indication of bit positions;
j indicates flipping bit with respect to indication bit i;
m_x represents the bit to flip in a message m.

Compression function, $Cmp(m')$: For each pair of bits of the ciphertext (m'), beginning with the MSB, m'_0, until the LSB, m'_{2l-1} [$2l$ represents the length of the expanded ciphertext], the compression function, $Cmp(m')$, is derived as follows. We will consider sublocks of m' of size 8. If $XOR(m_i, m_{i+3})$ is equal to m_{i+1}, m_{i+3} is equal to m_{i+7}, $XOR(m_{i+4}, m_{i+7})$ is equal to m_{i+5}, and if m_{i+6} is equal to m_{i+3}, then, the received cipher text is assumed not to have been modified (i.e. the message integrity is verified) and the decrypted message bits will be m_i, m_{i+3}, m_{i+4}, m_{i+7}. Figure 3 illustrates the decryption process.

Fig. 3 Overview of decryption process

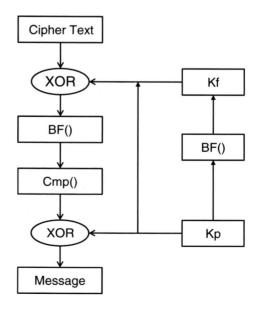

4   Security and Performance Evaluation

An RFID system is a resource-constrained system that cannot perform complex operations. The number of gates available within the RFID circuit is limited to implement advanced cryptographic algorithms. This leaves the system vulnerable to different attacks. In this section, we present a qualitative analysis of our algorithm, discussing its performance with respect to various attacks and also present a cryptanalysis. Further, we include a comparison of the logical gate estimation between our approach and the RBS approach [5].

4.1   Performance Against Known Attacks

A replay attack involves eavesdropping current information by an unauthorized entity to use at a later time to pretend to be an authorized entity [11]. As we change the key for every communication, eavesdropped information won't be useful for the next communication, which protects our approach from the replay attack.

A denial of service attack, on the other hand, is to stop the service between both communication parties, either temporarily or permanently. In our approach, the tag can easily identify the denial of service attack while attempting the next request from the reader before getting acknowledgment from the server for the current request.

In our approach, even if we assume that the attacker successfully retrieves the current key, it will not be useful, as the next key generated is based on n-bit nonce, which prevents the attacker from being able to generate the next or the previous keys.

4.2   Performance Evaluation and Discussion

Table 1 presented is a summary of the estimate of GE (Gate Equivalence) used in our approach and other algorithms RBS [5], AES [12], PRESENT [13] and Grain [14]. It can be seen that our algorithm is more efficient since it requires less logic gates, and hence, less operations to be performed.

To implement our algorithm, we use four 32-bit XOR gates, one NOT gate and one 2-bit comparator that is implemented using two 1-bit comparator circuits. Each

Table 1 Gate estimation comparison table

	Our approach	RBS [5]	AES [12]	PRESENT [13]	Grain [14]
# of gates	527 GE	1920 GE	3200 GE	2332 GE	1857 GE
Block size	64–128 bits	132 bits	128 bits	64 bits	1 bit

1-bit comparator is implemented using six 2 input NAND gates, and combined using two additional NAND gates. Each 2-bit XOR is considered to be implemented using four NAND gates. Approximately, 527 two-input NAND gates in our implementation.

It is our contentions that by combining the integrity check as part of the encryption processes, our algorithm will reduce the overall computational power required by the cryptosystem, thereby increasing the computational efficiency of the system.

Our protocol does not employ hashing in any way, which means that the overall computational cost of our protocol will be less than the computational cost of performing encryption with hashing, which we can summarize as follows:

Cost(EncryptionincludingIntegrityCheck) < Cost(Signature + Encryption) ≪ [Cost(Signature) + Cost(Encryption)] [15]

5   Conclusion and Future Work

In this paper, we have proposed a new low cost RFID algorithm which assures the confidentiality. It assures also the integrity check of messages which is included as part of the encryption process for a SKE using expansion function. Thereby eliminating the need for a separate hashing process in addition to encryption. Comparing with other protocols, our protocol will be good solutions for low-cost RFID systems that require good operational, efficiency and security properties. Our future work would involve writing this algorithm on VHDL code in order to ensure the functionality and also optimize the number of logic gates.

References

1. Juels, A.: Rfid security and privacy: a research survey. IEEE J. Sel. Areas Commun. **24**(2), 381–394 (2006)
2. Molnar, D., Wagner, D.: Privacy and security in library rfid: issues, practices, and architectures. In: Proceedings of the 11th ACM Conference on Computer and Communications Security, pp. 210–219. ACM (2004)
3. Batina, L., Guajardo, J., Kerins, T., Mentens, N., Tuyls, P., Verbauwhede, I.: Public-key cryptography for rfid-tags. In: Fifth Annual IEEE International Conference on Pervasive Computing and Communications Workshops. PerCom Workshops' 07, pp. 217–222. IEEE (2007)
4. Engels, D., Fan, X., Gong, G., Hu, H., Smith, E.M.: Hummingbird: ultra-lightweight cryptography for resource-constrained devices. In: Financial Cryptography and Data Security, pp. 3–18. Springer, Berlin (2010)
5. Jeddi, Z., Amini, E., Bayoumi, M.: Rbs: redundant bit security algorithm for rfid systems. In: 21st International Conference on Computer Communications and Networks (ICCCN), pp. 1–5. IEEE (2012)
6. Chen, X., Makki, K., Yen, K., Pissinou, N.: Sensor network security: a survey. Commun. Surveys Tutorials. IEEE **11**(2), 52–73, (2009)

7. Kumar, Y., Munjal, R., Sharma, H.: Comparison of symmetric and asymmetric cryptography with existing vulnerabilities and countermeasures. Int. J. Comput. Sci. Manag. Stud. **11**(03) (2011)
8. Dodis, Y., An, J.H.: Concealment and its applications to authenticated encryption. In: Advances in Cryptology EUROCRYPT, pp. 312–329. Springer, Berlin (2003)
9. Kutuboddin, J., Krupa, R.: Efficient implementation of hummingbird cryptographic algorithm 285 on a reconfigurable platform. Int. J. Eng. **2**(7) (2013)
10. Li, X., Zhang, W., Wang, X., Li, M.: Novel convertible authenticated encryption schemes without using hash functions. In: 2012 IEEE International Conference on Computer Science and Automation Engineering (CSAE), vol. 1, pp. 504–508. IEEE (2012)
11. Van Deursen, T., Radomirovic, S.: Attacks on rfid protocols. IACR Cryptol. ePrint Archive **2008**, 310 (2008)
12. Feldhofer, M., Wolkerstorfer, J., Rijmen, V.: Aes implementation on a grain of sand. IEEE Proc. Inf. Secur. **152**(1), 13–20 (2005)
13. Poschmann, A.Y.: Lightweight cryptography: cryptographic engineering for a pervasive world. Ph.D. dissertation, Ruhr-University Bochum, Germany (2009)
14. Good, T., Benaissa, M.: Hardware results for selected stream cipher candidates. State of the Art of Stream Ciphers, Workshop, pp. 191–204 (2007)
15. Zheng, Y.: Digital signcryption or how to achieve cost (signature & encryption) cost (signature) + cost (encryption). In: Advances in Cryptology CRYPTO'97, pp. 165–179. Springer, Berlin (1997)

Complex Design of Monitoring System for Small Animals by the Use of Micro PC and RFID Technology

Zoltán Balogh and Milan Turčáni

Abstract The paper describes a complex design of monitoring system handling created by the use of microcomputer Raspberry PI and LF RFID technology for the needs of monitoring and exploring the behavior of small terrestrial mammals and of selected species of birds. Similar systems already exist, but they are demanding and financially not affordable. To create such a system, low-cost products have been used which are easily modifiable, modular and expandable by measuring other activities. The paper concentrates on RFID/PIT technology. PIT tags are a useful tool to identify and follow individuals within a large population to monitor movement and behavior. Tags transmit a unique identifying number that can be read at a short distance that depends on the tag size and antenna design. Passive RFID tags are inductively charged by the reader and do not have a battery. Tags can remain operational for decades.

Keywords Raspberry Pi · Monitoring station · RFID technology · PIT tags · Web application

1 Introduction

The paper deals with the possibility of monitoring animals. While monitoring animals, it is necessary to do a few compromises considering the size of the animals, the price and number of individuals. The ideal transmitter should be light so that the animal can wear it safely. It should be adequately cheap so it can be possible to use it for other individuals. It should be also energy efficient so that the animal does not have wear a battery. It should be able to send the location with

Z. Balogh (✉) · M. Turčáni
Department of Informatics, Faculty of Natural Sciences, Constantine
the Philosopher, University in Nitra, Tr. a. Hlinku 1, 949 74 Nitra, Slovakia
e-mail: zbalogh@ukf.sk

M. Turčáni
e-mail: mturcani@ukf.sk

© Springer International Publishing Switzerland 2016
A. El Oualkadi et al. (eds.), *Proceedings of the Mediterranean Conference
on Information & Communication Technologies 2015*, Lecture Notes
in Electrical Engineering 380, DOI 10.1007/978-3-319-30301-7_7

great accurateness. As an example of methods which are currently supported by a great database about the animal movement, there is MoveBank [1]. Those are for example GPS (Global Positioning System), Argos Doppler system, VHF (Very High Frequency), ringing or tagging, GLS (Global Locating System) and of course RFID (Radio-frequency identification)/PIT (Passive Integrated Transponder) chips.

In the last thirty years, a great number of biological traits of animals have been explored due to PIT tags. It enabled animal growth rates assessment, movement patterns, and survivorship for many species by injecting the markers right into the animals. It was possible in a more reliable approach of external identification marking of animals. The devices have been also used for confirming the animal identity of the zoo, animal identity of pets or protected species illegally carried away from their natural habitat. New approaches bring progress in physiology and conservation biology and encourage greater understanding of social interactions in a population. The devices have their restrictions such as high purchase cost, low detection distance, or potential tag loss. Nevertheless, PIT tags reveal all mysteries of animals which could not have been named effectively [2].

Hedgepeth [3] pointed out that RFID is a wireless data collection technology that uses electronic tags the size of a grain of rice, for allowing an automatic trace and track of individual goods throughout the logistics network. Some researchers even expect that applications of RFID technology will become widespread as a means of tracking and identifying all kinds of animals [4]. So, by its fast, reliable, multi-readable, and non-line-of-sight capabilities [5], the dynamic animal information, like medical records and location, can be monitored and hence used to facilitate diagnosis and animals' behavior patterning.

Boarman et al. [6] used automated reader to demonstrate that the endangered species of desert tortoise (*Gopherus agassizii*) migrated through the sluices built under the highways.

The use of RFID technology in animal monitoring is one of the established methods in many countries in the world. RFID technology is used when monitoring domesticated, wild and farm animals. In the United States, USDA (U.S. Department of Agriculture) monitors deer and elk with RFID chips in order to find out how these animals react to CWD (chronic wasting disease), a severe neurological disorder [7].

By technology improvement, PIT chips are getting more appealing for researchers who want to observe individuals in long-term. In 2014 one chip costs a few cents and the price of the readers is about 700–2000€ and more. It is considered to be expensive but it is still cheaper than telemetry equipment. Automatic scanning is really time efficient in comparison to ringing and tagging. The problem is that if two animals tagged with PIT chip get closer to the reader in the same moment, the reader will be not able to read the chips. This is not a problem in satellite telemetry. Telemetry functions on greater distance while PIT readers are able to scan to a distance of about 100 cm.

In the majority of European countries for tagging animals with PIT chips, certain standards are fulfilled in order to keep compatibility of the chips and scanners. These ISO standards have great international recognition including Europe and

Canada. The International Organisation of Standardization assumes responsibility for the two standards ISO 11784 and ISO 11785. For the needs of international norms, they contain a code of three digits which indicates the country, the producer and also the serial number of the chip.

The paper deals with description, design and creation of monitoring device for automatic animal scanning by the use of microcomputer Raspberry Pi and RFID technology.

2 Material and Methods

The aim of the research was to design a system which was able to automate data collection from spatial activity (for data measurements and collection microcomputer Raspberry Pi was used) and consequently create an application which is able to observe and record the activity of small animals. There was a presumption that the resulting system will be cheaper than the accessible monitoring stations and that the resulting system would have wider use than the usual RFID monitoring stations.

2.1 The Proposed Monitoring Stations

As the controller, micro PC Raspberry Pi was used and its model B. Raspberry provides wide possibilities of use which are proved by various scientific projects based on this micro PC. The Raspberry Pi is a credit card-sized single-board computer developed in the UK by the Raspberry Pi Foundation with the intention of promoting teaching of basic computer science in schools [8]. The Raspberry Pi is based on a system on a chip (SoC) called the Broadcom BCM2835. It contains an ARM1176JZF-S 700 MHz processor, VideoCore IV GPU. First it was conveyed with 256 MB of RAM, which was later upgraded (Model B and Model B+) to 512 MB. The system has Secure Digital (SD) or MicroSD (Model A+ and B+) plugs for boot media and permanent storage [9]. The Foundation provides Debian and Arch Linux ARM distributions for download. Tools are available for Python as the main programming language, with support for BBC BASIC (via the RISC OS image or the Brandy Basic clone for Linux), C, C++, Java, Perl and Ruby. At present, its price ranges around 30–35€. Raspberry Pi (Fig. 1) enables sensors connection through 8 GPIO (General Purpose Input/Output) pins as the other devices through USB, UART, I2C or SPI. Thanks to its size, price and connection possibilities of low-level peripherals, it was a good choice for monitoring system controller.

Monitoring station consists of a controller—micro PC Raspberry Pi, RFIDRW-E-232 reader, signalling bicolour LED, 1 kΩ resistor, WiFi adaptor

Fig. 1 The Raspberry Pi packs a lot of power into a credit-card size package [10]

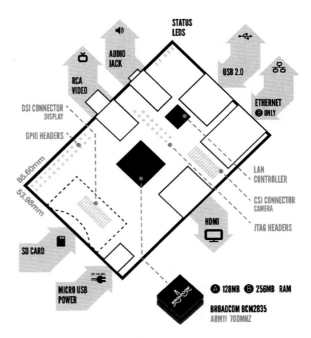

(TL-WN725NV2), and a translator from TTL logic to RS232 logic (Fig. 2). The communication takes place by means of two UART pins:

– RXD—pin 10 on Raspberry Pi
– TXD—pin 7 on Raspberry Pi

UART pins and Raspberry Pi work on 3.3 V logic and RFIDRW-E-232 reader works on 5 V logic. Because of this difference, communication had to take place through translator from TTL (3.3 V) to RS232 (−25 to 25 V). Communication between RXD and TXD should have been crossed so the pin RXD on Raspberry Pi

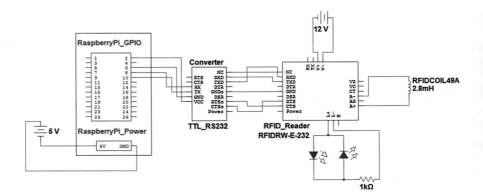

Fig. 2 Schematic diagram of the monitoring station—own design

would be linked with the pin TXD on the reader the pin TXD on Raspberry Pi would be linked to pin RXD on the reader.

The power supply on Raspberry Pi is 5 V and the recommended current is 1A. The converter needed power supply of 5 V which was consequently provided by raspberry Pi from pin number 2, was possible to link the ground to GPIO pin number 6 on Raspberry Pi. The reader needed power supply of 12 V and at least 38 mA. It was possible to link the reader to the bi-coloured LED diode which signalizes chip reading and writing. The diode was connected to pins L+ L−. It was also possible to link a buzzer on pin B.

The principles of communication RFID chip with RFID reader (RFIDRW-E reader) is generated magnetic field by an external antenna, usually on a frequency of 125 kHz or 134 kHz. Passive chips had integrated antenna which was tuned to the same frequency. When the chip was nearby to the magnetic field transmitted by the external antenna of the reader, the chip was able to draw enough energy from the electromagnetic field for the power supply of its own electronics. In the moment the chip was connected, it was able to modulate the magnetic field which was captured by the reader. This way it was possible for the chips to transfer data to the reader. Many types of chips exist which may work on various frequencies. By choosing the right antenna and capacity setting, can be the RFIDRW-E reader tuned to the same frequency as the scanned chip.

2.2 The Proposed Monitoring System

While monitoring, researchers dealing with exploring the behavior of small terrestrial mammals and of selected species of birds have to be "present in person" and manually record the behavior of these species. The designed monitoring system consists of two parts, a monitoring station and a server on which affordable micro PCs Raspberry PI were used with attachment and RFID reader.

The designed monitoring system (Fig. 3) consists of two parts, a monitoring station and a server on which affordable micro PCs Raspberry Pi was used. Monitoring station preserved transmitter scanning by a program created in Java and

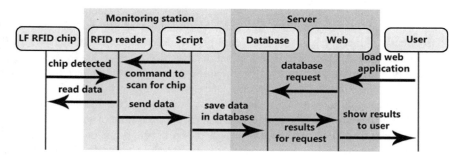

Fig. 3 Communication of monitoring system—own design

RFID chip reader and sending entries into relational database. The server provided WiFi net for all monitoring stations in the monitored zone on which they were connected. The server contained a web server where the web application was placed to access the entries in the relational database. The server contained a DHCP server in order not to set the IP address on each monitoring station separately. It also contained DNS server to simplify the access to web application. DNS server refers to public DNS server from Google for Internet access. Loaded entries were saved by monitoring stations into MySQL database established on servers.

Consequently, it would be possible to get to entries by accessing into WiFi net transmitted by the server. Web application would allow viewing the recordings saved in the database, their export or filtration.

3 Software for Monitoring System

Software for monitoring system contains two parts. The first preserved communication with RFID chip reader and loaded data entry into relational database. The second part was a web application which provided access to data saved in the database.

3.1 Software for Monitoring Stations

The main class that provided communication with RFID reader. Communication with the reader took place by the use of sending chains ended with special character (CR—cursor shift at the beginning of the line) through serial port. Communication through GPIO and serial port was provided by library *pi4j*. Communication through serial port was initiated by the line:

```
serial.open("/dev/ttyAMA0", 9600);
```

Serial port was placed on the path/dev/ttyAMA0 and the number 9600 was given by baudrate (speed of communication).

After opening the serial port, the application tried to establish a connection with the database which was placed on the server. If there is no connection, the application will wait 10 s and will try to establish connection again until it is active. Consequently, a listener was added to serial port in order to record the data which were sent back by RFID reader. The method dataReceived (SerialDataEvent event) provided processing of all data which came through serial port. Data accuracy was examined by checking its length because of possible errors while transfer. If the data are correct, chip identity and reader identity (which read the PIT chip) will be saved into database. While saving data into database, actual time of saving was also added to the record. The method run provided regular querying on RFID reader whether there was a RFID chip around. Any errors while program running were written into log file logy.log. The class Trap provided creation of log file and launch of RFID reader class. While launching, Trap waited 15 s in order to prevent unsuccessful experiments of database connection and to have enough time to launch the database.

3.2 Software for Server

The software for server is a web application which enabled access to data. It was possible to browse data, and export them into format CSV and XML. It was possible to export data according to a given date or according to the given last *x* records. It was also possible to process exported data in any program which supports data import in given formats. Web application also enabled to filtrate the shown results according to date, monitoring station identifier and chip identifier, and to show in what period the chip was found around the monitoring station.

ConnectionFactory is what took care of signing-on with the database. Every constant used by the application was saved in the WebConstants class, which enabled simple change of main program settings. Communication with the database was enabled by DataBean. This class provided data retrieval for each part of the web application.

Web application contained also data display according to holding near the monitoring station. This function displayed data in an extent in which the chip was present near the monitoring station.

4 Result and Discussion

The system took place in laboratory conditions. For testing, the PIT microchip was put into two mice (Mus musculus) which were placed into a big aquarium. Monitoring system consisted of a server and a monitoring station. The created sensor was represented by a plastic tube (Fig. 4) which was opened on one side.

Fig. 4 System with the RFID coil antenna—own design

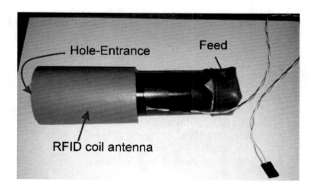

Here was a RFID coil antenna to load RFID/PIT chips. On the other side of the trap, there was feed for mice.

Testing and verification of the system lasted during 7 days of continuous system operation which was connected through electrical system. Microchips were put into two mice which moved freely around the aquarium. In this testing, 35,744 records have been captured into relational database and the system has shown to be reliable. The system was set so that it could record with the highest frequency of record which represented every 100 ms. The aim was to test system functionality and to obtain the most data. In some cases, when the mouse was under RFID coil antenna, the system scanned it 3× in 1 s. By the use of obtained data, it was necessary to find out system validity according to various statistical methods [11], in order to reduce the frequency of RFID/PIT chip inscription so that no important data would be lost. This way, a lot of energy has been saved which was the assumption for autonomous operation of the monitoring station. The next was testing power monitoring station with a battery of 9800 mAh. The battery powered Raspberry Pi (5 V) and the reader of RFID microchip (12 V), TTL on RS232 converter was consequently powered by Raspberry Pi. In these conditions of consumption ~800 mA, the battery had stamina of 8 h. In the future there is an assumption of lowering the consumption of monitoring system and testing solar power in combination with high capacity batteries so that the system could work autonomously. To optimize such a system would be possible by various modelling techniques and consequential simulations which would be used in the future [12, 13].

5   Conclusion

The research aim was to design and create a complex autonomous monitoring system which would monitor record and evaluate results of spatial activity. There are assumptions that the resultant system would be cheaper and application-flexible than the accessible monitoring systems. The resultant monitoring system would be spread by other possibilities in order to obtain as much information about spatial activities of terrestrial mammals and of selected species of birds. On the basis of created system there would be space to spread our research task in the field of ecology and other fields such as ambient intelligence [14].

References

1. MoveBank: movebank.org (2013). Available via https://www.movebank.org/node/857. Accessed 11 Oct 2014
2. Gibbons, J.W., Andrews, K.M.: PIT tagging: simple technology at its best. Bioscience 54(5), 447–454 (2004)
3. Hedgepeth, W.O.: RFID metrics: decision making tools for today's supply chains. CRC Press, Boca Raton (2007)

4. Jansen, M.B., Eradus, W.: Future developments on devices for animal radiofrequency identification. Comput. Electron. Agric. **24**(1-2), 109–117 (1999)
5. Lahiri, S.: RFID sourcebook. IBM Press (2006). Available via http://worldcat.org, http://www.books24x7.com/marc.asp?bookid=12240. Accessed 10 Sept. 2014
6. Boarman, W.I., Beigel, M.L., Goodlett, G.C., Sazaki, M.: A passive integrated transponder system for tracking animal movements. Wildl. Soc. Bull. **26**(4), 886–891 (1999)
7. Yan, L., Zhang, Y., Yang, L.T., Ning, H.: The Internet of Things: From RFID to the Next-Generation Pervasive Networked Systems. Auerbach Publications, India (2008)
8. Bush, S.: Dongle computer lets kids discover programming on a TV ElectronicsWeekly.com. http://www.electronicsweekly.com/news/design/embedded-systems/dongle-computer-lets-kids-discover-programming-on-a-2011-05/. Accessed 25 May 2011
9. Bradbury, A.: Open source ARM userland. http://www.raspberrypi.org/open-source-arm-userspace/. Accessed 7 Nov 2014
10. Halfacree, G.: Raspberry Pi interview: Eben Upton reveals all. http://www.linuxuser.co.uk/features/raspberry-pi-interview-eban-upton-reveals-all. Accessed 12 Dec 2014
11. Munk, M., Kapusta, J., Švec, P.: Data preprocessing evaluation for web log mining: Reconstruction of activities of a web visitor. In: 10th International Conference on Computational Science, ICCS 2010, pp. 2273–2280. Amsterdam (2010)
12. Skorpil, V., Stastny, J.: Comparison of learning algorithms. In: 24th Biennial Symposium on Communications, BSC 2008, pp. 231–234. Kingston (2008)
13. Hubalovsky, S.: Modeling and computer simulation of real process—solution of Mastermind board game. Int. J. Math. Comput. Simul. **6**(1), 107–118 (2012)
14. Mikulecky, P.: A need for knowledge management in intelligent outdoor spaces. In: 23rd International Business Information Management Association Conference, IBIMA 2014, pp. 527–531. Valencia (2014)

Performance Investigation of Radio Over Fiber Link with High-Speed Interface of IEEE 802.15.3c

Moussa El Yahyaoui, Mohammed Amine Azza and Ali Elmoussati

Abstract 60 GHz frequency band is candidate for future broadband wireless networks for the last mile because of high available of unlicensed bandwidth worldwide. However, 60 GHz frequency band has limited coverage due to the high free-space losses and the waves do not penetrate the walls. Radio over Fiber (RoF) technology combining both, the high capacity of optical communication and the flexibility of wireless access, can help in extending 60 GHz radio coverage in indoor environment. This paper presents a study of RoF link employing IEEE 802.15.3c Physical Layer (PHY) for 60 GHz frequency band. A simulation of 802.15.3c High-speed interface (HSI) mode in radio over fiber system using the co-simulation technique between OptiSystem and Simulink have been realized. Finally, a BER performance of IEEE 802.15.3c HSI mode in Radio over Fiber system versus the Multi-mode Fiber (MMF) length have been calculated.

Keywords Radio over fiber · IEEE 802.15.3c · 60 GHz · BER · MMF

1 Introduction

The evolution of wired technologies such as xDSL and Fiber to Home connections provides high broadband access network which now allows the delivery of services in the order of Gb/s. Current wireless systems have limited capacity e.g. 802.11a provide up to 54 Mb/s and 802.11n up to 400 Mb/s [1]. However, new indoor

M.E. Yahyaoui (✉) · M.A. Azza · A. Elmoussati
Signals, Systems and Information Processing Team, National School
of Applied Sciences Oujda, Oujda, Morocco
e-mail: mselyahyaoui@gmail.com

A. Elmoussati
e-mail: a.elmoussati@ump.ma

© Springer International Publishing Switzerland 2016
A. El Oualkadi et al. (eds.), *Proceedings of the Mediterranean Conference
on Information & Communication Technologies 2015*, Lecture Notes
in Electrical Engineering 380, DOI 10.1007/978-3-319-30301-7_8

wireless applications need for higher data throughput, such the wireless multimedia applications (IPTV, HDTV, in-room gaming) and kiosk applications. Data rates required is in range of several GB/s to transmit signals like uncompressed video streaming signals [2]. To support high data, new radio standards, as ECMA-387, IEEE 802.15.3c and IEEE 802.11ad, providing up to 7Gbit/s throughput using unlicensed 5 GHz of spectrum at 60 GHz are emerging. Transmission at 60 GHz can cover limited area, around 10–15 m in indoor environment due to free space path loss (loss over 1 m at 60 GHz is 68 dB). We investigate the use of radio over fiber solution to cover huge spaces or even building and large rooms.

Radio-over-Fiber (RoF) technology is becoming the promising solution for the broadband access networks because it can increase the capacity, coverage, and mobility and it decreases the cost of the base station [3]. In RoF systems, radio signals optically transported between the central management networks Home Communication Controller (HCC) and Radio Access Point (RAP), signal generation and processing are centralized in the HCC. RoF technology for micro cell and multiservice is becoming one of important technologies for the future communication systems [4, 5].

Wireless Personal Area Network WPAN IEEE 802.15.3c have been proposed for the future high data rate operates in the 60 GHz millimeter-wave band. The IEEE 802.15.3c support uncompressed video streaming at 1.78 or 3.56 GB/s and fast downloading in Kiosks [6]. Recently, numerous studies of RoF in millimeter-wave have being published [7–9]. BER performances of ieee 802.15.3c PHY for AWGN channel have been calculated in [8]. The generation of millimeter-wave techniques have been demonstrated in [10, 11]. This paper investigates the performances of applying the PHY layer of IEEE 802.15.3c in Radio over Fiber system at 60 GHz based on MMF fiber and Optical Carrier Suppression (OCS).

This paper has been organized as follows. We present the architecture of Radio over Fiber system, and then we introduce the IEEE 802.15.3c standard. Finally, we provide numerical results of the performance of the IEEE 802.15.3c in RoF system.

2   Radio Over Fiber Architecture

RoF network architecture employing IEEE 802.15.3c PHY is shown in Fig. 1 [5]. Home Communication Controller (HCC) is the central management unit for in-home networks. HCC is responsible for radio access control, signal processing, and RoF signal generation. Every room has one or more RAPs, which forwards the packets to its destinations. RAP contains only opto-electronic interfaces and RF modules and is connected to the HCC via optical fiber. There is three main Schemes of Signal Transport over fiber as shown in Fig. 2. The first Scheme consists of transposing the electrical radio signal onto an optical carrier for distribution over optical fiber. the second scheme consists in transposing the radio signal at an

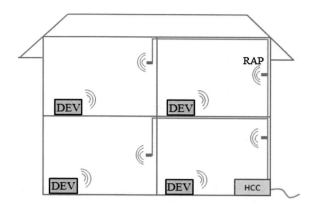

Fig. 1 RoF home network architecture

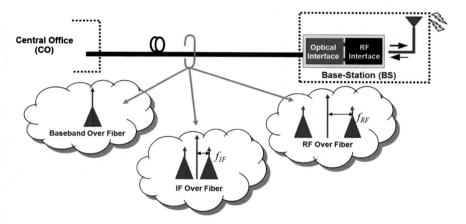

Fig. 2 Radio over fiber transport techniques

Intermediate Frequency (IF) before the optical transmission over fiber. The third scheme consist in transposing the base band signal onto an optical carrier for transmission over fiber.

2.1 RoF at Radio Frequency Architecture

RoF architecture realized in OptiSystem is shown in Fig. 3. The in-phase (I) and quadrature (Q) components are transposed to IF by mixing with an RF oscillator split in two paths with a 90 phase shift. The continuous-wavelength (CW) optical source of 850 nm wavelength is used to generate the optical carrier. the optical

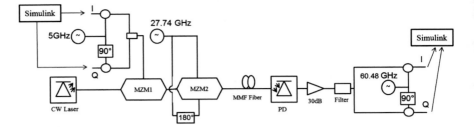

Fig. 3 Schematic diagrams RF architecture at 60.48 GHz

carrier is modulated by IF signal using Mach-Zehnder modulator MZM1, and then modulated by 27.75 GHz signal using dual arme Mach-Zehnder modulator MZM2 in order to delete the optical carrier [12]. The optical signal is transported by MMF fiber to the receiver. At the reception the PD convert the optical signal to electrical signal. The received signal amplified, filtered and demodulated to recover I and Q HSI OFDM signal.

2.2 802.15.3c HSI Model

The mmWave PHY is defined for the frequency band of 57.0–66.0 GHz, which consists of four channels. Three different PHYs modes are defined known as Single Carrier (SC PHY), High Speed Interface (HSI PHY), and Audio-Visual (AV PHY) modes. In our work, we have focused on HSI PHY. The HSI PHY is designed for NLOS operation and uses OFDM with an FEC based on Low-density Parity-check (LDPC) codes. In this usage model, all of the devices in the WPAN will have bidirectional, NLOS high speed, low-latency communication, which is provided for by the HSI PHY. OFDM technology is adopted in a large number of wireless standards as a mature technology. Figure 4 presents the block diagram of HSI PHY mode modeled in Simulink according to the standard 802.15.3c. The data bits

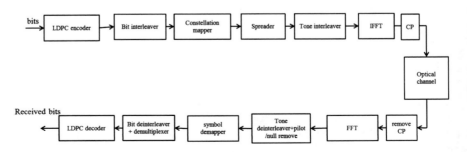

Fig. 4 HSI transmitter and receiver block diagram

Table 1 HSI OFDM PHY parameters

Description	Value
Reference sampling rate	2640 MHz
FFT size	512
Number of data subcarriers	336
Number of pilot subcarriers	16
Number of DC subcarriers	3
Nominal used bandwidth	1815 MHz
OFDM Symbol duration	218.18 ns

encoded by two LDPC encoders use 1/2 or 5/8 rate, and the output of LDPC encoder multiplexed to form a single data stream. After the data multiplexer, the bits are interleaved by a bloc interleaver of length of 2688 bits. The interleaved bits are mapped into a serial complex data using the QPSK, 16QAM or 64QAM modulation format. Each group of 336 complex numbers assigned to an OFDM symbol. The output data of the constellation mapper are then parallelized, and Pilots, dc and null tones are added up before the tone interleaver. Tone interleaver makes sure that neighboring symbols are not mapped into adjacent subcarriers. The interleaved tones are modulated by OFDM modulator that consists of 512-point IFFT. Finally, adding a cyclic prefix of 64 tones and transmitting the symbols in optical link.

The received signal from OptiSystem is removed from the cyclic prefix and then demodulated by OFDM that consist of 512-point FFT. After de-interleaving process, data carriers are identified and QPSK, 16QAM or 64QAM symbols are demodulated by the de-mapper block. The obtained bit stream is de-interleaved and demultiplixed into 2 bit streams to be decoded with 2 LDPC decoders. Table 1 presents the parameters of HSI OFDM system.

3 Results and Discussion

After the simulation of the complete system, IEEE 802.15.3c HSI mode with RoF link, we obtained results as presented below. HSI OFDM IF signal at 5 GHz is shown in Fig. 5. The 5 GHz IF signals were later transposed to optical carrier using MZM1 and MZM2 as shown in Fig. 6a. Figure 6b shows 60 GHz OFDM spectrum at receiver and Fig. 7 shows the recovered signal constellation diagram at the RF OFDM receiver, after propagating through 200 m MMF fiber length.

Figure 8 shows the BER performances of three MCS as shown in Table 2 versus MMF fiber length. As we can see in the graph, we have reached, 350, 295 and 210 m length of MMF fiber respectively for MCS1, MCS4 and MCS7 with BER below than 10^{-5}. Thinks to dual arm MZM enabling OCS that make able to reach long distances of MMF fiber.

Fig. 5 HSI OFDM spectrum at 5 GHz intermediate frequency

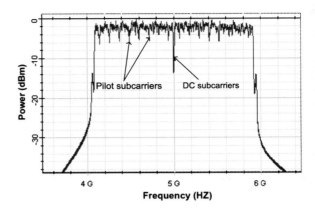

Fig. 6 Optical spectrum for double-sideband carrier suppression scheme after MZM2 (**a**). Electrical spectrum of HSI-OFDM signal at 60.48 GHz carrier frequency after 400 m MMF fiber Length at receiver unit (**b**)

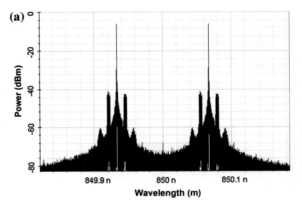

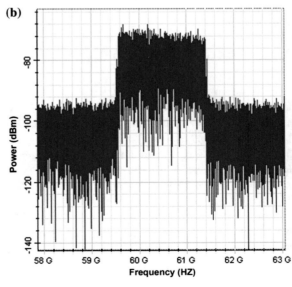

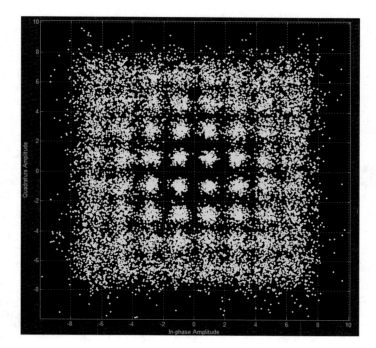

Fig. 7 Constellation diagram of 64QAM at the receiver with EVM = 14

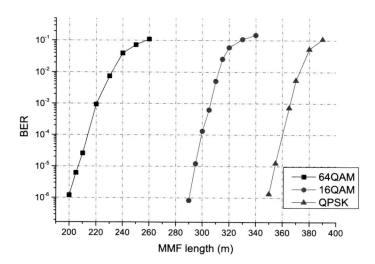

Fig. 8 BER performance in RF architecture

Table 2 HSI modulation and coding schemes parameters

MCS index	Data rate(Mb/s)	Modulation scheme	LDPC codes rate
1	1540	QPSK	1/2
4	3080	16QAM	1/2
7	5775	64QAM	5/8

4 Conclusion

In this paper, we established an ROF link with HSI PHY communication system in SIMULINK and OptiSystem. This have been done by ROF architecture based on MMF fiber and optical carrier suppression. We have calculated the BER performance of the system for various modulation schemes QPSK, 16QAM and 64QAM.

References

1. IEEE Standard for Information technology– local and metropolitan area networks—specific requirements—part 11: wireless LAN medium access control (MAC) and physical layer (PHY) specifications amendment 5: enhancements for higher throughput. IEEE Std 802.11n-2009 (Amendment to IEEE Std 802.11-2007 as amended by IEEE Std 802.11 k-2008, IEEE Std 802.11r-2008, IEEE Std 802.11y-2008, and IEEE Std 802.11w-2009), pp. 1565 (2009)
2. 802.11.ad-WLAN at 60 GHz, http://cdn.rohde-schwarz.com
3. Venkatesha Prasad, R., Quang, B., Chandra, K., An, X., Niemegeers, IGMM., Huong, N.: Analysing IEEE 802.15.3c protocol in Fi-Wi hybrid networks. In: Consumer Communications and Networking Conference (CCNC), 2013 IEEE, pp. 749–752 (2013)
4. Sampath Kumar, U., Saminadan, V., Williams, P.: Performance evaluation of Millimeter wave and UWB signals over Fiber radio networks. In: International Conference on Communications and Signal Processing (ICCSP), pp. 104–107 (2012)
5. Chandra, K., Venkatesha Prasad, R., Quang, B., Niemegeers, I.G.M.M., Abdur Rahim, M.D.: Analysis of Fi-Wi indoor network architecture based on IEEE 802.15.3c. In: 2014 IEEE 11th Consumer Communications and Networking Conference (CCNC) (CCNC 2014), pp. 113–118 (2014)
6. IEEE 802.15 TG3c Working Group: Part 15.3: Wireless Medium Access Control (MAC) and Physical Layer (PHY) Specifications for High Data Rate Wireless Personal Area Networks, (WPANs), http://www.ieee802.org/15/pub/TG3.html, 2009
7. Zhang, J.W., et al.: Experimental demonstration of 24-Gb/s CAP-64QAM radio-over-fiber system over 40-GHz mm-wave fiber-wireless transmission. Opt. Express **21**(22), 26888–26895 (2013)
8. Liso Nicolas, M., Jacob, M., Kurner, T.: Physical layer simulation results for IEEE 802.15.3c with different channel models. In: Advances in Radio Science, vol 9, pages 173–177 (2011)
9. Chandra, K., Venkatesha Prasad, R., Quang, B., Niemegeers, I.G.M.M., Abdur Rahim, M.D.: Analysis of Fi-Wi indoor network architecture based on IEEE 802.15.3c. In: 2014 IEEE 11th Consumer Communications and Networking Conference (CCNC)(CCNC 2014), pp. 113–118 (2014)

10. Stohr, A., Jger, D.: Photonic millimeter-wave and terahertz source technologies (invited paper). In: Jdger, D. (ed.) International Topical Meeting on Microwave Photonics, 2006. MWP '06, pp. 14 (2006)
11. Beas, J., et al.: Millimeter-wave frequency radio over fiber systems: a survey. Commun. Surv. Tutorials, IEEE **15**(4), 1593–1619 (2013)
12. Jia, Z., Yu, J., Chang, G.K.: A full-duplex radio-over fiber system based on optical carrier suppression and reuse. IEEE Photon. Technol. Lett. **18**(16), 1726–1728 (2006)

Impact of EIRP Constraint on MU-MIMO 802.11ac Capacity Gain in Home Networks

Khouloud Issiali, Valéry Guillet, Ghais El Zein
and Gheorghe Zaharia

Abstract In this paper, we evaluate a downlink Multi-User Multiple-Input Multiple-Output (MU-MIMO) scenario, in which a 802.11ac access point with multiple antennas (up to 10) is transmitting to two receivers, each one with two antennas. Block diagonalization (BD) method is investigated under the Equivalent Isotropic Radiated Power (*EIRP*) constraint. This study shows that scaling the transmitted power according to the *EIRP* constraint can improve the multi-user (MU) sum capacity to single-user (SU) capacity ratio compared to the gain achieved under the transmitted power constraint.

Keywords MU-MIMO capacity · IEEE 802.11ac · Home Networks · EIRP

1 Introduction

MU-MIMO techniques, proposed to increase the throughput, consist in applying a linear precoding to the transmitted (Tx) spatial streams. Thus, the antenna array pattern and gain are modified as functions of the user location and channel properties. This directly impacts the *EIRP*. In Europe, the *EIRP* is limited at 5 GHz band to 200 mW or 1 W depending on the propagation channels. This constraint may differ in other countries where it is rather based on the total Tx power.

K. Issiali · V. Guillet (✉)
Orange Labs, Belfort, France
e-mail: Valery.Guillet@orange.com

K. Issiali
e-mail: Khouloud.Issiali@orange.com

G.E. Zein · G. Zaharia
IETR—INSA, UMR 6164, Rennes, France
e-mail: Ghais.El-Zein@insa-rennes.fr

G. Zaharia
e-mail: Gheorghe.Zaharia@insa-rennes.fr

© Springer International Publishing Switzerland 2016
A. El Oualkadi et al. (eds.), *Proceedings of the Mediterranean Conference on Information & Communication Technologies 2015*, Lecture Notes in Electrical Engineering 380, DOI 10.1007/978-3-319-30301-7_9

The *EIRP* constraint is rarely evaluated for MIMO systems. Often, the packet error rate and the capacity value are evaluated based on the same total Tx power (P_{Tx}) which is a function of the Signal to Noise Ratio (*SNR*), commonly defined as the ratio of P_{Tx} to the average noise power. The propagation channel is usually normalized to have an average path loss of 0 dB. Few recent studies have focused on the capacity optimization problems under total Tx power constraint [1–3]. Sometimes, this optimization is performed on each subcarrier of the 802.11 OFDM signal [1]. In [4], a new *EIRP*-based solution for IEEE 802.11 power scaling is proposed. However, this study is dedicated to only one user system with a single spatial stream.

The MU-MIMO linear precoding, like BD [2, 5], modifies dynamically the antenna array pattern and gain. This changes the *EIRP* of the Tx antenna array if Tx power remains unchanged. MU-MIMO and Transmit Beamforming (TxBF) are commonly associated with a large number of Tx antennas used to improve the antenna array gain and performance, as previously stated for narrowband i.i.d. Rayleigh SISO channels forming the MIMO channel in [3, 5]. In the case of the *EIRP* constraint, it may not be evident that TxBF and MU-MIMO linear precoding still improve the system performance.

Therefore, this paper evaluates the impact of the *EIRP* constraint on 802.11ac MU-MIMO capacity gain using simulations. Two different power allocation schemes of the spatial streams are analyzed to optimize MU-MIMO capacity: equal and unequal power repartition under the same *EIRP* constraint. A typical indoor residential environment is evaluated based on the IEEE TGac correlated channels [6, 7]. Comparisons are given versus an i.i.d. Rayleigh channel.

The rest of this paper is organized as follows. Section 2 describes the system model and briefly presents the BD algorithm. The problem formulation to compute the sum capacity for MU-MIMO system under the *EIRP* constraint is given in Sect. 3. Section 4 describes the simulation process and presents the simulation results with analysis. Finally, the conclusion is drawn in Sect. 5.

Hereafter, superscripts $(.)^t$, $(.)^*$ and $(\bar{\cdot})$ denote transposition, transpose conjugate and complex conjugate, respectively. Expectation (ensemble averaging) is denoted by E(.). The Frobenius norm of a matrix is written as $|| \cdot ||$.

2   System Model

The studied 802.11ac MU-MIMO system is composed of K users connected to one Access Point (AP), as shown in Fig. 1. The AP has n_T antennas and each user k has n_{R_k} antennas.

We define $n_R = \sum_{k=1}^{K} n_{R_k}$. The $L_k \times 1$ (where L_k is the number of parallel symbols Tx simultaneously for the kth user) transmit symbol vector s_k is pre-processed at the AP before being transmitted.

For each 802.11ac OFDM subcarrier, the received signal at the kth receiver is:

Fig. 1 Diagram of
MU-MIMO system

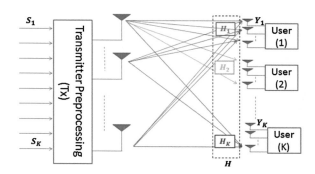

$$y_k = H_k W_k s_k + H_k \sum_{i=1,i\neq k}^{K} W_i s_i + n_k \qquad (1)$$

where H_k is the $n_{R_k} \times n_T$ channel matrix for the kth receiver, W_k is the $n_T \times L_k$ BD precoding matrix resulting in an $n_T \times L$ $(L = L_1 + L_2 + \cdots + L_K)$ precoding matrix $W = [W_1, \ldots, W_K]$, n_k is the Gaussian noise vector $(E(n_k n_k^*) = \sigma_n^2 I_{n_{R_k}})$.

BD [3, 5] decomposes the MU-MIMO downlink into K parallel independent SU-MIMO downlinks. The BD consists first of perfectly suppressing the inter-user interference IUI = $H_k \sum_{i=1,i\neq k}^{K} W_i s_i$ in order to have parallel SU-MIMO systems. Then, a classic TxBF is applied to optimize the capacity for each user [8]. In this study, $K = 2$ and perfect knowledge of channel state information is assumed at the transmitter. The channel model specified for the 802.11ac standard within the TGac task group [7] is selected. This model takes into account realisticTx and Rx correlations, contrary to an i.i.d. Rayleigh channel. It is based on a cluster model [6] amended by the TGac task group for the IEEE 802.11ac standard. The TGac modifications concern the power angular spectrum to allow MU-MIMO operation and are summarized as follows [7]:

- The TGn azimuth spread for each cluster remains the same for all users.
- For each user and for all taps, independent random offsets are introduced: between ±180° for the angles of arrival (AoA) and between ±30° for the angles of departure (AoD).

In this study, a typical home network is evaluated by using the channel model TGac-B (15 ns RMS delay spread) for the 5.25 GHz frequency band. Rayleigh fading is exhibited for each one of the 9 uncorrelated taps, except for the Line Of Sight (LOS) tap which follows a Rice fading with a 0 dB Rician factor. This study focuses on the TGac-B NLOS channel model. Similar results are obtained with the TGac-B LOS channel model as the 0 dB Rician factor does not display significantly different results from the TGac-B NLOS channel model. For each 802.11ac OFDM subcarrier, the channel matrix is computed through a discrete Fourier transformation (size: 56 subcarriers) of the tap delay representation. For comparison, an i.i.d.

Rayleigh channel is also evaluated. We apply the common normalization $E\left(||H||^2\right) = n_T n_R$ for each subcarrier which means an average propagation loss equal to 0 dB.

3 Problem Statement

3.1 Usual Definitions

The MU-MIMO system is decomposed into K independent SU-MIMO systems by applying the BD algorithm. For each one of the 802.11ac OFDM subcarriers, the MU-MIMO sum capacity is expressed as follows [3]:

$$C_{BD} = \sum_{k=1}^{K} \sum_{i=1}^{n_{R_k}} \log_2(1 + \frac{p_{ik}}{\sigma_n^2} \mu_{ik}^2) \tag{2}$$

where p_{ik} is the power dedicated to the ith antenna for the kth user, μ_{ik}^2 are the eigenvalues of the effective channel for the kth user after applying the IUI cancellation [2] and σ_n^2 is the noise power. The subcarrier index is not mentioned throughout this paper in order to simplify the notations, but since C_{BD} is related to H, C_{BD} depends on each OFDM subcarrier. For the corresponding SU-MIMO systems and for relevant comparisons with MU-MIMO, the number of antennas n_T and n_R remains unchanged. The considered SU-MIMO system applies a singular value decomposition and its capacity C_{SU} is computed as detailed in [8] for each OFDM subcarrier.

3.2 EIRP in Linear Precoding

For any receiver location, i.e. for any H matrix, the transmit antenna array pattern is modified by the W precoding matrix. We have used a linear array of omnidirectional 0 dBi gain antennas with regular spacing δ, typically $\delta = \lambda/2$. The transmitter antenna array manifold $a(\theta)$, is a function of the θ angle with the array axis:

$$a(\theta)^t = \left(1, e^{-2j\pi\delta\cos(\theta)/\lambda}, e^{-4j\pi\delta\cos(\theta)/\lambda}, \ldots, e^{-2j\pi(n_T-1)\delta\cos(\theta)/\lambda}\right). \tag{3}$$

Due to the used TGac channel model, the antenna array pattern is simplified to a 2D problem. The average radiated power $d(\theta)$ in any direction θ relative to the antenna array direction is expressed as a function of the input signals $x^t = (x_1, \ldots, x_{n_T})$:

$$d(\theta) = E\left(\left|a(\theta)^t x\right|^2\right). \tag{4}$$

With $x = Ws$, $d(\theta)$ is expressed as a function of $B = E(\overline{S}S^t) = diag(p_{ik})$:

$$d(\theta) = a(\theta)^* \overline{W} B W^t a(\theta). \tag{5}$$

Considering all the subcarriers of the system, the total radiated power $d(\theta)$ is:

$$d_{total}(\theta) = \sum_{subcarrier} d(\theta), \tag{6}$$

and the *EIRP* is:

$$EIRP = \max_{\theta} d_{total}(\theta) \tag{7}$$

If the total power P_{Tx} is equally shared among N_{SS} spatial streams, $p_{ik} = \frac{P_{Tx}}{NN_{SS}}$, where $N = 56$ is the number of subcarriers in 802.11ac. Thus:

$$d(\theta) = \frac{P_{Tx}}{NN_{SS}} a(\theta)^* \overline{W} \, B W^t a(\theta). \tag{8}$$

3.3 Optimization Problems

In order to find the optimal value of the Tx power p_{ik} compatible with the *EIRP* constraint, two power allocation schemes are evaluated: equal power allocation and unequal power allocation. This paper focused on the case where each subcarrier has the same allocated total Tx power. Furthermore, an unequal subcarrier power allocation may not have a favorable impact on the peak-to-average power ratio of the OFDM signal. The general optimization problem is thus expressed for each subcarrier as:

$$\max_{p_{ik}} \sum_{k=1}^{K} \sum_{i=1}^{n_{R_k}} \log_2 \left(1 + \frac{p_{ik}}{\sigma_n^2} \mu_{ik}^2\right) \text{ and } EIRP \leq 23 \text{ dBm.} \tag{9}$$

For the case with equally distributed powers, i.e. $p_{ik} = \frac{P_{Tx}}{NN_{SS}}$, the problem has only one variable P_{Tx}. It is simplified by seeking the maximum antenna array gain $\frac{d(\theta)}{P_{Tx}}$ and then scaling the power according to the *EIRP* limit. For the genaral case, the optimization is performed using a Matlab-based modeling for convex optimization namely CVX [9].The case $K = 1$ uses the same optimization method for computing the SU-MIMO capacity for both equal and unequal power allocation under *EIRP* constraint.

3.4   Evaluated Systems and SNR Considerations

802.11ac MU-MIMO systems based on BD schemes are evaluated. The results are presented in Sect. 4 and compared to SU-MIMO systems relying on the same antennas and total power or *EIRP* constraint. Three capacity optimization techniques are evaluated and compared. The first one is the usual MIMO system (denoted *basic*), with a constant Tx power P_{Tx} equally shared among the spatial streams. *BD-basic* and *SU-basic* denote the corresponding studied systems. For this case, the average signal to noise ratio is defined as $SNR = \frac{P_{Tx}}{N\sigma_n^2}$. This is the common *SNR* definition.

The second optimization labelled *eirp-equal* considers a 23 dBm *EIRP* constraint and a total Tx power equally shared among the spatial streams. A dynamic power scaling is applied, as a function of each channel matrix snapshot H. *SUeirp-equal* and *BDeirp-equal* denote respectively the corresponding SU and MU systems.

The third one (*eirp-unequal*) considers a 23 dBm *EIRP* constraint and a total Tx power unequally and dynamically shared among the spatial streams. *SUeirp-unequal* and *BDeirp-unequal* denote the corresponding systems applying this technique.

For *eirp-equal* and *eirp-unequal* systems, the common *SNR* definition is biased as P_{Tx} is no more constant, and depends on each channel matrix computation. Under *EIRP* constraint, we define $SNR_{\text{eirp}} = \frac{EIRP}{N\sigma_n^2}$ for *eirp-equal* and *eirp-unequal* systems. Note that the maximum antenna array gain is n_T. Since $SNR = \frac{P_{Tx}}{N\sigma_n^2}$ for a *basic* system, it implies that its corresponding SNR_{eirp} value is upper bounded by $\frac{n_T P_{Tx}}{N\sigma_n^2}$.

4   Simulation Results and Analysis

The simulated system is composed of one access point equipped with multiple antennas (linear array of 0 dBi omnidirectional and vertically polarized antennas), and two receivers. Each receiver has two 0 dBi omnidirectional antennas. The antenna spacing is 0.5λ. A Matlab source code [10] was used to compute the 802.11ac TGac-B channel samples over a 20 MHz bandwidth. To have representative results, 100 couples of users ($K = 2$) are randomly drawn around the access point following the IEEE TGac recommendations [9]. For each drawing, we use a simulation length equal to 55 coherence times of the MIMO channel to simulate the fading. By setting the "Fading Number of Iterations" in the Matlab channel model to 512, 488 interpolated channel samples are collected for each couple of users to simulate 10 fading periods.

4.1 Results for Equal Power Allocation

Figures 2 and 3 display the MU-MIMO to SU-MIMO capacity ratio for *basic* and *eirp-equal* systems. Average values, 10 and 90 % quantiles (q_{10} and q_{90}) are represented to estimate the confidence intervals. *SNR* = 20 dB and n_T varies from 4 to 10. They show that the MU to SU-MIMO capacity ratio increases with n_T for TGac-B and Rayleigh channels. The ratio changes from 1.2 to 1.77 for the *eirp-equal* system in a residential environment, which is more than 50 % of capacity gain. Note that the gain without the *EIRP* constraint is around 45 %.

It has been shown in [3, 5] that increasing n_T favorably impacts the capacity gain on an i.i.d. Rayleigh channel under *SNR* constraint. We have been able to prove that this result holds even under the *EIRP* constraint and with correlated channels as in TGac models. The difference $q_{90}-q_{10}$ decreses if n_T increases. This shows that fading has less impact on the capacity values.

The *basic* and *eirp-equal* comparisons are biased. In fact, a system relying on a total Tx power does not satisfy a constant *EIRP* constraint since it may have an increasing *EIRP* as n_T increases. We could expect that for a *basic* system, the MU-MIMO to SU-MIMO capacity ratio increases more rapidly in function of n_T than for an *eirp-equal* system, but simulations prove the opposite. SU-MIMO takes advantage of the power P_{Tx} when the system is not under *EIRP* constraint. For instance in our simulated case ($K = 2$) where $N_{SS} = 4$ for MU-MIMO and $N_{SS} = 2$ for each one of the single users, the *EIRP* reached by the MU-MIMO system

Fig. 2 MU to SU capacity ratio for an IEEE TGac-B channel (residential)

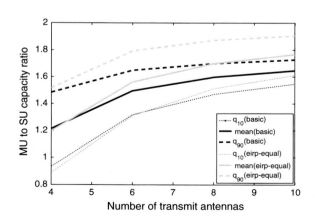

82 K. Issiali et al.

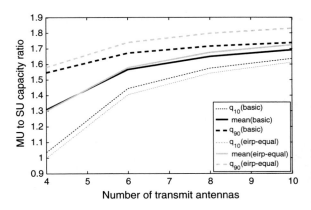

Fig. 3 MU to SU capacity ratio for an i.i.d. Rayleigh channel

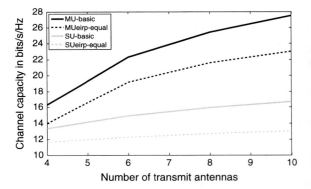

Fig. 4 Capacity value achieved by the *basic* and *eirp-equal* systems

$(EIRP_{MU})$ is expressed as $EIRP_{MU} = \frac{P_{Tx}}{NN_{SS}}[\max_{\theta}(a(\theta)^*\overline{W}W^t a(\theta))]$ and is upper bounded by $n_T \frac{P_{Tx}}{4}$.

Similarly, for the same P_{Tx}, the single user *EIRP* is upper bounded by $n_T \frac{P_{Tx}}{2}$. This means that under the same *EIRP* constraint, the allocated power tends to be lower for SU-MIMO than for MU-MIMO. For the *basic* system, the allocated power is the same for SU-MIMO and MU-MIMO. Figure 4 shows the average capacity value for MU and SU. It is well observed that the SU capacity increases rapidly with n_T.

4.2 Impact of Power Allocation Strategy

The probability when the MU-MIMO capacity is lower than the SU-MIMO capacity is illustrated in Fig. 5 for *eirp-equal* and *eirp-unequal* schemes versus n_T.

We also observe that the MU-MIMO capacity gain for unequal repartition is slightly greater than the one observed for a fair power distribution. Nevertheless, the

Fig. 5 $\text{Proba}(C_{BD}/C_{SU} \leq 1)$ versus the number of transmit antennas

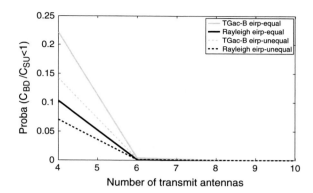

gain is not significant: we have around 3 % of capacity gain by contrast to high computational complexity. The probability is almost 0 ($\leq 1\%$) for $n_T = 6$ and is equal to 0 for $n_T \geq 8$. These results can be explained by examining the overall system: for $n_T = 4$, the MU-MIMO system is composed of 4 antennas in the transmit and the receive sides with 4 spatial streams. This gives no diversity possibilities, $N_{SS} < n_T$ when n_T increases. As a result, the system takes benefit from transmit diversity with probabilities which tend to 0.

5 Conclusion

This paper has analyzed the capacity optimization for 802.11ac MU-MIMO systems with multiple spatial streams for each user under the EIRP constraint. A typical home environment and correlated 802.11ac channels were considered. Two transmit power allocation methods have been evaluated: equal and unequal repartition based under the same EIRP constraint. These two strategies were compared with the more common MU-MIMO under the total Tx power constraint. It is shown under EIRP constraint, that the number of transmit antennas must be larger than the total number of spatial streams to guarantee a MU-MIMO capacity gain over SU-MIMO.

References

1. Saavedra, J.: Multidiffusion et diffusion dans les systèmes OFDM sans fil. *Thèse de doctorat d'état, Univ. Paris-Sud, Paris, France*, (2012)
2. Zhao, L., Wang, Y., Charge, P.: Efficient power allocation strategy in multiuser MIMO broadcast channels. PIMRC, 2591–2595, (2013)
3. Spencer, Q.H., Swindlehurst, A.L., Haardt, M.: Zero-forcing methods for downlink spatial multiplexing in multiuser MIMO channels. IEEE Trans. Sig. Proc. **52**(2), 461–471 (2004)

4. Kuzminskiy, A.M.: Downlink beamforming under EIRP constraint in WLAN OFDM systems. Sig. Process. **87**(5), 991–1002 (2007)
5. Choi, L.-U., Murch, R.D.: A transmit preprocessing technique for multiuser MIMO systems using a decomposition approach. IEEE Trans. Wireless Comm. **3**(1), 20–24 (2004)
6. Breit, G., Sampath, H., Vermani, S. et al.: TGac channel model addendum support material. Mentor IEEE, Doc IEEE 802.11-09/06/0569r0, 2009
7. Erceg, V., Schumacker, L., P. Kyritsi, P. et al.: TGn channel models. Mentor IEEE, Doc IEEE 802 11-03/940r4, 2004
8. Bouhlel, A., Guillet, V., El Zein, G., Zaharia, G.: Transmit beamforming analysis for MIMO systems in indoor residential environment based on 3D ray tracing. In: Springer Wireless Personal Communications, pp. 1–23, (2014)
9. Grant, M., Boyd, S.: CVX: Matlab software for disciplined convex programming, (2014)
10. Schumacher L., Dijkstra, B.: Description of a Matlab implementation of the indoor MIMO WLAN channel model proposed by the IEEE 802.11 TGn Channel Model Special Committee. Implementation note, version 3.2, (2004)

Joint Channel Maximization and Estimation in Multiuser Large-Scale MIMO Cognitive Radio Systems Using Particle Swarm Optimization

Mostafa Hefnawi

Abstract This paper investigates the use of particle swarm optimization (PSO) for joint channel maximization and estimation in large-scale multiuser multiple-input multiple-output (LS-MU-MIMO) cognitive networks with multiple primary users (PUs) and secondary users (SUs) sharing the same spectrum. The PSO algorithm in this paper is applied at two levels; the mobile station (MS) and the base station (BS). At the MS, PSO is used to seek iteratively the transmit beamforming weights that maximize the uplink MIMO channel capacity for each cognitive user, while controlling the interference levels to PUs without involving any gradient search. At the BS, PSO is used for channel estimation without involving any matrix inversion. The performance of the PSO-based capacity-aware (PSO-CA) cognitive system is compared to the one based on the gradient search scheme (GS-CA) and the results show that PSO-CA requires considerably less computational complexity while achieving essentially the same level of performance as the GS-CA.

1 Introduction

Recently, the use of multiuser large-scale multiple input multiple output (MU-LS-MIMO) systems have gained significant research attentions and have been considered as a promising candidate for the new 5th generation of wireless systems [1, 2]. This motivation is justified by their potential of realizing theoretically predicted MIMO benefits such as very high spectral efficiencies, increased reliability, and power efficiency. On the other hand, orthogonal frequency-division multiplexing (OFDM) based cognitive radio networks (CRNs) have been proposed as the essential technology needed to make significantly better use of available spectrum via "opportunistic use" of spectrum resources [3, 4]. However, opportunistic spectrum sharing may not be reliable and may limit the system capacity since it

M. Hefnawi (✉)
Electrical and Computer Engineering Department, Royal Military College of Canada,
Kingston, ON, Canada
e-mail: hefnawi@rmc.ca

© Springer International Publishing Switzerland 2016
A. El Oualkadi et al. (eds.), *Proceedings of the Mediterranean Conference on Information & Communication Technologies 2015*, Lecture Notes in Electrical Engineering 380, DOI 10.1007/978-3-319-30301-7_10

85

suffers from the interruptions imposed by the primary network (PN) on the secondary network (SN) that must leave the licensed channel when primary users (PUs) emerge. Also, with opportunistic spectrum sharing, secondary users (SUs) can still cause interference to PUs due to their imperfect spectrum sensing. One way to overcome these limitations is to incorporate LS-MU-MIMO into OFDM-based CRNs, which will help achieving higher spectral efficiency by multiplexing multiple users on the same time-frequency resources and allowing concurrent spectrum sharing instead of opportunistic sharing. OFDM-based MU-MIMO techniques has been successfully deployed in 3G/4G cellular systems based on traditional static spectrum access approach [5, 6] and a vast number of multi-user detection algorithms, such as maximum ratio combining (MRC), minimum mean-squared error (MMSE), are presently being tailored towards solving the MIMO processing in cognitive networks [7–10], where additional constraints to protect licensed users' QoS are imposed. Most of these techniques have used Lagrange multiplier and gradient search algorithm in order to solve the constrained optimization problem. Within this context, we have recently developed a capacity-aware (CA) based MU-MIMO-OFDM scheme [10] for cognitive radio that seeks iteratively the optimal transmit weight vector using the steepest ascent gradient of the cognitive MU-MIMO-OFDM channel capacity. We have shown that the proposed algorithm is able to enhance the overall system capacity for each SU and that under the assumption of large number of antennas and perfect channel state information (CSI), close to optimal performances can be achieved with the simplest forms of user detection and beamforming, i.e., MRC. These gradient-based methods, however, suffers local minima and require the constrained channel capacity to be differentiable, which might not be fulfilled in general. Also, when considering massive MIMO, traditional methods for channel estimation such as MMSE become challenging. These channel estimators basically involve matrix inversions, which become very computationally expensive in large-scale MIMO systems. Therefore, in this paper, we investigate the use of particle swarm optimization (PSO), which is an easy-to-use derivative-free algorithm and can perform global optimization. PSO was initially introduced by Kennedy and Eberhart in [11] and has received a lot of attention in recent years. It is an evolutionary computation technique inspired by swarm intelligence such as fish schooling and bird flocking looking for the best food spot (exploring the optimal solution) in the search space where a quality measure, fitness, can be evaluated without any a priori knowledge. This bio-inspired algorithm is particularly interesting because of its fast convergence, low complexity and simple implementation. The PSO algorithm in this paper will be applied at two levels; the mobile station (MS) and the base station (BS). At the MS, PSO is used to seek iteratively the transmit beamforming weights that maximize the UL MIMO channel capacity for each cognitive user, while controlling the interference levels to PUs without involving any gradient search. At the BS, PSO is used for channel estimation without involving any matrix inversion. The performance of the proposed PSO-based capacity-aware (PSO-CA) is compared to the one based on the gradient search scheme (GS-CA) in terms of channel capacity and symbol error rate.

2 System Model

We consider the UL multiuser access scenario shown in Fig. 1 where L_s SUs and one secondary base station (SBS) coexist with L_p PUs and one primary base station (PBS) via concurrent spectrum access. The users in both networks and the base stations are equipped with multiple antennas. It is also assumed that both the SBS and the PBS receivers detect independent OFDM data streams from multiple SUs and PUs simultaneously on the same time-frequency resources. Let $\mathbf{x}^s[k] = \left\{x_1^s, x_1^s, \ldots, x_{L_s}^s\right\}$ and $\mathbf{x}^p[k] = \left\{x_1^p, x_1^p, \ldots, x_{L_p}^p\right\}$ denote, respectively, the set of L_s SUs signals and L_p PUs signals transmitted on each subcarrier, $k = 1, \ldots, N_c$, where N_c denotes the number of subcarriers per OFDM symbol in the system. It is assumed that x_i^s and x_i^p are complex-valued random variables with unit power, i.e., $E\left[\|x_i^s\|^2\right] = E\left[\|x_i^p\|^2\right] = 1$.

The expression for the array output of the SBS in Fig. 1 can be written for each subcarrier as

$$\mathbf{y}_{SBS}[k] = \sum_{l_s=1}^{L_s} \mathbf{H}_{ss,l_s}[k]\mathbf{w}_{l_s}^t[k]\mathbf{x}_{l_s}^s[k] + \mathbf{n}[k] + \mathbf{I}_{PU}[k], \tag{1}$$

where $\mathbf{y}_{SBS}[k] = \left[y_1^s[k], y_2^s[k], \ldots, y_{N_s^r}^s[k]\right]^T$ is the $N_s^r \times 1$ vector containing the outputs of the N_s^r—element array at the SBS, with $(.)^T$ denoting the transpose

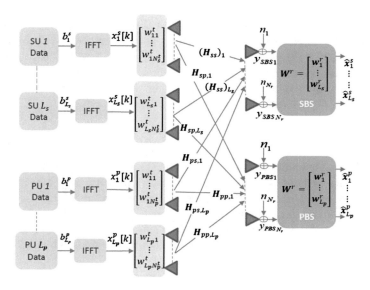

Fig. 1 MIMO-OFDM based cognitive radio system

operation, $\mathbf{H}_{ss,l_s}[k]$ is the $N_s^r \times N_s^t$ frequency-domain channel matrix representing the transfer functions from secondary user l_s's N_s^t—element antenna array to the SBS's N_s^r—element antenna array, $\mathbf{w}_{l_s}^t[k] = \left[[w_{l_s,1}^t[k], w_{l_s,2}^t[k]\ldots, w_{l_s,N_s^t}^t[k]] \right]^{\mathrm{T}}$ is the $N_s^t \times 1$ complex transmit weight vector for SU l_s, $l_s = 1,\ldots,L_s$, $\mathbf{n}[k] = [n_1[k], n_2[k]\ldots, n_{l_s}[k]]^{\mathrm{T}}$ is the $N_s^r \times 1$ complex additive white Gaussian noise vector, and $\mathbf{I}_{PU}[k]$ represents the interference introduced by PUs at the SBS, given by

$$\mathbf{I}_{PU}[\mathbf{k}] = \sum_{l_p=1}^{L_p} \mathbf{H}_{ps,l_p}[k]\mathbf{w}_{l_p}^t[k]\mathbf{x}_{l_p}^P[k], \tag{2}$$

where $\mathbf{H}_{ps,l_p}[k]$ is the $N_s^r \times N_p^t$ channel matrix representing the fading coefficients from PUs to the SBS's N_s^r—element antenna array. On the other hand, the interference power seen by the primary base station due to secondary transmission is given by

$$J_{sp}[k] = \sum_{l_s=1}^{L_s} \mathbf{H}_{sp,l_s}[k]\mathbf{w}_{l_s}^t[k]\mathbf{w}_{l_s}^{t,H}[k]\mathbf{H}_{sp,l_s}^H[k], \tag{3}$$

where $(.)^{\mathrm{H}}$ denotes the Hermitian transpose and $\mathbf{H}_{sp,l_s}[k]$ is the $N_p^r \times N_s^t$ channel matrix representing the fading coefficients from the l_s-th SU to the PBS's N_p^r—element antenna array. It is also assumed that $\mathbf{H}_{ss,l_s}[k] \in \mathbb{C}^{N_s^r \times N_s^t}$, $\mathbf{H}_{sp,l_s}[k] \in \mathbb{C}^{N_p^r \times N_s^t}$, $\mathbf{H}_{ps,l_p}[k] \in \mathbb{C}^{N_s^r \times N_p^t}$, and $\mathbf{H}_{pp,l_s}[k] \in \mathbb{C}^{N_p^r \times N_p^t}$ consist of iid (independent and identically distributed) complex Gaussian entries, with zero mean and unit variance. The transfer functions from the l_s-th SU device to the SBS antenna array (the cascade of $\mathbf{H}_{ss,l_s}[k]$ and $\mathbf{w}_{l_s}^t[k]$), result in a unique spatial signature for each SU, which can be exploited to effect the separation of the user data at the SBS using appropriate multiuser detection techniques. The SBS detects all L_s SUs, simultaneously at the multiuser detection module of the SDMA system, by multiplying the output of the array with the $N_s^r \times 1$ receiving weight vectors as follows

$$\hat{x}_{l_s}[\mathbf{k}] = \mathbf{w}_{l_s}^{r,H}[k]\mathbf{y}_{SBS}[k], \tag{4}$$

During the simulation, pilot-based channel estimation with time-division duplex (TDD) MIMO systems is assumed. It is also assumed that each MS employs an orthogonal pilot sequence, so that no pilot contamination (no interference between pilots) is present in the system. Pilot contamination is a well-known problem in pilot-aided channel estimation and it happens when other users in the system are reusing the same set of pilot signals due to the limitation imposed by the number of available orthogonal pilots [12].

3 PSO-Based Channel Maximization

Our objective is to find the optimal beamforming vector, $\left(\mathbf{w}_{l_s}^t[k]\right)_{opt}$, that maximizes the Ergodic capacity of the cognitive MU-MIMO-OFDM channel for each SU l_s of the SN imposing the following two sets of constraints: (1) each secondary user l_s has a limited maximum transmission power equal to $P_{max,l}^t$ and (2) the total maximum interference power at the PBS from the SUs does not exceed the maximum power constraint of $J_{sp}^{\max}$. In mathematical terms, these two constraints are expressed as follows [13]:

$$
\begin{array}{c}
\max_{\mathbf{w}_{l_s}^t[k]} \left[\mathrm{E}\left(\log_2\left\{ \mathbf{I} + \frac{\rho_{l_s}}{N_s^t} \left| B_{l_s}^{-1/2}[k]\tilde{\mathbf{H}}_{ss,l_s}^t[k] \right|^2 \right\} \right) \right] \\[2mm]
\text{Subject to:} \left\{ \begin{array}{c}
\mathbf{w}_{l_s}^{t,H}[k]\mathbf{w}_{l_s}^t[k] \leq P_{max,l_s}^t \\[2mm]
J_{sp} = \sum_{i=1}^{L_s} \tilde{\mathbf{H}}_{sp,l_s}^t[k]\tilde{\mathbf{H}}_{sp,l_s}^{t,H}[k] \leq J_{sp}^{\max}
\end{array} \right\},
\end{array}
\tag{5}
$$

where E [.] denotes the expectation operator,

$$
\tilde{\mathbf{H}}_{ss,l_s}^t[k] = \mathbf{H}_{ss,l_s}[k]\mathbf{w}_{l_s}^t[k], \quad \tilde{\mathbf{H}}_{sp,l_s}^t[k] = \mathbf{H}_{sp,l_s}[k]\mathbf{w}_{l_s}^t[k],
$$

$$
\tilde{\mathbf{H}}_{ps,l_p}^t[k] = \mathbf{H}_{ps,l_p}[k]\mathbf{w}_{l_p}^t[k], \quad \mathbf{B}_{l_s}[k] = \mathbf{B}_{ss}[k] + \mathbf{B}_{ps}[k] + \sigma_n^2\mathbf{I}_{N_s^r}[k],
$$

$$
\mathbf{B}_{ss}[k] = \sum_{i=1,i\neq l_s}^{L_s} \tilde{\mathbf{H}}_{ss,l_s}^t[k]\tilde{\mathbf{H}}_{ss,l_s}^{t,H}[k], \text{ and } \mathbf{B}_{ps}[k] = \sum_{l_p=1}^{L_p} \tilde{\mathbf{H}}_{ps,l_s}^t[k]\tilde{\mathbf{H}}_{ps,l_s}^{t,H}[k].
$$

This problem is a constrained optimization problem, which is highly non-convex and complicated to solve. However, a sub-optimal solution can be obtained by exploiting the method of Lagrange multipliers as follows:

$$
\begin{aligned}
\mathcal{L}\left(\mathbf{w}_{l_s}^t, v_{l_s}, \lambda_{l_s}\right) =& \mathrm{E}\left(\log_2\left\{ \mathbf{I} + \frac{\rho_{l_s}}{N_s^t} \left| \mathbf{B}_{l_s}[k]^{-\frac{1}{2}}\tilde{\mathbf{H}}_{ss,l_s}^t[k] \right|^2 \right\} \right) \\
& - v_{l_s}\left(\frac{\sum_{l_s=1}^{L_s} \tilde{\mathbf{H}}_{sp,l_s}^t[k]\, \tilde{\mathbf{H}}_{sp,l_s}^{t,H}[k]}{J_{sp}^{max}} - 1 \right) \\
& - \lambda_{l_s}\left(\frac{\mathbf{w}_{l_s}^{t,H}[k]\mathbf{w}_{l_s}^t[k]}{P_{max,l_s}^t} - 1 \right),
\end{aligned}
\tag{6}
$$

where v_{l_s} and λ_{l_s} are the Lagrange multipliers associated with the l_s-th SU transmission power and the PBS received interference, respectively. In the gradient search-based cognitive capacity-aware algorithm (GS-CA), the weight vector for user l_s is updated at each iteration n, according to

$$\mathbf{w}_{l_s}^t(n+1) = \mathbf{w}_{l_s}^t(n) + \mu \nabla_{\mathbf{w}_{l_s}^t} \mathcal{L}\left(\mathbf{w}_{l_s}^t, v_{l_s}, \lambda_{l_s}\right), \tag{7}$$

where $\nabla_{\mathbf{w}_{l_s}^t}$ is the gradient of $\mathcal{L}\left(\mathbf{w}_{l_s}^t, v_{l_s}, \lambda_{l_s}\right)$ w.r.t. to $\mathbf{w}_{l_s}^t$ and μ is an adaptation constant to be chosen relatively small in order to achieve convergence. Since the update is done separately on each subcarrier, we drop the frequency index $[k]$ and concentrate on the iteration index (n) in this recursion. This gradient-based equation involves the derivative of the Lagrange multiplier of Eq. (6), which might be difficult to achieve (highly non-linear and non-differentiable). As an alternative we propose, in this paper, to employ the PSO algorithm to optimize the transmit weight vector. PSO is a stochastic algorithm; where the birds or particles are mapped to the transmit beamforming weights, fly in the search space, aiming to optimize a given objective. In this paper, the beamforming weights are optimized towards maximizing the constrained channel capacity given by (6). First, the PSO generates B random particles for each secondary user (i.e., random weight vector $\mathbf{w}_{l_s}^{t,(b)}$, $b = 1$, ..., B of length $N_s^t \times 1$) to form an initial population set S (swarm). The algorithm computes the constrained channel capacity according to (6) for all particles $\mathbf{w}_{l_s}^{t,(b)}$ and then finds the particle that provides the global optimal channel capacity for this iteration, denoted $\mathbf{w}_{l_s}^{t,(b,gbest)}$. In addition, each particle b memorizes the position of its previous best performance, denoted $\mathbf{w}_{l_s}^{t,(b,pbest)}$. After finding these two best values, PSO updates its velocity $\mathbf{v}_{l_s}^{t,(b)}$ and its particle positions $\mathbf{w}_{l_s}^{t,(b)}$, respectively at each iteration n as follows:

$$\mathbf{v}_{l_s}^{t,(b)}(n+1) = \omega \mathbf{v}_{l_s}^{t,(b)}(n) + c_1 \varphi_1 \left(\mathbf{w}_{l_s}^{t,(b,pbest)}(n) - \mathbf{w}_{l_s}^{t,(b)}(n)\right)$$
$$+ c_2 \varphi_2 \left(\mathbf{w}_{l_s}^{t,(b,gbest)}(n) - \mathbf{w}_{l_s}^{t,(b)}(n)\right), \tag{8}$$

$$\mathbf{w}_{l_s}^{t,(b)}(n+1) = \mathbf{w}_{l_s}^{t,(b)}(n) + \mathbf{v}_{l_s}^{t,(b)}(n+1), \tag{9}$$

where c_1 and c_2 are acceleration coefficients towards the personal best position (*pbest*) and/or global best position (g *best*), respectively, φ_1 and φ_2 are two random positive numbers in the range of [0, 1], and ω is the inertia weight which is employed to control the exploration abilities of the swarm. Large inertia weights will allow the algorithm to explore the design space globally. Similarly, small inertia values will force the algorithms to concentrate in the nearby regions of the design space. This procedure is repeated until convergence (i.e., channel capacity remains constant for a several number of iterations or reaching maximum number of iterations). An optimum number of iterations is tuned and refined iteratively by evaluating the average number of iterations required for PSO convergence as a function of the target minimum square error for algorithm termination and as a function of the population size. Since random initialization does not guarantee a fast convergence, in our optimization procedure we consider that the initial value of $\mathbf{w}_{l_s}^{t,(b)}(n)$ at iteration index $n = 0$ is given by the eigen-beamforming (EBF) weight, i.e., $\mathbf{w}_{l_s}^{t,(b)}(0) = \mathbf{w}_{l_s}^{t,EBF} = \sqrt{P_{max,l_s}^t} \mathbf{u}_{max,l_s}$, where $\mathbf{u}_{max,l_s}$ denotes the eigenvector corresponding to λ_{max,l_s}, the maximum eigenvalue of $\mathbf{H}_{ss,l_s}^H \mathbf{H}_{ss,l_s}$. This initial guess enables the algorithm to reach a more refined solution iteratively by ensuring fast convergence and allows to compute the initial value of the received beamforming vector at iteration index n = 0. In our case we assume MRC at the receiving SBS; that is $\mathbf{w}_{l_s}^r(0) = (\mathbf{B}_{l_s}(0))^{-1} \tilde{\mathbf{H}}_{ss,l_s}^t(0)$.

4 PSO-Based Channel Estimation

At the receiver side, channel parameters are estimated using PSO-based MMSE, which does not require any matrix inversion. Once an estimate is obtained it is then used by the MRC receiver. To support the pilot-aided channel estimation, SUs transmit orthogonal training sequences $\mathbf{x}$ of length $l_s \times N_t$, subject to $\mathbf{x}\mathbf{x}^H = \mathbf{I}_{l_s \times N_t}$. The PSO fitness function for channel estimation is represented by the maximum-likelihood metric, which can be written as follows:

$$f(\mathbf{P}_i) = \sum_{k=1}^{l_s \times N_t} \|y(k) - \mathbf{P}_i \mathbf{x}(k)\|^2, \tag{10}$$

where the position of the ith particle $\mathbf{P}_i \in \mathbb{C}^{N_r \times N_t \times l_s}$ is used as a potential solution for the metric and represents the estimate of the channel matrix $\mathbf{H}_e$. In case of MU-MIMO channel estimation, $N_r \times N_t \times l_s$ channel coefficients are estimated assuming a quasi-invariant channel. The steps for the PSO-based MMSE for channel estimation are similar to the PSO-based capacity-aware (PSO-CA) for data beamforming where the fitness function is replaced by the maximum-likelihood metric and the transmit weight vector is replaced by the channel estimate. PSO-MMSE optimizes all channel coefficients with one swarm and PSO-CA optimizes the transmit beamforming weights with a separate swarm.

5 Simulation Results

In our simulation setups we consider a CR-based MU-MIMO-OFDM system with $N_s^t = N_p^t = 2$ transmit antennas. The number of antennas at the PBS and at the SBS is the same, $N_p^r = N_s^r$, and is equal to 16 and 24. $L_s = 4SUs$ and $L_p = 4PUs$. We

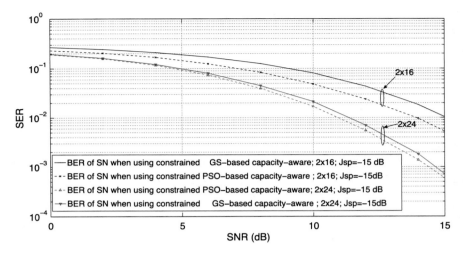

Fig. 2 SER performance of SUs with PSO-CA and GS-CA

assume BPSK modulation. We impose $P^t_{max,l} = 0dB$ and $J^{max}_{sp} - 15dB$ on the SUs. For OFDM configurations, we assume the 256-OFDM system ($N_c = 256$), which is widely deployed in broadband wireless access services. For the PN we assume an MU-MIMO system with non-constrained MIMO-MRC, i.e., EBF at the transmitter and MRC at the receiving PBS. For the PSO parameters of both swarms, the swarm size is 30, the maximum iteration number is 25 and the acceleration coefficients are $c_1 = c_2 = 2$. The inertia weight ω ranges from 0.9 to 0.4 and varies as the iteration goes on. Figure 2 compares the SER performance of SUs achieved by the proposed PSO-CA and the GS-CA schemes for $J^{max}_{sp} = -15$ dB and $N^r_s = 16$ and 24. It is observed from the results that for both cases of N^r_s, the SER of PSO-CA is out-performing the GS-CA. It is also noted that as N^r_s increases the performance gap between the two schemes is reduced.

6 Conclusion

This paper proposes a PSO-based adaptive beamforming and channel estimation algorithm for multiuser access in large-scale cognitive OFDM-SDMA systems. On one hand, the proposed algorithm iteratively seeks the optimal transmit weight vectors that maximize the channel capacity of each secondary user in the network while protecting PUs from SUs' interferences without requiring a gradient search of the channel capacity. On the other hand, channel estimation is performed at the base station using PSO-based MMSE without involving matrix inversion. It was shown that the proposed system is able to achieve a low computational complexity with the same level or better performance than the convectional gradient search for channel maximization and MMSE for channel estimation.

References

1. Marzetta, T.L.: Noncooperative cellular wireless with unlimited numbers of base station antennas. IEEE Tran. Wireless Comm. **9**(11), 3590–3600 (2010)
2. Hoydis, J., ten Brink, S., Debbah, M.: Massive MIMO in the UL/DL of cellular networks: How many antennas do we need? IEEE J. Sel. Areas Commun. **31**(2), 160–171 (2013)
3. Mitola, J., Maguire, G.Q.: Cognitive radio: Making software radios more personal. IEEE Pers. Commun. **6**(6), 13–18 (1999)
4. Haykin, S.: Cognitive radio: brain-empowered wireless communications. IEEE J. Select. Areas Commun. **23**(2), 201–220 (2005)
5. Jiang, M., Ng, S., Hanzo, L.: Hybrid iterative multiuser detection for channel coded space division multiple access OFDM systems. IEEE Trans. Veh. Technol. **55**(1), 115–127 (2006)
6. Sulyman, A.I., Hefnawi, M.: Performance evaluation of capacity-aware MIMO beamforming schemes in OFDM-SDMA systems. IEEE Trans. Commun. **58**(1), 79–83 (2010)
7. Yang L.-L., Wang L.-C.: Zero-forcing and minimum mean-square error multiuser detection in generalized multicarrier DS-CDMA systems for cognitive radio. EURASIP J. Wireless Commun. Netw., 1–13, (2008)
8. Yiu, S., Vu, M., Tarokh, V.: Interference reduction by beamforming in cognitive networks. In: IEEE GLOBECOM Telecommunications Conference, pp. 1–6 (2008)
9. Bixio, L., Oliveri, G., Ottonello, M., Raffetto, M., Regazzoni, C.: Cognitive radios with multiple antennas exploiting spatial opportunities. IEEE Trans. Sig. Process. **58**(8), 4453–4459 (2010)
10. Hefnawi, M.: SER performance of large scale OFDM-SDMA based cognitive radio networks. Int. J. Antennas Propag., pp. 1–8 (2014)
11. Kennedy, J., Eberhart, R.C.: Particle swarm optimization. In: Proceedings of the IEEE Conference on Neural Networks IV, Piscataway, NJ, 1942–1948 (1995)
12. Jose, J., Ashikhmin, A., Marzetta, T., Vishwanath, S.: Pilot contamination and precoding in multi-cell TDD systems. IEEE Trans. Wireless Commun. **10**(8), 2640–2651 (2011)
13. Smith, P.J., Roy, S., Shafi, M.: Capacity of MIMO systems with semicorrelated flat fading. IEEE Trans. Inform. Theory **49**(10), 2781–2788 (2003)

Part III
Signal Processing and Information Theory

Motion Detection Using Color Space-Time Interest Points

Insaf Bellamine and Hamid Tairi

Abstract Detecting moving objects in sequences is an essential step for video analysis. Among all the features which can be extracted from videos, we propose to use Space-Time Interest Points (STIP). STIP are particularly interesting because they are simple and robust low-level features providing an efficient characterization of moving objects within videos. In general, Space-Time Interest Points are based on luminance, and color has been largely ignored. However, the use of color increases the distinctiveness of Space-Time Interest Points. This paper mainly contributes to the Color Space-Time Interest Points (CSTIP) extraction and detection. To increase the robustness of CSTIP features extraction, we suggest a pre-processing step which is based on a Partial Differential Equation (PDE) and can decompose the input images into a color structure and texture components. Experimental results are obtained from very different types of videos, namely sport videos and animation movies.

Keywords Color space-time interest points · Color structure-texture image decomposition · Motion detection

1 Introduction

Detecting moving objects in dynamic scenes is an essential task in a number of applications such as video surveillance, traffic monitoring, video indexing, recognition of gestures, analysis of sport-events, sign language recognition, mobile robotics and the study of the objects' behavior (people, animals, vehicles, etc....). In the literature, there are many methods to detect moving objects, which are based

I. Bellamine (✉) · H. Tairi
Department of Computer Science, Sidi Mohamed Ben Abdellah University LIIAN,
Fez, Morocco
e-mail: insaf.bellamine@usmba.ac.ma

H. Tairi
e-mail: hamid.tairi@usmba.ac.ma

© Springer International Publishing Switzerland 2016
A. El Oualkadi et al. (eds.), *Proceedings of the Mediterranean Conference
on Information & Communication Technologies 2015*, Lecture Notes
in Electrical Engineering 380, DOI 10.1007/978-3-319-30301-7_11

on: optical flow [1], difference of consecutive images [2, 3], Space-Time Interest Points [4] and modeling of the background (local, semi-local and global) [5].

For grayscale sequences, the notion of Space-Time Interest Points (STIP) is especially interesting because they focus information initially contained in thousands of pixels on a few specific points which can be related to spatiotemporal events in an image. Laptev and Lindeberg were the first who proposed STIP for action recognition [4], by introducing a space-time extension of the popular Harris detector [6]. They detect regions having high intensity variation in both space and time as spatio-temporal corners. The STIP detector usually suffers from sparse STIP detection [4]. Later, several other methods for detecting STIP have been reported. Dollar et al. [7] improved the sparse STIP detector by applying temporal Gabor filters and selecting regions of high responses. Dense and scale-invariant Space-Time Interest Points were proposed by Willems et al. [8]. An evaluation of these approaches has been proposed in [9].

For color sequences, we propose a color version of Space-Time Interest Points extension of the Color Harris detector [10] to detect what they call "Color Space-Time Interest Points detector" (CSTIP). To increase the robustness of CSTIP features extraction, we propose a color version of Structure-Texture image decomposition extension of the Structure-Texture image decomposition technique [11].

The Color Structure-Texture image decomposition technique is essential for understanding and analyzing images depending on their content. This one decomposes the color image f into a color structure component u and a color texture component v, where $f = u + v$. The use of the color structure-texture image decomposition method enhances the performance and the quality of the motion detection.

Our contribution is twofold: First, after giving two consecutive images, we split each one into two components (Structure, Texture). Second, we compute the Color Space Time Interest Points (CSTIP) associated to the color structure (respectively texture) components by using the proposed algorithm of the detection of Color Space- Time Interest Points. The fusion procedure is implemented to compute the final Color Space- Time Interest Points. One of the aims of the present paper is to propose a new parallel algorithm so as to generate good results of moving objects.

This paper is organized as follows: Sect. 2 presents the materials and methods, Sect. 3 presents our proposed approach, and finally, Sect. 4 shows our experimental results.

2 Materials and Methods

2.1 Color Structure-Texture Image Decomposition

Let f be an observed image which contains texture and/or noise. Texture is characterized as repeated and meaningful structure of small patterns. Noise is

Fig. 1 The color
structure-texture image
decomposition algorithm

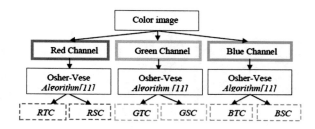

characterized as uncorrelated random patterns. The rest of an image, which is called cartoon, contains object hues and sharp edges (boundaries). Thus an image f can be decomposed as f = u + v, where u represents image cartoon and v is texture and/or noise. In recent years, several models based on total variation, which are inspired by the ROF model, were created [12]. In the literature there is also another model called Mayer [13, 14] that is more efficient than the ROF model. Many algorithms have been proposed to solve numerically this model. In the following, we represent the most popular algorithm, Osher-Vese algorithm [11] for grayscale sequences:

This algorithm is based on the decomposition model as follows:

$$F^{OV}_{\lambda,\mu,p}(u, g) = J(u) + \lambda \|f - (u + \text{div}(g))\|^2_{L^2} + \mu \left\| \sqrt{g_1^2 + g_2^2} \right\|_{L^p} \tag{1}$$

The additional term $\|f - (u + \text{div}(g))\|^2_{L^2}$. ensures that we have the constraint f = u + v where f is the original image that consists of two components: a component u containing all the objects, a component v: is the sum of textures and noise.

In this paper, we propose an adapted decomposition Osher-Vese algorithm for color images shown in Fig. 1. where RTC, RSC, GTC, GSC, BTC and BSC are respectively the red texture component, red structure component, green texture component, green structure component, blue texture component and blue structure component .The color texture component of the color image will be calculated by the combination between RTC, GTC and BTC. The color structure component of the color image will be calculated by the combination between RSC, GSC and BSC.

The Color Aujol Structure-Texture decomposition method has been applied on the karate's fight image of size 352 × 288. The results of the decomposition are shown in Fig. 2.

2.2   *Color Space-Time Interest Points*

The idea of color interest points [10] in the spatial domain can be extended into the color spatio-temporal domain by requiring the image values in space-time to have large variations in both the spatial and the temporal dimensions. For grayscale

(a)   **(b)**   **(c)**

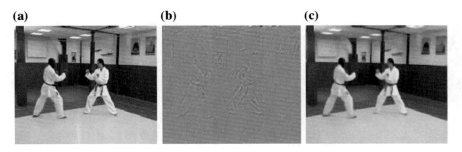

Fig. 2 The color structure-texture image decomposition algorithm: **a** original images (karate's fight image), **b** the color texture component of karate's fight image, **c** the color structure component of karate's fight image

sequences, Laptev et al. [4], proposed a spatio-temporal extension of the Harris detector to detect what they call "Space-Time Interest Points", denoted STIP in the following. Detection of space-time interest points is performed by using the Hessian-Laplace matrix H, which is defined by:

$$H(x,y,t) = g(x,y,t;\sigma_s^2,\sigma_t^2) \otimes \begin{pmatrix} \frac{\partial^2 I(x,y,t)}{\partial x^2} & \frac{\partial^2 I(x,y,t)}{\partial x \partial y} & \frac{\partial^2 I(x,y,t)}{\partial x \partial t} \\ \frac{\partial^2 I(x,y,t)}{\partial x \partial y} & \frac{\partial^2 I(x,y,t)}{\partial y^2} & \frac{\partial^2 I(x,y,t)}{\partial y \partial t} \\ \frac{\partial^2 I(x,y,t)}{\partial x \partial t} & \frac{\partial^2 I(x,y,t)}{\partial y \partial t} & \frac{\partial^2 I(x,y,t)}{\partial t^2} \end{pmatrix} \quad (2)$$

where $I(x, y, t)$ is the intensity of the pixel s (x, y) at time t denotes the convolution. $g(x, y, t; \sigma_s^2, \sigma_t^2)$ is the Gaussian smoothing (see Eq. (4)) and the two parameters (σ_s and σ_t) control the spatial and temporal scale.

For color sequences, we propose a color version of space-time interest points extension of the Color Harris detector [10] to detect what they call "Color Space-Time Interest Points", denoted CSTIP in the following. The information given by the three RGB color channels: Red, Green and Blue.

Color plays a very important role in the stages of feature detection. Color provides extra information which allows the distinctiveness between various reasons of color variations, such as change due to shadows, light source reflections and object reflectance variations.

The H matrix can be computed by a transformation in the RGB space [15]. The first step is to determine the gradients of each component of the RGB color space. The gradients are then transformed into desired color space. By multiplying and summing of the transformed gradients, all components of the matrix are computed as follows:

$$\frac{\partial^2 I(x,y,t)}{\partial x \partial t} = \frac{\partial^2 R(x,y,t)}{\partial x \partial t} + \frac{\partial^2 G(x,y,t)}{\partial x \partial t} + \frac{\partial^2 B(x,y,t)}{\partial x \partial t}$$

$$\frac{\partial^2 I(x,y,t)}{\partial x^2} = \frac{\partial^2 R(x,y,t)}{\partial x^2} + \frac{\partial^2 G(x,y,t)}{\partial x^2} + \frac{\partial^2 B(x,y,t)}{\partial x^2}$$

$$\frac{\partial^2 I(x,y,t)}{\partial y^2} = \frac{\partial^2 R(x,y,t)}{\partial y^2} + \frac{\partial^2 G(x,y,t)}{\partial y^2} + \frac{\partial^2 B(x,y,t)}{\partial y^2}$$

$$\frac{\partial^2 I(x,y,t)}{\partial t^2} = \frac{\partial^2 R(x,y,t)}{\partial t^2} + \frac{\partial^2 G(x,y,t)}{\partial t^2} + \frac{\partial^2 B(x,y,t)}{\partial t^2} \qquad (3)$$

$$\frac{\partial^2 I(x,y,t)}{\partial y \partial t} = \frac{\partial^2 R(x,y,t)}{\partial y \partial t} + \frac{\partial^2 G(x,y,t)}{\partial y \partial t} + \frac{\partial^2 B(x,y,t)}{\partial y \partial t}$$

$$\frac{\partial^2 I(x,y,t)}{\partial x \partial y} = \frac{\partial^2 R(x,y,t)}{\partial x \partial y} + \frac{\partial^2 G(x,y,t)}{\partial x \partial y} + \frac{\partial^2 B(x,y,t)}{\partial x \partial y}$$

where $R(x,y,t)$, $G(x,y,t)$ and $B(x,y,t)$ are respectively the red, green and blue components of the color pixel s (x, y) at time t.

As with the Color Harris detector, a Gaussian smoothing is applied both in spatial domain (2D filter) and temporal domain (1D filter).

$$g(x,y,t;\sigma_s^2,\sigma_t^2) = \frac{\exp(-\frac{x^2+y^2}{2\sigma_s^2} - \frac{t^2}{2\sigma_t^2})}{\sqrt{(2\pi)^3 \sigma_s^4 \sigma_t^2}} \qquad (4)$$

The two parameters (σs and σt) control the spatial and temporal scale. As in [4], the color spatio-temporal extension of the Color Harris corner function, entitled "salience function", is defined by:

$$M(x,y,t) = \det(H(x,y,t)) - k \times \operatorname{trace}(H(x,y,t))^3 \qquad (5)$$

where k is a parameter empirically adjusted at 0.04, det is the determinant of the matrix H and trace is the trace of the same matrix.

3　Proposed Approach

The most famous algorithm to detect Space-Time Interest Points is that of Laptev [4]; however we can reveal four major problems when a local method is used:

- Texture, Background, lighting changes and Objects that may influence the results;
- Noisy datasets such as the KTH dataset [16], which is featured with low resolution, strong shadows, and camera movement that renders clean silhouette extraction impossible;

Fig. 3 The adapted Laptev
algorithm

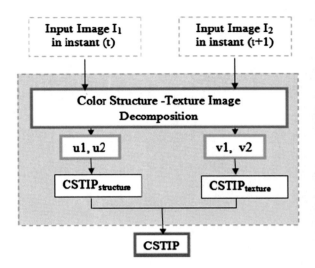

- Features extracted are unable to capture smooth and fast motions, also they are sparse. This also explains why they generate poor results;
- RGB video must be converted into grayscale video.

However, to overcome the four problems, we propose a new technique based on the Color Space-Time Interest Points and Color Structure-Texture Image Decomposition algorithm (Fig. 1).

Let CSTIP denote the final extracted color space-time interest points; $CSTIP_{texture}$ and $CSTIP_{structure}$ denote respectively the extracted Space-Time Interest Points fields on texture components and structure components. The Color Structure-Texture Image Decomposition decomposes each input image into two components (u1, v1) for the first image I1 and (u2, v2) for the second image I2, where u1 and u2 denote the color structure components, v1 and v2 are the color texture components.

The proposed approach detects the color space-time interest points by given a pair of consecutive images in several stages; it will help to have a good detection of moving objects and even reduce the execution time by proposing a parallel algorithm (see Fig. 3). Our new Space-Time Interest Points will be calculated as the following:

$$CSTIP = CSTIP_{structure} \cup CSTIP_{texture} \qquad (6)$$

4   Results and Discussion

This section presents several experimental results on different kinds of sequences.

In Fig. 4, we represent some examples of clouds of Color Space-Time Interest Points detected in the sequence (sport video: "karate's fight" lasts for 2 min and

Fig. 4 Examples of clouds of color space-time interest points, in each frame (Image (t = 44)/ Image (t = 45)), we use the parameters (σt = 1.5, k = 0.04 and σs = 1.5)

49 s with 200 images and the size of each image frame is 400 by 300 pixels.). The results illustrated in Fig. 4, show that the objects moving (the two players) are detected with our approach (Fig. 3). The red points represent the extracted Color Space-Time Interest Points with our proposed approach (Fig. 3).

In order to correctly gauge performance of our algorithm (Fig. 3), we will proceed with a comparative study to use the mask of the moving object and the precision.

For each moving object, we have a number of the color space-time interest points detected in the moving object (NTP) and a number of the color space-time interest points extracted off the moving object (NFP).

The test is performed many examples of sequences [17] and gives in the following results:

The proposed approach is much less sensitive to noise, and the reconstruction of the image, it also extracts the densest features (see Fig. 5).

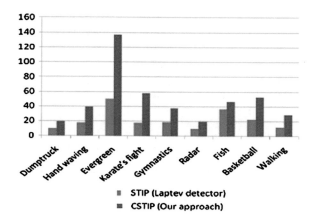

Fig. 5 Number of color space-time interest points extracted in each frame

Table 1 Compared results

Videos	Laptev detector [4]	Our approach (Fig. 3)
	Precision (%)	Precision (%)
Radar	80	84
Fish	94	96
Basketball	91	94
Walking	90	92
Dumptruck	78	89
Hand waving	92	93
Evergreen	96	98
Karate's fight	88	96
Gymnastics	89	92

The results, illustrated in Table 1, show that our approach allows a good detection of moving objects.

5 Conclusion

This paper mainly contributes to the Color Space-Temporal Interest Points extraction and detection. To increase the robustness of CSTIP features extraction, our approach suggests a pre-processing step which can decompose the input images into a color structure component and a color texture component. The experiment results show that the proposed method outperforms the original STIP feature detector with promising results.

References

1. Simac, A.: Optical-flow based on an edge-avoidance procedure. Comput. Vis. Image Underst. **113**(2009), 511–531 (2008)
2. Galmar, E., Huet, B.: Analysis of vector space model and spatiotemporal segmentation for video indexing and retrieval. In: CIVR van Leeuwen, J. (ed.) Computer Science Today. Recent Trends and Developments. Lecture Notes in Computer Science, vol. 1000. Springer, Berlin Heidelberg New York (1995)
3. Zhou, B.: A phase discrepancy analysis of object motion, ACCV 2010
4. Laptev, I.: On space-time interest points. Int. J. Comput. Vis. **64**(2/3):107–123 (2005)
5. Nicolas, V.: Suivi d'objets en mouvement dans une séquence vidéo. Doctoral thesis, Paris Descartes university (2007)
6. Harris, C., Stephens, M.J.: A combined corner and edge detector. In: Alvey Vision Conférence (1988)
7. Dollar, P., Rabaud, V., Cottrell, G., Belongie, S.: Behavior recognition via sparse spatio-temporal features. In: VS-PETS (2005)
8. Willems, G., Tuytelaars, T., Van Gool, L.: An efficient dense and scale-invariant spatio-temporal interest point detector. Eur. Conf. Comput. Vis. **5303**(2), 650–663 (2008)

9. Wang, H.: Evaluation of Local Spatio-Temporal Features for Action Recognition. BMVC '09 London (2009)
10. Stöttinger, J., Hanbury, A., Sebe, N.: Sparse color interest points for image retrieval and object categorization IEEE Trans. Image Process. **21**(5), (2012)
11. Vese, L., Osher, S.: Modeling textures with total variation minimization and oscillating patterns in image processing. J. Sci. Comput. **19**(1–3), 553–572 (2002)
12. Gilles, J.: Décomposition et détection de structures géométriques en imagerie. Doctoral thesis, Ecole Normale Supérieure de Cachan (2006)
13. Meyer, Y.: Oscillating Patterns in Image Processing and in Some Nonlinear Evolution Equations. The Fifteenth Dean Jacquelines B. Lewis Memorial Lectures, American Mathematical Society (2001)
14. Chambolle, A.: An algorithm for total variation minimization and application. J. Math. Imaging vis. **20**(1–2), 89–97 (2004)
15. van de Weijer, J., Gevers, T.: Edge and corner detection by photometric quasi-invariants. IEEE Trans. Pattern Anal. Mach. Intell. **27**(4), 625–630 (2005)
16. Laptev, I., Caputo, B.: Recognizing human actions: a local SVM approach. ICPR **3**, 32–36 (2004)
17. Baker, et al.: A database and evaluation methodology for optical flow. Int. J. Comput. Vision **92**(1), 1–31 (2011)

Time-Frequency Analysis of GPR Signal for Cavities Detection Application

Houda Harkat, Saad Bennani Dosse and Abdellatif Slimani

Abstract In a dispersive medium, GPR signal is classified as a nonstationary signal, and which is widely attenuated, as so, some reflected echoes become none visible neither in the time or the frequency representation of the signal. In fact, we are aimed to calculate the travel time inside a cavity, i.e. the time between two transition that are highly attenuated, in order to identify the nature of the dielectric inside this cavity, which become impossible due to the attenuation phenomena. In this outlook, we proposed to analysis this signal using a time-frequency representation. The continuous wavelet transform is the alternative approach to the Fourier transform due to the fact that the spectrogram is limited in the resolution by the width of the window. Besides the Stockwell transform, in addition to the Hilbert Huang transform are widely used for the analyzed of the electrical, biomedical (ECGs), GPR signals, and seismic sections.

Keywords Dispersive medium · GPR signal · Time-frequency representation · Continuous wavelet transform · Fourier transform · Hilbert Huang transform · Stockwell transform · Electrical and biomedical signals

1 Introduction

GPR signals is a nonstationary signal affected by several phenomena, such as noise and clutter signal caused principally by little diffracting objects localized in the subsurface layers. As so, the signal is widely attenuated, and the transitions echo's

H. Harkat (✉) · S. Bennani Dosse · A. Slimani
Faculty of Science and Technology, Laboratory Renewable Energy and Intelligent Systems (LERSI), Road Imouzzer Fez, PO Box 2202, FEZ, Morocco
e-mail: harkat.houda@gmail.com

S. Bennani Dosse
e-mail: bennani.saad.ensaf@gmail.com

A. Slimani
e-mail: slimani.abdellatif.ma@gmail.com

© Springer International Publishing Switzerland 2016
A. El Oualkadi et al. (eds.), *Proceedings of the Mediterranean Conference on Information & Communication Technologies 2015*, Lecture Notes in Electrical Engineering 380, DOI 10.1007/978-3-319-30301-7_12

107

become undistinguished. In a dispersive medium this phenomena becomes more important, as so, some reflected echoes become none visible neither in the time or the frequency representation of the signal.

In fact, in geophysical data analysis, the concept of a stationary time series is a mathematical idealization that is never realized and is not particularly useful in the detection of signal arrivals. Although the Fourier transform of the entire time series does contain information about the spectral components in a time series, for a large class of practical applications, this information is inadequate. The spectral components of such a time series clearly have a strong dependence on time. It would be desirable to have a joint time-frequency representation (TFR).

Besides, in this study, we must calculate the travel time inside a cavity, i.e. the time between two transition that are highly attenuated, in order to identify the nature of the dielectric inside this cavity.

For time-frequency signal analysis, different techniques can be used. Three techniques that have notably better properties in the analysis of nonstationary signals are Hilbert Huang Transform (HHT), Continuous Wavelet Transform (CWT), and Stockwell Transform.

In fact, Short Time Fourier Transform (STFT) is commonly used. However, the limitation of this technique is that it has a constant resolution in time and in frequency, where the width of the windowing function defines the resolution.

Continuous Wavelet Transform (CWT), is a technique suitable for time localization of frequency content of the signal, has drawn significant attention in the field, especially in seismic trace analyzing, signal denoising and enhancement.

The Stockwell transform is a combination of short-time Fourier (STFT) and wavelet transforms, since it employs a variable window length and the Fourier kernel. The advantage of S-transform is that it preserves the phase information of signal, and also provides a variable resolution similar to the wavelet transform. In addition, the S-transform is a linear transform that can be used as both an analysis and a synthesis tool, which is not the case with some of the bilinear transforms such as Wigner–Ville distribution. However, the S-transform suffers from poor energy concentration at higher frequency and hence poor frequency localization.

However, Hilbert Huang Transform (HHT) is a recent technique for time-frequency analysis, based on the empirical mode decomposition concept, which simultaneously provides excellent resolution in time and frequency [1]. Its essential feature is the use of an adaptive time-frequency decomposition that does not impose a fixed basis set on the data, and therefore it is not limited by the time-frequency uncertainty relation characteristic of Fourier or Wavelet analysis.

HHT has been applied for target signal enhancement in depth resolution problem for high frequency range. In fact, Hilbert-Huang transform is widely studied and applied in the many fields such as system simulation, spectral data preprocessing, geophysics and the like through being developed for more than ten years. Besides, Feng et al. [2] applied Hilbert transform to convert ground penetrating radar real signal into complex signals. The instantaneous amplitude, instantaneous phase and instantaneous frequency waveform Figs were extracted, and independent profiles of three parameters were formed, thereby improving the accuracy of radar interpretation.

Based on Hilbert transform, the signals must be narrowband when Hilbert transform is used for calculating instantaneous parameters of signals. The GPR data often adopt broadband, and error physical interpretation can be caused through directly adopting Hilbert for calculating instantaneous parameters, such as negative frequency and the like.

This paper gives comparative survey of the three time-frequencies approaches with the intention of giving the guidelines for deciding which of these techniques will be chosen for the analysis of signals obtained from GPR survey, considering the desired outcomes of the analysis, and the specific application for the resolution of the problem of undistinguished reflections in dispersive medium, due to the attenuation phenomena. Besides brief mathematical foundations of the transforms, the paper illustrates their utilization using simulated examples.

2 Time-Frequency Analysis

2.1 Wavelet Transform

A wavelet function, is a function $\psi \in L^2(\mathbb{R})$ with zero average, normalized (i.e. $\int_{\mathbb{R}} \psi = 0$), and centered in the neighborhood of $t = 0$. Scaling ψ by a positive quantity s, and translating it by $u \in \mathbb{R}$, we define a family of time-frequency atoms, $\psi_{u,s}$, as:

$$\psi_{u,s}(t) = \frac{1}{\sqrt{s}} \psi\left(\frac{t - u}{s}\right), u \in \mathbb{R}, s > 0 \tag{1}$$

Given a signal sig $\in L^2(\mathbb{R})$, the continuous wavelet transform (CWT) of sig at time u and scale s is defined as:

$$W \, sig(u, s) = \langle sig, \psi_{u,s} \rangle = \int\limits_{-\infty}^{+\infty} sig(t) \psi_{u,s}^*(t) dt \tag{2}$$

It provides the frequency component (or details) of sig corresponding to the scale s and time location t.

The revolution of wavelet theory comes precisely from this fact: the two parameters (time u and scale s) of the CWT make possible the study of a signal in both time and frequency domains simultaneously, with a resolution that depends on the scale of interest. According to these considerations, the CWT provides a time-frequency decomposition of sig in the so called time-frequency plane. This method, as it is discussed previously, is more accurate and efficient than the windowed Fourier transform (WFT) basically.

The scalogram of sig is defined by the function :

$$S(s) = \| \, Wsig(s,u) \, \| = \sqrt{\int_{-\infty}^{+\infty} |W \, sig \, (s,u)|^2 du} \tag{3}$$

Representing the energy of (W sig) at a scale s. Obviously, $S(s) \geq 0$ for all scales s, and if $s > 0$ we will say that the signal sig has details at scale s. Thus, the scalogram allows the detection of the most representative scales (or frequencies) of a signal, that is, the scales that contribute the most to the total energy of the signal.

Versatile as the wavelet analysis is, the problem with the most commonly used Morlet wavelet is its leakage generated by the limited length of the basic wavelet function, which makes the quantitative definition of the energy-frequency-time distribution difficult. Sometimes, the interpretation of the wavelet can also be counterintuitive. In fact, to define a change occurring locally, one must look for the result in the high-frequency range, for the higher the frequency the more localized the basic wavelet will be. If a local event occurs only in the low-frequency range, one will still be forced to look for its effects in the high-frequency range. Such interpretation will be difficult if it is possible at all. Another difficulty of the wavelet analysis is its non adaptive nature. Once the basic wavelet is selected, one will have to use it to analyze all the data. Since the most commonly used Morlet wavelet is Fourier based, it also suffers the many shortcomings of Fourier spectral analysis, it can only give a physically meaningful interpretation to linear phenomena. In spite of all these problems, wavelet analysis is still the best available non-stationary data analysis method so far [2].

2.2 Stockwell Transform

There are several methods of arriving at the S transform. We consider it illuminating to derive the S transform as the phase correction of the CWT [3].

The S transform of a function sig(t) is defined as a CWT with a specific mother wavelet multiplied by the phase factor :

$$S(r,f) = e^{i2\pi fu} W \, sig(u,s) \tag{4}$$

where W sig (u,s) is the CWT given by (2), and the mother wavelet is defined as :

$$\psi(t,f) = \frac{|sig|}{k\sqrt{2\pi}} e^{-\frac{t^2 f^2}{2k^2}} e^{i2\pi ft}, \quad \forall k > 0 \tag{5}$$

where k is the scaling factor which controls the time-frequency resolution. The wavelet is generally chosen to be positive and normalized Gaussian. In particular, the transform with a Gaussian window can be rewritten in terms of a Morlet wavelet transform [4].

In fact, the S-transform provides a useful extension to the wavelets by having a frequency dependent progressive resolution as opposed to the arbitrary dilations used in wavelets. The kernel of the S-transform does not translate with the localizing window function, in contrast to the wavelet counterpart. For that reason the S-transform retains both the amplitude and absolutely referenced phase information [3]. Absolutely referenced phase information means that the phase information given by the S-transform is always referenced to time. This is the same meaning of the phase given by the Fourier transform, and is true for each S-transform sample of the time-frequency space. The continuous wavelet transform, in comparison can only localize the amplitude or power spectrum, not the absolute phase. There is, in addition, an easy and direct relation between the S-transform and the Fourier transform.

2.3 Hilbert Huang Transform

Hilbert-Huang transform is a new method for analyzing nonlinear and non-stationary signals, which was proposed by Huang in 1998. It mainly includes empirical mode decomposition (EMD) and Hilbert spectral analysis, which firstly decomposes signals into a number of intrinsic mode functions (IMF) through utilizing EMD method, then acts Hilbert transform on every IMF, and obtains corresponding Hilbert instantaneous spectrum, and the multi-scale oscillation change characteristics of original signals are revealed through analyzing each component and its Hilbert spectrum [5].

2.3.1 EMD Decomposition

In fact, we define as IMF any function having the same number of zero-crossings and extrema, and also having symmetric envelopes defined by the local maxima and minima respectively. Since IMFs admit well-behaved Hilbert transforms, the second stage of the algorithm is to use the Hilbert transform to provide instantaneous frequencies as a function of time for each one of the IMF components. Depending on the application, only the first stage of the Hilbert-Huang Transform may be used.

For a signal x(t) the EMD starts by defining the envelopes of its maxima and minima using cubic splines interpolation. Then, the mean of the two envelopes (E_{max} and E_{min}) is calculated:

$$m_1(t) = (E_{max}(t) + E_{min}(t))/2 \qquad (6)$$

Accordingly, the mean $m_1(t)$ is then subtracted from the original signal $x(t)$:

$$h_1(t) = x(t) - m_1(t) \qquad (7)$$

And the residual $h_1(t)$ is examined for the IMF criteria of completeness. If it is an IMF then the procedure stops and the new signal under examination is expressed as:

$$x_1(t) = x(t) - h_1(t) \qquad (8)$$

However, if $h_1(t)$ if is not an IMF, the procedure, also known as sifting, is continued k times until the first IMF is realized. Thus:

$$h_{11}(t) = h_1(t) - m_{11}(t) \qquad (9)$$

where the second subscript index corresponds to sifting number, and finally:

$$IMF_1(t) = h_{1k}(t) = h_{k-1}(t) - m_{1k}(t) \qquad (10)$$

In fact, the sifting process is continued until the last residual is either a monotonic function or a constant. It should be mentioned that as the sifting process evolves, the number of the extrema from one residual to the next drops, thus guaranteeing that complete decomposition is achieved in a finite number of steps. The signal given by:

$$x(t) = \sum_{i=1}^{n} IMF_i(t) + r(t) \qquad (11)$$

where k is the total number of the IMF components and $r(t)$ is the residual [5].

The final product is a wavelet-like decomposition going from higher to lower oscillation frequencies, with the frequency content of each mode decreasing as the order of the IMF increases. The big difference however, with the wavelet analysis is that while modes and residuals can intuitively be given a spectral interpretation in the general case, their high versus low frequency discrimination applies only locally and corresponds in no way to a predetermined sub-band filtering. Selection of modes instead, corresponds to an automatic and adaptive time variant filtering.

2.3.2 Hilbert Spectrum

For given decomposed signal, sig(t), the Hilbert transform, HT (t), is defined as:

$$HT(t) = \frac{1}{\pi} \int\limits_{-\infty}^{+\infty} \frac{sig(\tau)}{t - \tau} d\tau \qquad (12)$$

HT (t) can be combined to form analytical signal A (t), given by:

$$A(t) = sig(t) + iHT(t) = a(t)e^{\theta(t)} \qquad (13)$$

From the polar coordinate expression, the instantaneous frequency can be defined

$$\omega = \frac{d\theta}{dt} \qquad (14)$$

Applying the Hilbert transform to the n IMF components, the signal sig(t) can be written as :

$$sig(t) = R \sum_{j=1}^{n} a_j(t)e^{i\theta \int \omega_j(t)dt} \qquad (15)$$

where R is the real part of the value to be calculated and a_j the analytic signal associated with the jth IMF.

The above Equation is written in terms of amplitude and instantaneous frequency associated with each component as functions of time. The time-dependent amplitude and instantaneous frequency might not only improve the flexibility of the expansion, but also enable the expansion to accommodate nonstationary data [5].

The frequency-time distribution of the amplitude is designated as the Hilbert amplitude spectrum, $H(\omega, t)$, or Hilbert spectrum simply, defined as:

$$H(\omega, t) = \sum_{j=1}^{n} a_{jd}(t)dt \qquad (16)$$

The marginal spectrum, $h(\omega)$, is defined as :

$$h(\omega) = \sum_{j=1}^{n} \int_{0}^{T} a_{jd}(t)dt \qquad (17)$$

It is provides a measure of total amplitude contribution from each frequency value, in which T denotes the time duration of data.

3   Simulation

3.1   *Example of Simulation*

An example of simulation for a prospection type surface-surface, is realized for a cavity embedded in a dispersive clay (Fig. 1) using Matlab. The received signal is analyzed with the three time-frequency techniques discussed bellow.

3.2   *Results and Discussion*

We present the result of the GPR signal analyzed by wavelet transform (Fig. 2), Stockwell transform (Fig. 3), and Hilbert Huang Transform (Fig. 4).

CWT (Fig. 2) shows excellent time localization for different frequencies, but it is limited by the choice of the mother wavelet. Nevertheless, although CWT is able to provide sharper time localization of appearance of different frequencies than HHT (Fig. 4), it is not capable to determine the instantaneous frequency of the signal. But, otherwise, the Hilbert transform, with EMD algorithm, is unstable with noisy data (Fig. 5), which is the case for real data.

The S-transform (Fig. 3) improves the short-time Fourier transform and the continuous wavelet transform by merging the multi-resolution and frequency-dependent analysis properties of wavelet transform with the absolute phase retaining of Fourier transform.

The common S-transform applies a Gaussian window to provide appropriate time and frequency resolution and minimizes the product of these resolutions. However, the Gaussian S-transform is unable to obtain uniform time and frequency

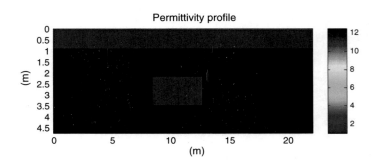

Fig. 1 Simulation model: permittivity profile

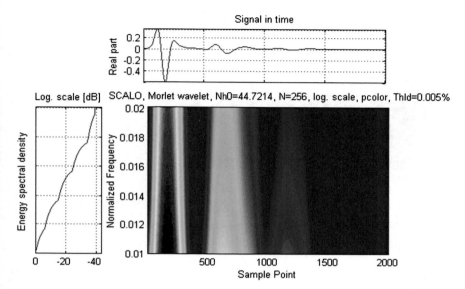

Fig. 2 Time-frequency representation using continuous wavelet transform

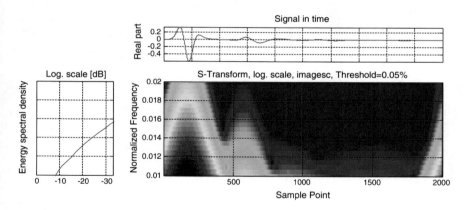

Fig. 3 Time-frequency representation using Stockwell transform

resolution for all frequency components. In fact, the S-transform suffers from inherently poor frequency resolution, particularly at the high frequencies, in addition to the misleading interference terms.

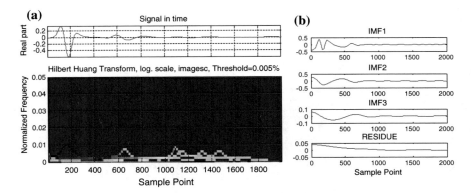

Fig. 4 Time-frequency representation using Hilbert Huang transform: **a** Time-frequency spectrum, **b** EMD decomposition which show the three IMF and the residue

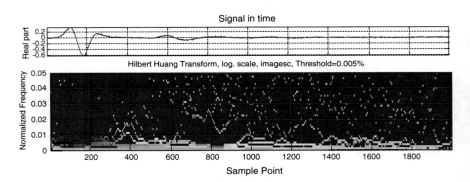

Fig. 5 Time-frequency representation using Hilbert Huang Transform for noisy data: SNR = −46.05 dB

4 Conclusion

This research aims to calculate the travel time inside a cavity, i.e. the time between two transition that are highly attenuated, in order to identify the nature of the dielectric inside this cavity, which become impossible due to the attenuation phenomena. Accordingly to the comparison study done between the three time-frequency representations algorithms explained, it was clear that the CWT gives good time-frequency resolution of the events (not affected by noise or misleading terms).

However the Hilbert Huang technique could be modified in the phase of EMD decomposition and IMF selection to perform the desired task. Future works will be concentrated on this points.

References

1. Jakovljević, Ž.B.: Comparative analysis of Hilbert Huang and discrete wavelet transform in processing of signals obtained from the cutting process: an intermittent turning example. FME Trans. **41**(4), 342–348 (2013)
2. Feng, D.S., Chen, C.S., Yu, K.: Signal enhancement and complex signal analysis of GPR based on Hilbert-Huang transform. In: Electrical Power Systems and Computers. LNEE 99, International Conference on Electric and Electronics (EEIC), 375–384 (2011)
3. Stockwell, R.G., Mansinha, L., Lowe, L.: Localization of the complex spectrum: the S-transform. Sig. Process. IEEE Trans. **44**(4), 998–1001 (1996)
4. Ventosa, S., Simon, C., Schimmel, M., Danobeitia, J.J., Manuel, A.: The S -transform from a wavelet point of view. Sig. Process. IEEE Trans. **56**(7), 2771–2780 (2008)
5. Huang, N.E., Shen, Z., Long, S.R.: The Empirical mode decomposition and the Hilbert spectrum for nonlinear and non-stationary time series analysis. Proc. R. Soc. Lond. A **454**, 903–995 (1998)

A New Scale and Orientation Adaptive Object Tracking System Using Kalman Filter and Expected Likelihood Kernel

Hamd Ait Abdelali, Fedwa Essannouni, Leila Essannouni and Driss Aboutajdine

Abstract This paper presents a new scale and orientation adaptive object tracking system using Kalman filter in a video sequence. This object tracking is an important task in many vision applications. The main steps in video analysis are two: detection of interesting moving objects and tracking of such objects from frame to frame. We use an efficient local search scheme (based on expected likelihood kernel) to find the image region with a histogram most similar to the histogram of the tracked object. In this paper, we address the problem of scale adaptation. The proposed approach tracker with scale selection is compared with recent state-of-the-art algorithms. Experimental results have been presented to show the effectiveness of our proposed system.

Keywords Object tracking · Computer vision · Integral image · Expected likelihood kernel · Kalman filter

1 Introduction

Real-time object tracking is a critical task in computer vision, and many algorithms have been proposed to overcome the difficulties arising from noise, occlusions, clutters, pose, and changes in the foreground object and/or background environment. Many different algorithms [1–3], have been proposed for object tracking, including mean-shift tracking [4], optical flow, and feature matching. Each algorithm has strengths in certain environments and weaknesses in others. This project aims to combine several such algorithms as inputs or "measurements" to a single Kalman filter [5], for robust object tracking. The filter can favor the algorithm that is most applicable to the current environment by decreasing its measurement noise

H. Ait Abdelali (✉) · F. Essannouni · L. Essannouni · D. Aboutajdine
Faculty of Sciences Rabat GSCM-LRIT Laboratory Associate, Unit to CNRST (URAC 29),
Mohammed V University, B.P. 1014, Rabat, Morocco
e-mail: hamd.abdelali@gmail.com

© Springer International Publishing Switzerland 2016
A. El Oualkadi et al. (eds.), *Proceedings of the Mediterranean Conference
on Information & Communication Technologies 2015*, Lecture Notes
in Electrical Engineering 380, DOI 10.1007/978-3-319-30301-7_13

variance, and similarly ignore less suitable algorithms by increasing their measurement variances.

Two major components can be distinguished in a typical visual tracker. Target Representation and Localization is mostly a bottom-up process which has also to cope with the changes in the appearance of the target. Filtering and Data Association is mostly a top-down process dealing with the dynamics of the tracked object, learning of scene priors, and evaluation of different hypotheses [4, 6, 7]. The way the two components are combined [8, 9], and weighted is application dependent and plays a decisive role in the robustness and efficiency of the tracker. In real-time applications, only a small percentage of the system resources can be allocated for tracking, the rest being required for the preprocessing stages or to high-level tasks such as recognition, trajectory interpretation, and reasoning. Therefore, it is desirable to keep the computational complexity of a tracker as low as possible.

The goal of this paper is dedicated to improve the similarity measure for the target representation in the Kalman filter. We derive a similarity measure by combining between the expected likelihood kernel [10–12], and the integral image [13], as a similarity measure between target and estimated scale/shape regions in the frames of video sequence. In this paper we analyzes and compares between our system with: Firstly in [7], the efficient local search framework for real-time tracking of complex non-rigid objects. The shape of the object is approximated by an ellipse and its appearance by histogram based features derived from local image properties based on mean-shift and Kalman filter. Secondly the Kalman filter is used as in [4]. Where the shape of the tracked object is approximated by an ellipse and the appearance within the ellipse is described by a histogram based model. The obvious advantage of such a model is its simplicity and general applicability. Another advantage, that made this observation model rather popular, is the existence of efficient local search schemes to find the image region with a histogram most similar to the histogram of the tracked object. Experimental results show that the proposed approach has superior discriminative power and achieves good tracking performance.

The rest of the paper is organized as follows: Sect. 2, introduces basic Kalman filter for object tracking. Section 3, present the expected likelihood kernel. Section 4, present scale estimation. And then Sect. 5 the proposed approach. Section 6 the experiment result. Section 7 concludes the paper.

2 Kalman Filter

The Kalman filter is a framework for predicting a process state, and using measurements to correct or "update" these predictions.

2.1 State Prediction

For each time step k, a Kalman filter first makes a prediction $\hat{s}_k$ of the state at this time step:

$$\hat{s}_k = A \times s_{k-1} \tag{1}$$

where s_{k-1} is a vector representing process state at time $k - 1$ and A is a process transition matrix. The Kalman filter concludes the state prediction steps by projecting estimate error covariance P_k^- forward one time step:

$$P_k^- = A \times P_{k-1} \times A' + W \tag{2}$$

where P_{k-1} is a matrix representing error covariance in the state prediction at time $k - 1$, and W is the process noise covariance.

2.2 State Correction

After predicting the state $\hat{s}_k$ (and its error covariance) at time k using the state prediction steps, the Kalman filter next uses measurements to "correct" its prediction during the measurement update steps. First, the Kalman filter computes a Kalman gain K_k, which is later used to correct the state estimate $\hat{s}_k$:

$$K_k = P_k^- \times (P_k^- + R_k)^{-1} \tag{3}$$

where R is measurement noise covariance. Determining R_k for a set of measurements is often difficult. In our implementations we calculated R dynamically from the measurement algorithms state. Using Kalman gain K_k and measurements z_k from time step k, we can update the state estimate:

$$\hat{s}_k = \hat{s}_k + K_k \times (z_k - \hat{s}_k) \tag{4}$$

Conventionally, the measurements Z_k are often derived from sensors. In our approach, measurements Z_k are instead the output of various tracking algorithm given the same input: one frame of a streaming video, and the most likely x and y coordinates of the target object in this frame (taken the first two dimensions of $\hat{s}_k$).

The final step of the Kalman filter iteration is to update the error covariance P_k^- into P_k:

$$P_k = (I - K_k) \times P_k^- \tag{5}$$

The updated error covariance will be significantly decreased if the measurements are accurate (some entries in R_k are low), or only slightly decreased if the measurements are noise (all of R_k is high). For more details, see [8, 9, 5].

3   Expected Likelihood Kernel

Let p and q be probability distributions on a space χ and ρ be a positive constant. In this work, we are using the probability product kernels ($K_\rho : \chi \times \chi \rightarrow R$ on the space of normalized discrete distributions over some indexs set Ω) as the similarity measures for comparing two discrete distributions $p_1, p_2, p_3, \ldots, p_N \in \chi$ and $q_1, q_2, q_3, \ldots, q_N \in \chi$. The probability product kernel between distributions $\{p\}_{1..N} \varepsilon \chi$ and $\{q\}_{1..N} \in \chi$ is defined as:

$$K_\rho(p, q) = \sum_{k=1}^{N} p(k)^\rho q(k)^\rho \tag{6}$$

It is easy to show that such a similarity measure is a valid kernel, since for any $p_1, p_2, p_3, \ldots, p_N \in \chi$, the Gram matrix K consisting of elements $K_{ij} = K_\rho(p_i, q_j)$ is positive semi-definite:

$$\sum_i \sum_j \alpha_i \alpha_j K_\rho(p_i, p_j) = \sum_k (\sum_i \alpha_i p_i(k)^\rho)^2 \geq 0 \tag{7}$$

for $\alpha_1, \alpha_2, \alpha_3, \ldots, \alpha_N \in IR$. Different ρ values are corresponded to different types of probability product kernels. For $\rho = 1$, we have:

$$K_1(p, q) = \sum_k p(k)q(k) = IE_p[q(k)] = IE_q[p(k)] \tag{8}$$

We call this the Expected Likelihood Kernel, is defined by $K(p, q) = \sum_k p(k)q(k)$. We denote the histogram of Target of object tracking T as h_T, and the number of pixels inside T as $|T|$, which is also equal to the sum over bins, $|T| = \sum_k h_T(k)$. Let q be the normalized version of h_T given by $q = \frac{h_T}{|T|}$, so we can consider q as a discrete distribution, with $\sum_k q(k) = 1$. Let p the normalized histogram obtained in the farms of video sequence. For the k-bin of h_T, its value is obtained by counting the pixels that are mapped to the index k:

$$h_T(k) = \sum_{x \varepsilon T} \delta[b(x) - k] \tag{9}$$

where $\delta[t]$ is the Kronecker delta, with $\delta[t] = 1$ if $t = 0$, and $\delta[t] = 0$ otherwise. The mapping function $b(x)$ maps a pixel x to its corresponding bin index. The computation of the expected likelihood kernel can be expressed as:

$$K(p,q) = \sum_k p(k)q(k) = \sum_k p(k)\left(\frac{1}{|T|}\sum_{x\varepsilon R}\delta[b(x)-k]\right)$$
$$= \frac{1}{|T|}\sum_{x\varepsilon T}\sum_k p(k)\delta[b(x)-k] = \frac{1}{|T|}\sum_{x\varepsilon T}p(b(x))$$

(10)

Therefore, the computation of the expected likelihood kernel can be done by taking the sum of values $p(b(x))$ within candidate target T. The output of the following algorithm is a support map using integral image to compute the similarity measure between target and candidate region from each frame of the video sequence.

4 Adaptive Scale of Target Model

A target is represented by an ellipsoidal region in the image. To eliminate the influence of different target dimensions. Let x_i denote the pixel locations of target model and i are all the pixels that belong to the object tracker, and θ is the location of the center of the object tracker in the frame to frame in sequence video. Suppose we are given an arbitrary shape S in an image specified by a set of pixel locations x_i, i.e., $S = \{x_i\}$. The original shape S we have been initially selected manually. The covariance matrix can be used to approximate the shape of the object:

$$\theta = \frac{1}{N_s}\sum_{x_i\varepsilon S}x_i, \quad and, \quad V = \frac{1}{N_s}\sum_{x_i\varepsilon S}(x_i-\theta)(x_i-\theta)'$$

(11)

where N_s pixels that belong to the object of interest, and V describe an arbitrary elliptical region. We use here the following parametrization $s = [\theta', scale_x, scale_y, skew]'$ where $scale_x$ and $scale_y$ are the scaling and $skew$ is the skew transformation obtained from V using the unique Cholesky factorization:

$$V = \begin{bmatrix} scale_x & skew \\ 0 & scale_y \end{bmatrix}'\begin{bmatrix} scale_x & skew \\ 0 & scale_y \end{bmatrix}$$

(12)

We will refer to the state S as $s = (\theta, V)$ to explicitly highlight the dependence on θ and V. Similarly, $S(s)$ will denote the elliptical shape defined by s.

The appearance of an object is described by a set of k scalar features $r_1, \ldots, r_k$ that are extracted from the local area of an image I defined by $S(s)$. We view each r_k as a "bin" of a histogram. Let Γ be the set of pixel values $I(x_i)$, for example $\Gamma = [0, 255]^3$ for RGB images. We define a quantization function $b : \Gamma \rightarrow [1, \ldots, k]$, that associates with each observed pixel value a particular bin index k. The value r_k of the k-th bin is calculated from the elliptical image region $S(s = (\theta, V))$ using:

$$r_k(I, s) = |V|^{\frac{\gamma}{2}} \sum_{x_i \varepsilon S(s)} N(x_i; \theta, V)\delta[b(I(x_i)) - k] \qquad (13)$$

where δ is the Kronecker delta function. The kernel function N is chosen such that pixels in the middle of the object have higher weights than pixels at the borders of the objects. A natural choice is a Gaussian kernel defined by:

$$N(x; \theta, V) = \frac{1}{|2\pi V|} \exp(-\frac{1}{2}(x - \theta)^t V^{-1}(x - \theta)) \qquad (14)$$

The prefactor $|V|^{\frac{\gamma}{2}}$ in (13) discounts for the fact that in practice we use only the N_s pixels from a finite neighborhood of the kernel center. We disregard samples further than 1.6-sigma and it is easy to show that one should use $\gamma \approx 1.6$ in this case. The smooth kernel function will suppress the influence of the (arguably less reliable) pixels near the borders.

5   Proposed Approach

To ensure good organization the progress of work, we used the benefits of modular design in our approach implemented using MATLAB. The goal of an object tracking is to generate the trajectory of an object over time by discovering its exact position in every frame of the video sequence. We have implemented several object tracking algorithms (Kalman filter, Expected likelihood kernel, Adaptive scale) with different processing methods. The step of object tracking system are shown in Fig. 1.

The proposed approach for object tracking is composed of four blocks named as: Block processing, Block prediction, Block Tracking, Block correction and Block result. The functions of these blocks are as follows:

Block Processing In block processing, we start video sequence and converting video into images processing for extracting color information of images and target of object tracking.

Block Prediction Block Prediction step attempts to evaluate how the state of the target will change by feeding it through a state prediction of Kalman filter. The state prediction serves two purposes: The time update equations are responsible for projecting forward (in time) the current state and error covariance estimates to obtain the a priori estimate for the next time step.

Block Tracking In this block we combine between the expected likelihood kernel, and the integral image to compute similarity measure, and the histograms of all possible target regions of object tracking in video sequence. And we based of state predicted to estimate shape and orientation of object tracker.

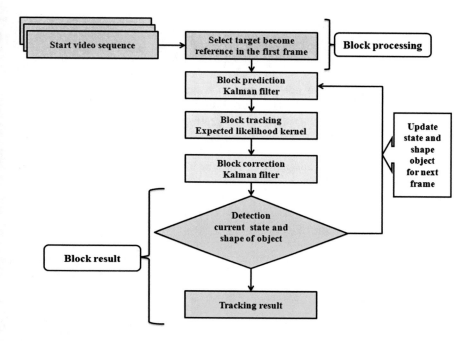

Fig. 1 Basic block diagram for proposed algorithm

Block Correction The Block correction update equations are responsible for the feedback. That is used for incorporating a new measurement into the a priori estimate to obtain an improved a posteriori estimate. The time update equations can also be thought of as predictor equations, while the measurement update equations can be thought of as corrector equations based on Block Tracking.

Block Result Tracking trajectory of object is done on the basis of the region properties of the object such as, shape, centroid, etc.

The Algorithm of the proposed approach can be explained as follows:
1. Start video sequence and select the target of object tracker in the first frame
2. Prediction using State prediction of Kalman filter to estimate how the state of an target will change by feeding it through a the current state and error covariance estimates to obtain the a priori estimate for the next time step. Using Eqs. (1) and (2)
3. Calculate similarity measure between target model and candidate regions and estimate shape and orientation of object tracker using Eqs. (10) and (13)
4. Correction and update equations into the a priori estimate to obtain an improved a posteriori estimate, using Eqs. (3)–(5), and state of similarity measure, which calculates the new position of the object
5. Draw trajectory by line joining each stored position has been drawn in every frame which shows the trajectory of the selected moving object. And go to step 2 in the next frame

6 Experiment Result

To verify the efficiency of the proposed approach (PA), we compared our system with two existing algorithms MKF [7] and MS [4], the experimental results show that, the PA system achieves good estimation accuracy of the scale and orientation of object in the sequences videos. We used different sequences, each has its own characteristics but the use of a single object in movement is a commonality between these different sequences, and we set up experiments to listed the estimated width, height, trajectory, and orientation of object. In this work, we selected RGB color space as the feature space and it was quantised into $16 \times 16 \times 16$ bins for a fair comparison between different algorithms. One synthetic video sequence and two real videos sequences are used in the experiments:

We first use a Synthetic Ellipse sequence to verify the efficiency of the proposed approach. As shown in Fig. 2. The external ellipses represent the target candidate regions, which are used to estimate the real targets, that is, the inner ellipses. The experimental results show that the proposed approach could reliably track the trajectory of ellipse with scale and orientation changes. Meanwhile, the experimental results by the MKF and MS are not good because of significant scale and orientation changes of the object.

The second video is a Occlusion sequence is on a more complex sequence. As can be seen in Fig. 3, both proposed approach and MKF [7], algorithm can track the target over the whole sequence, and MS [4], does not estimate the trajectory of target orientation change and has bad tracking results. However, the proposed approach system works much better in estimating the scale and orientation of the target, especially when occlusion occurs.

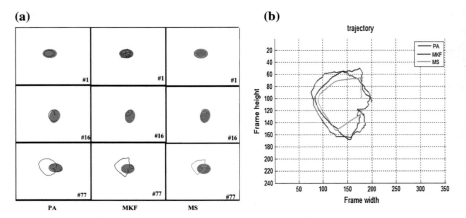

Fig. 2 **a** Tracking results of the synthetic ellipse sequence by different tracking algorithms. The frames 1, 16 and 77 are displayed. **b** Trajectory results of the synthetic ellipse video sequence by different tracking algorithms

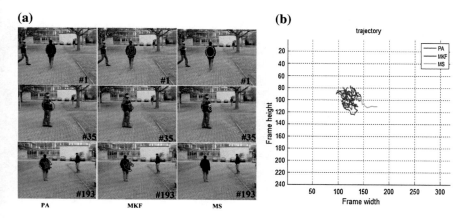

Fig. 3 **a** Tracking results of the occlusion sequence by different tracking algorithms. The frames 1, 35 and 193 are displayed. **b** Trajectory results of the occlusion video sequence by different tracking algorithms

The last video is a Player sequence where the scale of the object increases gradually as shown Fig. 4. The experimental results show that the proposed approach estimates more accurately the scale changes and good trajectory of target region than the MKF and MS algorithms.

Table 1 lists the average time by different methods on the videos sequences. We notice that our proposed approach (PA) has an average time of execution better than MKF and MS algorithms.

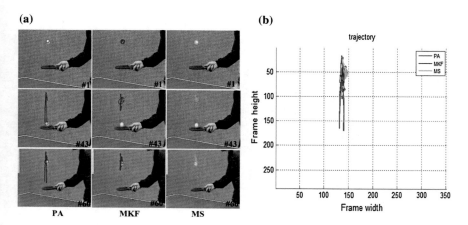

Fig. 4 **a** Tracking results of the player sequence by different tracking algorithms. The frames 1, 43 and 60 are displayed. **b** Trajectory results of the player video sequence by different tracking algorithms

Table 1 The average time by different methods on the videos sequences

Methods/sequences	PA (s)	MKF (s) [7]	MS (s) [4]
Ellipse sequence (77 frames)	0.07	0.48	0.19
Occlusion sequence (193 frames)	0.08	0.56	0.17
Player sequence (60 frames)	0.08	0.40	0.15

The experimental results demonstrate that the proposed approach is robust to track the trajectory of objects in different situations (scale variation, pose, rotation, and occlusion). It can be seen that the proposed approach achieves good estimation accuracy in real-time of the scale and orientation of the target.

7 Conclusion

In this paper, the proposed approach has been presented for tracking a single moving object in the video sequence using color information. In this approach we combine between Kalman filter and expected likelihood kernel as a similarity measure using integral image to compute the histograms of all possible target regions of object tracking in video sequence. The newly proposed approach has been compared with the state-of-the-art algorithms on a very large dataset of tracking sequences and it outperforms in the processing speed. The extensive experiments are performed to testify the proposed approach and validate its robustness to the scale and orientation changes of the target in real-time. This implemented system can be applied to any computer vision application for moving object detection and tracking.

References

1. Ning, J., Zhang, L., Zhang, D., Wu, C.: Scale and orientation adaptive mean shift tracking. IET Comput. Vis. **6**(1), 52–61 (2012)
2. Klein, D.A., Schulz, D., Frintrop, S., Cremers, A.B.: Adaptive real-time video-tracking for arbitrary objects. In: International Conference on Intelligent Robots and Systems (IROS), IEEE/RSJ, pp. 772–777 (2010)
3. Vojir, T., Noskova, J., Matas, J.: Robust scale-adaptive mean-shift for tracking. Pattern Recogn. Lett. **49**, 250–258 (2014)
4. Comaniciu, D., Ramesh, V., Meer, P.: Kernel-based object tracking. IEEE Trans. Pattern Anal. Mach. Intell. **25**(5), 564–577 (2003)
5. Kalman, R.E.: A new approach to linear filtering and prediction problems. J. Fluids Eng. **82**(1), 35–45 (1960)
6. Zivkovic, Z., Krose, B.: An EM-like algorithm for color-histogram-based object tracking. In: Proceedings of the 2004 IEEE Computer Society Conference on Computer Vision and Pattern Recognition, CVPR, vol. 1, pp. I–798 (2004)
7. Zivkovic, Z., Cemgil, A.T., Krose, B.: Approximate Bayesian methods for kernel-based object tracking. Comput. Vis. Image Underst. **113**(6), 743–749 (2009)

8. Xu, S., Chang, A.: Robust Object Tracking Using Kalman Filters with Dynamic Covariance. Cornell University
9. Salhi, A., Jammoussi, A.Y.: Object tracking system using camshift, meanshift and Kalman filter. World Acad. Sci. Eng. Technol. **64**, 674–679 (2012)
10. Jebara, T., Kondor, R., Howard, A.: Probability product kernels. J. Mach. Learn. Res. **5**, 819–844 (2004)
11. Jebara, T., Kondor, R.: Bhattacharyya and expected likelihood kernels. Learning Theory and Kernel Machines, pp. 57–71. Springer, Berlin (2003)
12. Chang, H.W., Chen, H.T.: A square-root sampling approach to fast histogram-based search. In: IEEE Conference on Computer Vision and Pattern Recognition (CVPR), pp. 3043–3049 (2010)
13. Viola, P., Jones, M.J.: Robust real-time face detection. Int. J. Comput. Vision **57**(2), 137–154 (2004)

A Lightweight System for Correction of Arabic Derived Words

Mohammed Nejja and Abdellah Yousfi

Abstract In this paper, we address the lexicon insufficient problem used in the automatic spell checker. In order to solve that deficiency, we developed an approach that aims to correct the derived words, considering that the most Arabic words are derived ones by adjusting the Levenshtein algorithm to our need. This method is based on a corpus constituted of surface patterns and roots characterized by a scaled down size compared to conventional approaches The proposed method reduced the execution time while maintaining the highest correction rate.

Keyword ANLP (Arabic Natural Language Processing) · Misspelled word · Spell checker tool · Edit distance · Surface patterns

1 Introduction

According to Lunsford and Lunsford [1], the spelling mistakes are considered among the most common mistakes when writing a text. In fact, approximately 6.5 % of all errors detected in a US national sample of college composition essays were identified as misspellings.

Arabic is one of the languages in which the spelling mistakes are frequently detected. In fact, by comparing the rate of committing a misspelling in Arabic, French and English, the Arabic language has been identified as having the highest rate. This result is due to the fact that Arabic words are much closer to each other with an average number of related forms of 26.5 while English words and French words have an average number of related forms of 3 and 3.5 respectively [2].

M. Nejja (✉)
TSE Team, ENSIAS Mohammed V University in Rabat, Morocco
e-mail: mohammed.nejja@gmail.com

A. Yousfi
ERADIASS Team, FSJES Mohammed V University in Rabat, Morocco
e-mail: yousfi240ma@yahoo.fr

© Springer International Publishing Switzerland 2016
A. El Oualkadi et al. (eds.), *Proceedings of the Mediterranean Conference on Information & Communication Technologies 2015*, Lecture Notes in Electrical Engineering 380, DOI 10.1007/978-3-319-30301-7_14

131

Automatic spelling correction is an active area where several studies have been developed to provide effective solutions for the gaps in this research area. As an example, Kukich [3] and Mitton [4] have presented a classical approach based on research in dictionaries. This approach aims at assessing whether the input string appears or not in the list of valid words. In case the string is missing from the dictionary then it is called erroneous chain. A second approach based on the calculation of the edit distance was presented by Drameau [5] and Levenshtein [6]. The main objective of this approach is to calculate the minimum number of operations required to move from a string to another. Pollock and Zamora [7] have also developed another approach that associate each dictionary word to a skeleton key. This technique is used to correct character duplication errors, deletion and insertion of some occurrences of character as well as accent mistakes.

For Arabic language, several correction techniques and studies have emerged and are available for exploitation, namely:

- Gueddah [8] suggested a new approach so as to improve planning solutions of an erroneous word in Arabic documents by integrating frequency editing errors matrices in the Levenshtein algorithm.
- A new approach has been advised by Bakkali [9] based on the use of a dictionary of the stems of Buckwalter to integrate morphological analysis in the Levenshtein algorithm.

In this paper, we present an improved extension of the approach proposed by Nejja and Yousfi [10]. The approach exposed in a previous work is based on using surface patterns to overcome the lexicon insufficient problem. Thereby, this approach aims at improving the identification accuracy of the surface pattern nearest to the wrong word so as to properly classify surface patterns with the same edit distance.

2 Correction by the Levenshtein Distance

The Levenshtein algorithm (also called Edit-Distance) calculates the least number of edit operations that are necessary to modify one string to obtain another string.
Elementary editing operations considered by Levenshtein are:

- Permutation (لعب [laåiba: to play] → لعت[laåita])
- Insertion (سمع [samiåa: to hear] → شسمع[šasamiåa])
- Deletion (جمع [jamaåa: to collect] → جع[jaå])

The Levenshtein algorithm uses the matrix of $(N + 1) * (P + 1)$ (where N and P are the lengths of the strings to compare T, S) that allows calculating recursively the distance between the strings T, S. The matrix can be filled from the upper left to the lower right corner. Each jump horizontally or vertically corresponds to an insert or a delete, respectively. The cost is normally set to 1 for each of the operations. The diagonal jump can cost either one, if the two characters in the row and column do not match or 0, if they do. The calculation of the cell M[N, P] equals to the minimum between the elementary operations:

$$M(i,j) = Min \begin{cases} M(i-1,j)+1 \\ M(i,j-1)+1 \\ M(i-1,j-1)+Cost(i-1,j-1) \end{cases} \quad (1)$$

where

$$Cost(i,j) = \begin{cases} 0 & \text{if } T(i) = S(j) \\ 1 & \text{if } T(i) \neq S(j) \end{cases} \quad (2)$$

3 The Surface Pattern

Arabic pattern essentially aims at identifying the structure of most of the words. The patterns allow producing Stems from a root or conversely extracting the root of a word. The Patterns are variations of the word فعل [faåala] which are obtained by using diacritics or adding of affixes.

The surface pattern [11] is a way to present the morphological variations of words that are not submitted by the classical scheme. Example: The conjugation of the verb رَعَى [raåa] to the active participle in the 1st person singular is رَاعٍ [râåin]; therefore, the surface pattern of the root رَعَى [raåa] is فَعَى [faåa] and فَاع [fâåin] is the surface pattern of رَاعٍ [râåin]. The surface pattern of آجِرٌ [ajiron] is آفِعٌ [afiåon] and of آجِرَاتٌ ['ajirâton'] is آفِعَاتٌ [afiåâton] [11].

4 The Morphological Correction by Surface Patterns in the Levenshtein Algorithm

An automatic spelling correction system is a tool that allows analyzing and eventually correcting spelling mistakes. To this end, that system uses a dictionary to compare the text's word to the dictionary's words.

However, the dictionary's size is considered as a major concern in the automatic spelling correction. In order to have an efficient automatic spelling correction system, those dictionaries need to contain all the words of the processed language as well as linguistic information of each word.

Some techniques are expected to use modules to calculate edition distances. Other techniques are meant for exploiting the morphological analysis. The objective of these techniques is to remedy the deficiency of the used dictionary.

In the same context, and in order to solve that deficiency, we developed an approach that aims to correct the derived words, considering that the most Arabic words are derived ones.

This approach consists in finding first the surface pattern that is the nearest lexically to the misspelled word. Then, the word is corrected through this surface pattern and using an assigned method. To identify the nearest surface scheme to the misspelled word, we adopted the Levenshtein algorithm to the Arabic language, and extended it in a way to select the nearest surface scheme to the word input.

We note by:

- $A = \{A_1, A_2, ..., A_n\}$: all patterns.
- W_{err}: erroneous word
- $\beta = \{\text{'ل'},\text{'ع'},\text{'ف'}\}$: the basic letter.

Therefore, the Levenshtein algorithm adapted to extract the correct surface patterns is defined (for all W_{err}, A_n) by:

$$M(k,p) = Min \begin{cases} M(k-1,p)+1 \\ M(k,p-1)+1 \\ M(k-1,p-1)+Cost(k-1,t-1) \end{cases} \tag{3}$$

where

$$Cost(k,p) = \begin{cases} 1 & \text{if } A(k) \sim\; = B(p) \text{ and } A(k) \notin \beta \\ 0 & \text{if } A(k) = B(p) \text{ or } A(k) \in \beta \end{cases} \tag{4}$$

We denote by $U(k)$ the letter of the word U at position k.

4.1 Approach 1

This approach consists in finding the nearest surface pattern lexically to the misspelled word using the formula 3, then correcting the word through the identified surface pattern. For example, for the misspelled word شتلعيون [šatalåayûna] the nearest surface pattern is ستفعلون [satafåalûna] so the corrected word is ستلعيون [satalåayûna]. As soon as we have the corrected word, we extract the potential root based on the letters $\beta = \{\text{'ل'},\text{'ع'},\text{'ف'}\}$ of the identified surface pattern. For our example ستلعيون [satalåayûna], the potential root is لعي [laåaya]. We then compare that potential root with the roots in our base to get the nearest one in such a way that the root's size is equal to the size of β (the surface pattern تفعلل [tafaålala] has a root's size equal to 4 because the size of $\beta = \{\text{'ف'},\text{'ع'},\text{'ل'},\text{'ل'}\}$ is 4). For our example, the correct root is لعب [laåiba]. At the end, we gather that information to construct the correct word, which is in our example ستلعبون [satalåabûna].

4.2 Approach 2

This new approach was developed to remedy the gap of the previous one. Indeed, the previous approach showed an issue when the deleted characters were characters of the root word. For example, if we use the previous approach with the misspelled word سضيرب [saDayribo] and the surface pattern سيفعل [sayafåalo] the character ض [D] will be deleted. Due to this gap, we improved the first approach in such a way that the deleted characters belong to both the surface pattern and the misspelled word and take up the same position in both of them. For the word شتتبون [šatatibûna] and the surface pattern ستفعلون [satafåalûna] the characters ون [ûna] will be deleted. That way, we make sure that we only deleted the characters that belong to the affixes of the concerned surface pattern.

That approach has proven its efficiency. Besides, and considering that the obtained results were satisfactory, we changed the formula (4) to improve the classifying of the selected surface pattern [12].

$$
M(k,p) = Min \begin{cases} M(k-1,p)+1 \\ M(k,p-1)+1 \\ M(k-1,p-1)+cost(k-1,t-1) \end{cases} \tag{5}
$$

where

$$
cost(k,p) = \begin{cases} 1 \; if \; A(k) \neq B(p) \; and \; A(k) \notin \{ل ع ف\} \\ 0 \; if \; A(k) \in \{ل ع ف\} \\ 0{,}2 \; if \; A(k) = B(p) \; and \; A(k) \notin \{ل ع ف\} \; and \; p = k \\ 0{,}5 \; if \; A(k) = B(p) \; and \; A(k) \notin \{ل ع ف\} \; and \; p \neq k \end{cases} \tag{6}
$$

The rectifications we introduced to the formula helped us improve the precision of identifying the surface pattern having the same Edit-Distance while displaying first the most adaptable solution (Fig. 1).

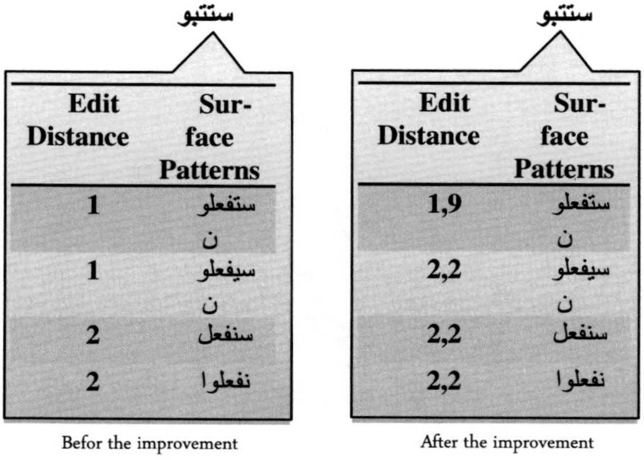

Fig. 1 An example of the result provided by our improvement

Edit Distance	Surface Patterns
1	ستفعلو ن
1	سيفعلو ن
2	سنفعل
2	نفعلوا

Befor the improvement

Edit Distance	Surface Patterns
1,9	ستفعلو ن
2,2	سيفعلو ن
2,2	سنفعل
2,2	نفعلوا

After the improvement

5 Test and Result

In order to evaluate the automatic spelling correction system efficiently, the ranking of the correct word in comparison with other candidates should be identified. And to achieve this, we have chosen, in a first place, to display the first 10 solutions for each erroneous word in order to obtain satisfactory results.

To test our method, we have performed a comparison between our approach and that of Levenshtein. This approach has been evaluated on 10000 erroneous words. For this reason, we considered:

- A training corpus containing 290 words for our approach (40 of surface patterns and 250 of root).
- A training corpus containing 10,000 words for the Levenshtein algorithm.
- Ever since the tests have been done, we have obtained a set of results:
- The characteristics of the used machine are:
- **System**: *Win XP.*
- **Memory**: *1G.*
- **Processor**: *Intel® Pentium® Dual CPU* 1.46 GHz. (Table 1).

We notice that our approach has reduced, considerably, the execution time which may be due to the lexicon size adopted by our method. In fact, the lexicon size used in this study is reduced compared to conventional approaches that are based on the edit distance.

Thereby, thanks to the improvement contributed to approach 2, we were able to increase classification accuracy of selected words while keeping a high rate of correction.

Table 1 Comparative table between our approach and method of Levenshtein

10,000 W_{err}	Method of Levenshtein	Approach 1	Approach 2
Nbr of solution in position 1 (%)	48.13	50.1	54.32
Nbr of solution in position 2 (%)	13.04	15.48	16.81
Nbr of solution in position 3 (%)	6.59	4.38	7.98
Nbr of solution in other position (%)	2.16	7.84	9.17
Correction rate (%)	69.72	77.8	84.84
Time/nbr W_{err} (ms)	0.14128	0.03178	0.03188

6 Conclusion

The automatic spell checker is a system that allows correcting spelling mistakes committed in a text, by using a set of methods often based on dictionaries. In fact, if the word processed by the system belongs to the adopted dictionary, it will be accepted as a word of the language; otherwise, the correction system will report it as a misspelled word and will suggest a set of similar words in it.

Inadequate vocabularies used in dictionaries are the major problem that hinders the most existing spell-checkers, consequently that requires a very large size to form a dictionary which can contain all possible terminologies.

Today, many works in ANLP were been done in order to focus on the development of independent methods of vocabulary's correction, by including morphological analysis, syntax, context, etc.

To remedy this problem, we were interested in this article in reducing the size of lexicon used. Thus, our proposed method deals with a particular case of derived words correction. This is due to the fact that most Arabic words are derived ones. Therefore, we have started by describing the way in which the Levenshtein algorithm was adapted to extract the nearest surface pattern of the erroneous word pattern. Then we have proposed a new solution for word's correction.

Thanks to our new approach, we were able to reduce the size of the dictionary, which reflects positively on the performance of our system while maintaining a higher coverage.

Among the performance criteria such spelling correction systems, we include the number of candidates proposed for a misspelled word. It is in this context that our next work is progressing. Actually, we aim to extend this study in order to reduce the number of words proposed candidates that have the same frequency of occurrence for a misspelled word.

References

1. Lunsford, A.A., Lunsford, K.J.: Mistakes are a fact of life: a national comparative study. Coll. Compos. Commun. **59**(4), 781–806 (2008)
2. Ben Othmane Zribi, C., Zribi, A.: Algorithmes pour la correction orthographique en arabe. TALN, 12–17 (1999)
3. Kukich, K.: Techniques for automatically correcting words in text. ACM Comput. Surv. **24**, 377–439 (1992)
4. Mitton, R.: English Spelling and the Computer. Longman Group (1996)
5. Damerau, F.J.: A technique for computer detection and correction of spelling errors. Commun. Assoc. Comput. Mach. (1964)
6. Levenshtein, V.: Binary codes capable of correcting deletions, insertions and reversals. SOL Phys Dokl, 707–710 (1966)
7. Pollock, J.J., Zamora, A.: Automatic spelling correction in scientific and scholarly text. Commun. ACM **27**(4), 358–368 (1984)
8. Gueddah, H., Yousfi, A., Belkasmi, M. :Introduction of the weight edition errors in the Levenshtein distance. Int. J. Adv. Res. Artif. Int., 30–32 (2012)

9. Bakkali, H., Yousfi, A., Gueddah, H., Belkasmi, M.: For an independent spell-checking system from the Arabic language vocabulary. Int. J. Adv. Comput. Sci. Appl. **5** (2014)
10. Nejja, M., Yousfi, A.: Correction of the Arabic derived words using surface patterns. In: Workshop on Codes, Cryptography and Communication Systems (2014)
11. Yousfi, A.: The morphological analysis of Arabic verbs by using the surface patterns. Int. J. Comput. Sci. Issues, **7** (2010)
12. Nejja, M., Yousfi, A.: Classification of the solutions proposed in the correction of the Arabic words derived using the use of surface patterns. In: Workshop on Software Engineering and Systems Architecture (2014)

A New Multi-Carrier Audio Watermarking System

Mohammed Khalil and Abdellah Adib

Abstract This paper presents a new audio watermarking system based on multi-carrier modulation. The proposed system improves the performances of spread spectrum modulations widely used in the existing state of the art watermarking systems. The embedded data is transmitted in the phase of the modulated signal and the amplitude of the carrier signal is computed according to the inaudibility constraint. The new system allows an embedding data rate equivalent to spread spectrum ones with almost no loss of data transparency and with an acceptable detection reliability.

Keywords Audio watermarking · Multi-carrier · Spread spectrum · Modulation

1 Introduction

Audio watermarking, primarily proposed as a potential solution for copyright protection, can be also used for hidden data transmission [1]. The problem can be viewed as a realization of a basic communications system, where the audio signal represents the channel noise and the watermark is the transmitted information [1]. However, watermarking schemes present major differences with communication ones. This is due to audio noise features, illustrated in Table 1, that makes the watermark extraction very difficult.

Audio watermarking related to data transmission interested several researchers [2]. Spread Spectrum (SS) methods remain among the most widely used techniques [3], because of the two following reasons (i) Transmission of the watermark signal with low power and hence a satisfaction of the inaudibility constraint,

M. Khalil (✉) · A. Adib
LIM@II-FSTM, B.P.146, 20650 Mohammedia, Morocco
e-mail: medkhalil87@gmail.com

A. Adib
e-mail: adib@fstm.ac.ma

© Springer International Publishing Switzerland 2016
A. El Oualkadi et al. (eds.), *Proceedings of the Mediterranean Conference on Information & Communication Technologies 2015*, Lecture Notes in Electrical Engineering 380, DOI 10.1007/978-3-319-30301-7_15

Table 1 Noise in communication and watermarking systems

Digital communication	Audio watermarking
White	Colored
Stationary	Not stationary
Gaussian	Not Gaussian
Weak power	High power
Uncorrelated with the channel response	Correlated with the channel response

(ii) Robustness against ISI (Inter-Symbol Interference) and hence an improved transmission reliability.

On the other side, bandpass modulations, widely used in telecommunications, have acquired considerably less amount of attention in audio watermarking. Indeed, few works have employed carrier modulations in audio watermarking [4–6]. In [5], authors propose to use frequency modulation to embed low bit-rate data imperceptibly. Another method is proposed in [6], where multiple watermarks are embedded in an audio content using frequency division multiplexing based on phase modulation. In that work, three kinds of watermarks are embedded with low-bit rates and two of them are not extracted blindly. In [4], the proposed method combines SS and phase modulations to ensure the inaudibility of the embedded watermark and to increase the robustness against acoustic path transmission.

In this paper we propose a new approach that will be analyzed thoroughly from various aspects, including not only the inaudibility but also the capacity and the robustness of the embedded information. The main contribution of this paper consists of a new system based on multi-carrier modulations that improves performances of standard SS watermarking algorithms. The new system is implemented by a specific control of different modulated signal parameters. The amplitude of the modulated signal is controlled by the inaudibility constraint to ensure the perceptual transparency of the embedded watermark. Then, the hidden data will be transmitted in the phase of the modulated signal.

The rest of this paper is outlined as follows: In Sect. 2 we present the spread spectrum watermarking system. The new multi-carrier watermarking system is presented in Sect. 3. Finally, performance of the new watermarking system is evaluated by a number of experiments in Sect. 4.

2   SS-Based Watermarking System

The information contained in the watermark is represented as an independent and identically distributed binary sequence $b_i = \{0, 1\}$. We group the sequence b_i to form a symbol sequence a_l (each symbol a_l is chosen from an alphabet of $M = 2^n$ symbols, where n represents the number of bit per symbol) [7]. The modulation step is characterized by spread spectrum waveforms (pseudo-random codes), where each one corresponds to a given symbol a_l [3]. The modulated signal $v(t)$ will be given then by:

$$v(t) = \sum_l a_l u(t - lT_s) \tag{1}$$

where $u(t)$ represents the spreading waveforms and T_s is the symbol time. To satisfy the inaudibility constraint, the embedded signal power should be controlled by filtering the modulated signal $v(t)$ by a shaping filter $H(f)$; so that the resulting watermark signal $w(t)$ will have the nearest PSD (Power Spectral Density) $S_w(f)$ to the masking threshold $S_{mask}(f)$ [8]. Then, the watermarked signal $y(t)$ is written as:

$$\begin{aligned} y(t) &= \alpha w(t) + x(t) \\ &= v(t) * h(t) + x(t) \end{aligned} \tag{2}$$

where $h(t)$ is the impulse response of $H(f)$, α is a scale factor and $*$ denotes the convolution product.

At the receiver, the signal $y(t)$ is passed through a zero-forcing equalizer $G(f) = \hat{H}^{-1}(f)$ [8]. However, since the audio signal $x(t)$ used to derive $H(f)$ is not available at the receiver, an estimate $\frac{1}{\hat{H}(f)}$ is computed by psychoacoustic modeling of the watermarked signal $y(t)$. The estimated signal $\hat{v}(t)$ is submitted initially to a demodulation step that assesses the similarity between $\hat{v}(t)$ and the waveforms known at the receiver. Finally, a decision step determines the estimated binary sequence $\hat{b}_i$.

3 Multi-Carrier Audio Watermarking

The principle of MC (Multi-Carrier) modulation is to convert a serial high-rate bits into N_c low-rate sub-streams. Each of N_c symbols from serial-to-parallel (S/P) conversion is carried out by different sub-carriers. Since the symbol rate on each sub-carrier is much less than the initial serial data symbol rate, the ISI effects, significantly decrease. Due to the S/P conversion, the duration of transmission time for N_c symbols will be extended to $N_c T_s$, which forms a single MC symbol with a length of T_{sym} (i.e., $T_{sym} = N_c T_s$). The complex envelope of the modulated MC signal has the form:

$$v_{CE}(t) = \frac{1}{N_c} \sum_{i=0}^{N_c-1} s_i \exp(j2\pi f_i t) \quad 0 \le t < T_{sym} \tag{3}$$

where s_i are the N_c parallel modulated source symbols and the complex exponential signals $\{\exp(j2\pi f_i t)\}_{i=0}^{N_c-1}$ represent different sub-carriers at the corresponding f_i in the MC signal.

The most important requirement of the watermarking system is the inaudibility constraint. In [9], a shaping filter $H(f)$ is designed by a Psychoacoustic Auditory Model (PAM) in such a way that the modulated signal $v(t)$ has an unit power,

$$\sigma_v^2 = 1 \tag{4}$$

In order to use the shaping filter $H(f)$ in the proposed scheme, we must adapt the amplitude of the MC signal while satisfying (4). Each sub-carrier component of the MC symbol with the effective duration T_{sym}, can be regarded as a single-tone signal multiplied by a rectangular window of length T_{sym}. The modulated signal $v(t)$ can be written as the sum of N_c sinusodal random phase [10] with constellation points positioned at uniform angular spacing around a circle.

$$v(t, \varphi) = \sum_{i=0}^{N_c-1} A \cos(2\pi f_i t + \varphi) \tag{5}$$

where A is the amplitude of each single-tone signal and the phase φ is a random variable that is uniformly distributed over the interval $[-\pi, \pi]$ with a probability density [10]:

$$p(\varphi) = \frac{1}{2\pi} \quad with \quad \varphi \in [-\pi, \pi] \tag{6}$$

According to (5) and (6), we compute the statistical moments of the random variable $v(t, \varphi)$ at a given time t_k,

$$
\begin{aligned}
E[v(t_k, \varphi)] &= E[v(t_k)] \\
&= \int_{-\infty}^{\infty} p(\varphi) \sum_{i=0}^{N_c-1} A \cos(2\pi f_i t_k + \varphi) d\varphi \\
&= \frac{1}{2\pi} \sum_{i=0}^{N_c-1} A \int_{-\pi}^{\pi} \cos(2\pi f_i t_k + \varphi) d\varphi \\
&= 0
\end{aligned}
\tag{7}
$$

and

$$
\begin{aligned}
E[v^2(t_k, \varphi)] &= E[v^2(t_k)] \\
&= \int_{-\pi}^{\pi} p(\varphi) \sum_{i=0}^{N_c-1} A^2 \cos^2(2\pi f_i t_k + \varphi) d\varphi \\
&= N_c \frac{A^2}{2}
\end{aligned}
\tag{8}
$$

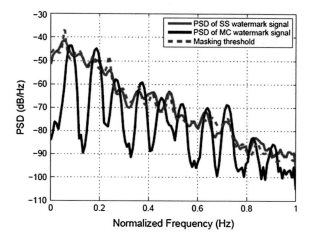

Fig. 1 Comparison of PSD watermark signal (in the case of MC and SS techniques) with the masking threshold

Hence, the modulated signal $v(t)$ is zero mean with a variance $\sigma_v^2 = N_c \frac{A^2}{2}$. In order to achieve the inaudibility constraint (4), we have to choose $A = \sqrt{2/N_c}$. Then, the modulated signal $v(t)$ passed through the shaping filter $H(f)$ so that the resulting watermark signal $w(t)$ will have the nearest PSD $S_w(f)$ to the masking threshold $S_{mask}(f)$. Figure 1 shows the PSD of watermark signal $w(t)$ using MC and SS modulations compared to the masking threshold signal in the spectral domain. It illustrates that the MC watermark signal power is considered as the sum of the frequency shifted *sinc* functions in the frequency domain.

4   Simulation Results

Several experiments are carried out to test the performances of the proposed audio watermarking system. In this experiment, the well-known SQAM[1] audio signals are selected as test samples to analyze the performances of the proposed method. A total of 30 host audio signals with different styles (single instruments, vocal, solo instruments, orchestra and pop music) and sampled at 44.1 kHz are used. The performance of the proposed watermarking system is assessed according to the four following constraints:(1) the inaudibility of the watermark, (2) the detection reliability and (3) the robustness against disturbances. All of these constraints are evaluated according to (4) the embedded data rate.

The watermark transparency is the first constraint imposed to any watermarking system. In our study, the choice was made on the Perceptual Evaluation of Audio Quality (PEAQ) algorithm that provides an ODG values describing the perceptual

---

[1]Available at http://soundexpert.org/sound-samples.

Fig. 2 Comparison of ODG
values between MC and SS
watermarking systems

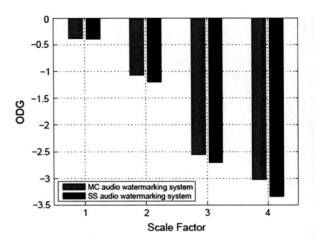

difference, from imperceptible (when the ODG is 0) to very annoying (when the
ODG is −4) [11].

Figure 2 presents the obtained ODGs for SS and MC modulations according to
the scale factor α. We have changed the value of α in order to investigate the effect
of watermark power on the inaudibility. By increasing α, lower quality of the
watermarked signal is achieved. The obtained ODG values confirm that by using
the adaptive regulation amplitude of MC modulation, the inaudibility constraint is
ensured almost like SS method. Although tones[2] are more audible and notable than
random noise,[3] the transparency of the watermark in the MC system, is ensured
because of the satisfaction of the constraint defined by the Eq. (4).

Once, the inaudibility constraint is satisfied, it will be necessary to evaluate the
detection reliability. We vary the embedded data rate from 0 to 500 b/s and we
compare the performances. Figure 3 shows a small improvement of MC method
compared to SS one, in terms of detection reliability, when the channel is free from
perturbations.

Finally, the watermarked signal has been submitted to AWGN (Additif White
Gaussian Noise) with a SNR (Signal to Noise Ratio) varying from −10 dB to 30 dB
in order to evaluate the robustness of the proposed method. The choice of this
perturbation stems from previous analytic works that have assumed the effect of
various distortions on the overall watermarked signal as a stationary AWGN [12].
The watermark bit rate has been fixed to R = 200 b/s. Figure 4 compares the
robustness of MC et SS systems against AWGN. We notice that the robustness is
improved when we increase the SNR. On the other side, MC audio watermarking
system presents an equivalent performance to SS one with a little improvement of
the MC system.

---

[2]Used in MC watermarking system.

[3]Used in SS watermarking system.

Fig. 3 Detection reliability of different modulations: MC and SS

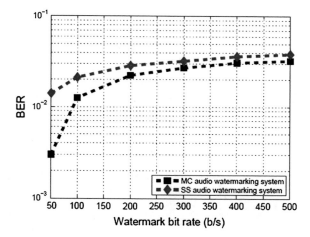

Fig. 4 Robustness against AWGN

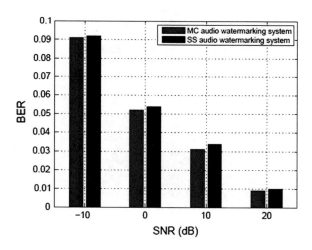

5 Conclusion

In this paper, we have proposed a new method based on MC modulations in audio watermarking. The experimental results have shown some advantages of MC modulations compared to SS ones. One particularity of the proposed method is the judicious choice of MC parameters that have been adapted to the watermarking context in order to improve the inaudibility of the watermark. The next stage of our work is to increase the robustness of the proposed method against various perturbations like MPEG compression.

References

1. Cox, I.J., Miller, M.L., McKellips, A.L.: Watermarking as communications with side information. Proc. IEEE. **87**(7), 1127–1141 (1999)
2. Khalil, M., Adib, A.: Audio watermarking with high embedding capacity based on multiple access techniques. Digit. Signal Proc. **34**, 116–125 (2014)
3. Khalil, M., Adib, A.: Audio watermarking system based on frequency hopping modulation. In: International Conference on Multimedia Computing and Systems (ICMCS), pp. 211–216 (2014)
4. Arnold, M., Baum, P.G., VoeÃŸing, W: A phase modulation audio watermarking technique. Lecture Notes in Computer Science. vol. 5806, pp 102–116 (2009)
5. Girin, L., Marchand, S.: Watermarking of speech signals using the sinusoidal model and frequency modulation of the partials. In: IEEE International Conference on Acoustics, Speech, and Signal Processing (ICASSP) (2004)
6. Takahashi, A., Nishimura, R.,Suzuki, Y.: Multiple watermarks for stereo audio signals using phase-modulation techniques. IEEE Trans. Signal Process. **53**(2) (2005)
7. Haykin, S.: Digital Communications. 4th edn. John Wiley and Sons (2001)
8. Baras, C., Dymarski, P., Moreau, N.: An audio watermarking scheme based on an embedding strategy with maximized robustness to perturbations. In: IEEE International Conference on Acoustics, Speech, and Signal Processing (ICASSP) (2004)
9. Baras, C., Moreau, N., Dymarski, P.: Controlling the inaudibility and maximizing the robustness in an audio annotation watermarking system. IEEE Trans. Audio Speech Lang. Process. **14**(5), 1772–1782 (2006)
10. Khalil, M., Adib, A.: Informed audio watermarking based on adaptive carrier modulation. Multimedia Tools Appl. **74**(15), 5973–5993 (2015)
11. Kondo, K.: On the use of objective quality measures to estimate watermarked audio quality. In: International Conference on Intelligent Information Hiding and Multimedia Signal Processing (2012)
12. Kundur, D., Hatzinakos, D.: Diversity and attack characterization for improved robust watermarking. IEEE Trans. Signal Process. **49**(10), 2383–2396 (2001)

Efficient Implementation of Givens QR Decomposition on VLIW DSP Architecture for Orthogonal Matching Pursuit Image Reconstruction

Mohamed Najoui, Anas Hatim, Mounir Bahtat and Said Belkouch

Abstract Orthogonal Matching Pursuit (OMP) is one of the most used image reconstruction algorithm in compressed sensing technique (CS). This algorithm can be divided into two main stages: optimization problem and least square problem (LSP). The most complex and time consuming step of OMP is the LSP resolution. QR decomposition is one of the most used techniques to solve the LSP in a reduced processing time. In this paper, an efficient and optimized implementation of QR decomposition on TMS320C6678 floating point DSP is introduced. A parallel Givens algorithm is designed to make better use of the 2-way set associative cache. A special data arrangement was adopted to avoid cache misses and allow the use of some intrinsic functions. Our implementation reduces significantly the processing time; it is 6.7 times faster than the state of the art implementations. We have achieved a 1-core performance of 1.51 GFLOPS with speedups of up to x20 compared to Standard Givens Rotations (GR) algorithm.

Keywords Orthogonal matching pursuit · Image reconstruction · Compressed sensing · QR decomposition · Givens rotations · TMS320C6678

M. Najoui (✉) · A. Hatim · M. Bahtat · S. Belkouch
LGECOS Lab ENSA-Marrakech, University of Cadi Ayyad,
Marrakech, Morocco
e-mail: m.najoui@uca.ma

A. Hatim
e-mail: hatim.anas@gmail.com

M. Bahtat
e-mail: mnr.bahtat@gmail.com

S. Belkouch
e-mail: s.belkouch@uca.ma

© Springer International Publishing Switzerland 2016
A. El Oualkadi et al. (eds.), *Proceedings of the Mediterranean Conference on Information & Communication Technologies 2015*, Lecture Notes in Electrical Engineering 380, DOI 10.1007/978-3-319-30301-7_16

1 Introduction

Recently, compressed sensing (CS) [1] has received a considerable attention in many applications such as communication [2], radar and biomedical imaging [3]. In image processing, CS opens a radically new path to reconstruct images from a number of samples which is far smaller than the desired resolution of the images. In radar imaging applications [4, 5] CS is used due to its fast and efficient signal processing ability while in Medical Resonance Imaging [6] it is used to reconstruct the image from sparse measurements and reduces scan time, which is proportional to the number of acquired samples. CS can reconstruct the original signal by multiplying a measurement matrix, which shall be known in advance, by the compressively sampled input signal. The reconstruction process consists of finding the best solution to an underdetermined system of linear equations given by (1).

$$y = \Phi x \tag{1}$$

Where the measurement matrix $\Phi \in R^{MxN}$ and the measured signal $y \in R^M$ are known.

The CS reconstruction is very complex and requires high computational intensive algorithms. Several reconstruction algorithms have been proposed and most of them are computationally intensive. OMP [7] is one of the most widely used algorithms in CS reconstruction. The main stage in OMP algorithm is the least square problem which is based on matrix inversion and multiplication operations. When matrix lengths are important these operations are complex and need huge resources to be executed. The least square problem can be resolved by QR decomposition (QRD) algorithms. However, QRD is considered a computationally expensive process, and its sequential implementations fail to meet the requirements of many time-sensitive applications. There are three basic algorithms for computing QR decomposition: the Modified Gram-Schmidt (MGS) algorithm, Householder reflections and Givens Rotations (GR). All these algorithms require multiplications, divisions and square-root operations, resulting in high hardware complexity and computation latency. The Householder and MGS algorithms lend themselves to efficient sequential implementation; however their inherent data dependencies complicate parallelization. On the other hand, the structure of Givens rotations provides many opportunities for parallelism due to its matrix updates independencies, but it is typically limited by the availability of computing resources. In addition, it was proven that Givens rotations algorithm is stable and robust numerically [8, 9].

In this paper, we introduce an efficient parallel implementation scheme of GR QRD that fits well on VLIW architecture. The data level parallelism (DLP) and thread level parallelism (TLP) of the TMS320C6678 VLIW DSP were exploited. All the useless arithmetic operations were avoided. We took advantage of the memory hierarchy to make efficient data access using 2-way set associative caches. This will allow us to make parallel processing with lower delays and cycles. The results of our optimized implementation prove that our algorithm speedups the conventional implementation by a factor of 20x with a 1-core peak performance of 1.51 GFLOPS.

This paper is organized as follows. In Sect. 2, a comparison between the state of the art works is presented. The Standard GR Algorithm and some improvements are described in Sect. 3. Section 4 presents the optimized implementation, the experimental results and the discussion. The conclusion and the future work are outlined in Sect. 5.

2 Related Work

Nikolic et al. [10] compared real time performance between floating point and fixed point implementations for five linear algebra algorithms including Householder QR decomposition. They have introduced an advanced code optimization based on inline function expansion that saves the overhead of functions call like inverse square root and division and allow optimizations. An implementation by DSP-specific fixed point C code generation was also presented. Compiler-driven optimizations were used to achieve more performance. They have implemented their algorithm on two floating point targets: TMS320C6713 and TMS320C6727 DSPs and on the fixed point DSP TMS320C6455; these targets runs at 200 MHz and 300 MHz and 700 MHz, respectively. These targets execute the QR algorithm in 8006 and 3201 and 5893 cycles respectively for 5×5 matrix size. We mention that in this study both levels of the CPU cache were enabled and all data and program were kept in the internal memory. While in [11] a new algorithm called Modified Gram-Schmidt Cholesky Decomposition was proposed. The algorithm combines the low execution time advantage of Cholesky decomposition and the robustness against the round-off error propagation of the modified Gram-Schmidt QR factorization. The algorithm was implemented on the fixed point TMS320C6474 running at 1 GHz clock rate. The algorithm takes more than 100 μs to be executed for 10×16 input matrix size. Table 1 summarizes the presented algorithms and results. We note that these two works are the only DSP implementations of QRD available on the state of the art.

Table 1 Comparison between state of the art QRD implementations on DSP

Works	Target	Performances				
		Clock	# of cores	Input size	# of cycles	Processing time (μs)
[11]	TMS320C6474	1 GHz	–	10×16	–	100
[10]	TMS320C6713	200 MHz	–	5×5	8006	40.03
	TMS320C6727	300 MHz	–	5×5	3201	10.67
	TMS320C6455	700 MHz	–	5×5	5893	8.41
This work	TMS320C6678	1 GHz	1	5×5	1599	1.6
	TMS320C6678	1 GHz	1	10×16	8159	8.16

3 Givens Rotations Algorithm

3.1 Standard Givens Rotations (SGR) Algorithm

Given a matrix $A \in RMxN$ where $M < N$, using Givens QR decomposition A can be shown as:

$$A = QR \tag{2}$$

Where $R \in R^{MxN}$ is an upper triangle matrix and $Q \in R^{MxM}$ is an orthogonal matrix that satisfies (3):

$$Q^T Q = I \tag{3}$$

The Givens rotation (GR) zeros one element of the matrix at a time by two-dimensional rotation [12]. The pseudo-code of the standard Givens Rotations (SGR) algorithm is given as follows:

```
01 Q = I;
02 R = A;
03 For each (j = 1 : +1 : N)
04 Begin
05     For each (i = M : -1 : j+1)
06     Begin
07         α = sqrt((R[i-1][j])² + (R[i][j])²);
08         c = R[i-1][j] / α;
09         s = R[i][j] / α;
10         For each (k=j : +1 : N)
11         Begin
12             x = R[i-1][k];
13             y = R[i][k];
14             R[i-1][k] = c*x + s*y;
15             R[i][k] = -s*x + c*y;
16         End
17         For each (k=0 : +1 : M)
18         Begin
19             x = Q[k][i-1];
20             y = Q[k][i];
21             Q[k][i-1] = c*x  + s*y;
22             Q[k][i] = -s*x  + c*y;
23         End
24     End
25 End
```

3.2   Our Enhanced SGR Algorithm

The SGR algorithm updates the lines R_{i-1} and R_i and the columns Q_{i-1} and Q_i in each *i*th iteration. Analyzing the SGR results, some optimizations must be done before DSP implementation. Firstly, the QRD block input matrix size is *MxN* where $N \gg M$. As we can see in the pseudo-code given above, in line 05 the loop index i is initialized by M and decremented at each iteration until j+1, while j is initialized by 0 and incremented at each iteration until N. Therefore, when j is equal to or greater than M, the second loop (cf line 05) is never executed but the comparison is still done at each iteration. Consequently, all j iterations from M to N are useless for QRD process. To avoid all these lost N-M cycles of comparison operations, we have reduced the number of loops on j from N to M-1 and therefore we made important savings on the algorithm. Secondly, when updating the columns of *Q* some elements remain unchangeable between the *(i-1)*th and the *i*th iteration. For example, in the first iteration there are just two elements per column that are changed, but all the other elements remain unchangeable. We have determined an efficient formula for k loop counter (4) that leads to a significant reduction of: (i) the number of required iteration to update Q columns, and (ii) the number floating point operations per iteration.

$$Cnt = j + M - i + 1 \qquad (4)$$

4   Implementation and Discussion

4.1   Optimized C Implementation of GR Algorithm

The implementation is done on TI TMS320C6678, which is a VLIW multicore fixed and floating point DSP. It integrates 8 CorePac cores running at 1.0 GHz. A single device performance is about 128 GFLOPS [13]. Each C66x integrates four execution units exist per side, denoted .L, .S, .M and .D. Each core can performs 8 floating/fixed-point 32-bit multiplications per cycle using .M units, and 8 floating/fixed-point 32-bit additions/substractions per cycle using .L or .S units. Loads/stores can be done using .D units. The classic canonical implementation of the QR decomposition does not make a good balance in terms of resources usage within the VLIW core. Indeed, within each iteration 4 non-contiguous single precision floats are accessed from/to memory, which occupies the load/store units for at least 2 cycles per iteration, furthermore, exploiting only a half of its available bandwidth (64-bit used instead of 128-bit available). Besides, the 4 needed multiplications can be achieved theoretically using only 1 ".M" unit, leaving other units non-used along the iteration width of 2 cycles. Our optimized

Fig. 1 The optimized givens algorithm kernel's flow of operations

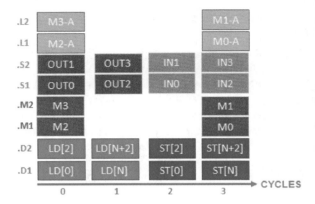

C implementation takes advantage of loop unrolling and software pipelining to guarantee the good use of resources, and to avoid cache-related penalties. To do so, during the two first cycles we load 128-bit in each one using both.D units; afterward both .M, .S and .L units are used to make the 8 required multiplications/cycle, data arrangement and addition operations, respectively. Since during each iteration: the R matrix is accessed along its lines, while the Q matrix is accessed along its columns, we aligned differently the 2 matrices in memory. The R matrix was aligned according to its lines and Q according to its columns. The cache memory will be efficiently exploited, the cache lines are well used and the cache misses are minimized due to the good alignment that we have done for the matrices to avoid the majority of memory jumps, which are the principal cause of cache misses. The kernel was built based on an unrolling value of 4. The kernel's algorithm is illustrated on Fig. 1. The 4 floating-point multiplications and 2 additions/subtractions needed per basic iteration can be expressed efficiently using a complex multiplication, which is supported in hardware as a fast intrinsic function.

To do so, some pre-arrangements and post-arrangements of registers have to be made, which are represented as INx and OUTx operations respectively. Post single precision additions (Mx-A) are necessary to complete the processing of complex multiplications. The root-square operation '1/sqrt(x)' is avoided by the use of an available intrinsic algorithm approximating its expression; therefore 2 additional multiplications are performed to get the C and S values. The pseudo-code of our optimized algorithm is the following:

```
For (n=0:M-2) do
   For (m=M-1:-1:n+1) do
      a:=R[m-1][n]; b:=R[m][n];
      r:=sqrt_reciprocal_intrinsic(a²+b²);
      sc:={ b*r, a*r }; padd:=N; loop1_cnt:=N-n ;
      loop2_cnt:=M+n-m+1 ; loop1_rst:=loop1_cnt%4;
      loop2_rst:=loop2_cnt%4;
      ptr1:=&R[m-1][n]; ptr2:=&Q[m-1][m-1-n];
      loop_cnt:=loop1_cnt/4 + loop2_cnt/4;
      loop_rst:=loop1_rst + loop2_rst;
      For (k=1:loop_cnt) do
         x0_x1:=load8(&ptr[0]); x2_x3:=load8(&ptr[2]);
         y0_y1:=load8(&ptr[padd]);
         y2_y3:=load8(&ptr[padd+2]);
         x0_y0:={ low_r(y0_y1) , low_r(x0_x1) };
         x1_y1:={ high_r(y0_y1) , high_r(x0_x1) };
         x2_y2:={ low_r(y2_y3) , low_r(x2_x3) };
         x3_y3:={ high_r(y2_y3) , high_r(x2_x3) };
         nx0_ny0:=complex_mpy(sc , x0_y0);
         nx1_ny1:=complex_mpy(sc , x1_y1);
         nx2_ny2:=complex_mpy(sc , x2_y2);
         nx3_ny3:=complex_mpy(sc , x3_y3);
         store8(&ptr[0]):={low_r(nx1_ny1),low_r(nx0_ny0)};
         store8(&ptr[2]):={low_r(nx3_ny3),low_r(nx2_ny2)};
   store8(&ptr[padd]):={ high_r(nx1_ny1),high_r(nx0_ny0)};
store8(&ptr[padd+2]):={high_r(nx3_ny3),high_r(nx2_ny2)};
         ptr:=ptr + 4;
         If (k==loop1_cnt/4) do
            padd:=M; ptr1_s:=ptr; ptr:=ptr2;
         End If
      End For
      For (k=1:loop_rst) do
         x_y:={ptr[padd],ptr[0]};
         nx_ny :=complex_mpy(sc, x_y);
         ptr[0]:=low_r(nx_ny);ptr[padd]:=high_r(nx_ny);
         ptr++;
         If (k==loop2_rst) do
            padd:=N ; ptr:=ptr1_s;
         End If
      End For
   End For
End For
```

4.2 *Experimental Results and Synthesis*

To evaluate our Optimized GR (OGR) implementation we put all the inputs and outputs data in DDR3 which is the very slow level of memory in our EVM. L1 was fully configured as cache. The number of CPU cycles against the input matrix size is illustrated in Fig. 2. The presented OGR algorithm outperforms the SGR and reduces the execution time significantly. Our OGR implementation is faster up to 20 times than the SGR. Table 2, presents the number of GFLOPS obtained for both algorithms SGR and OGR. The performances speedup (GFLOPS) achieved by our optimization are comprised between x6 and x23. The clock savings made are explained by: (i) the complexity reduction done on the algorithm, (ii) the use of low level instructions that exploit the full performances of the processor, (iii) the kernel of the presented algorithm is well scheduled and allows taking benefit of the efficiency of the architecture and (iv) the memory hierarchy was well exploited to enhance the efficiency of data access to avoid the cache misses. The comparison shown in Table 1 between state of the art DSP implementations of QR factorization proves that our optimized implementation is 6.7 times faster and 5 times faster than the floating point and fixed point implementations of work [10], respectively. It is also 12 times faster than the implementation presented in [11].

Fig. 2 Performance (# of cycles) of OGR and SGR implementations

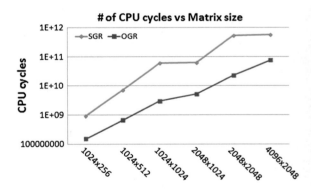

Table 2 Performance in GFLOPS

Matrix size NxM	# of GFLOPS		Speedup
	SGR	OGR	
1024 × 256	0,226644	1,404417	x6
1024 × 512	0,129347	1,398249	x10
1024 × 1024	0,068537	1,451392	x21
2048 × 1024	0,116645	1,461124	x12
2048 × 2048	0,063598	1,512419	x23
4096 × 2048	0,104313	0,794585	x7

5 Conclusion and Future Work

In this paper, we implemented a high-throughput Givens QR decomposition module for OMP image reconstruction. The number of iterations in SGR was reduced by eliminating all useless loops and arithmetic operations. Based on the SGR structure and the VLIW architecture, we have developed an optimized version of GR algorithm which fits well with our C6678 DSP. An efficient GR kernel was built to exploit the DLP and TLP levels of the processor parallelism. Also the memory cache levels were exploited to enhance the efficiency of the algorithm. We have shown that the proposed implementation scheme is efficient and can achieve a 1-core peak performance of 1.51 GFLOPS. In addition, our implementation scheme is fully scalable and can be used for higher matrix sizes in OMP algorithm. The proposed implementation will accelerate the CS reconstruction algorithms which is the main concern of our future work.

References

1. Candes, E., Wakin, M.: An introduction to compressive sampling. Sig. Process. Mag. IEEE **25**(2), 21–30 (2010)
2. Yang, D., Li, H., Peterson, G.D., Fathy, A.: "Compressed sensing based UWB receiver: Hardware compressing and FPGA reconstruction", In: Proceedings of 43rd Annual Conference Information Science Systems (CISS), pp. 198–201 (2009)
3. Dixon, A.M.R., Allstot, E.G., Chen, A.Y., Gangopadhyay, D., Allstot, D.J.: "Compressed sensing reconstruction: Comparative study with applications to ECG bio-signals", In: Proceedings of IEEE International Symposium Circuits Systems (ISCAS), pp. 805–808 (2011)
4. Herman, M.A., Strohmer, T.: High-resolution radar via compressed sensing. Sig. Process. IEEE Trans. **57**(6), 2275–2284 (2009)
5. Yu, Y., Petropulu, A., Poor, H.:"Compressive sensing for mimo radar," in acoustics, speech and signal processing, In: ICASSP 2009 IEEE International Conference, pp. 3017–3020 (2009)
6. Lustig, M., Donoho, D., Santos, J., Pauly, J.: Compressed sensing MRI. Sig. Process. Mag. IEEE **25**(2), 72–82 (2008)
7. Tropp, J.A., Gilbert, A.C.: Signal recovery from random measurements via orthogonal matching pursuit. IEEE Trans. Inf. Theor. **53**(12), 4655–4666 (2007)
8. El-Amawy, A., Dharmarajan, K.R.: Parallel VLSI algorithm for stable inversion of dense matrices, IEEE Proc. **136**(6) (1989)
9. Heath, M.T.: Numerical methods for large sparse linear least squares problems, ORNL/CSD-114, Distribution category UC-32 (1983)
10. Nikolic, Z., Nguyen, H.T., Frantz, G.: Design and implementation of numerical linear algebra algorithms on fixed point DSPs, EURASIP J. Adv. Sig. Process. **2007**, p. 22 (2007). doi:10.1155/2007/87046
11. Maoudj, R., Fety, L., Alexandre, C.: Performance analysis of modified Gram-Schmidt Cholesky implementation on 16 bits-DSP-chip. Int. J. Comput. Digit. Syst. **2**, 21–27 (2013). doi:10.12785/ijcds/020103
12. Huang, Z.Y., Tsai, P.Y.: Efficient implementation of QR decomposition for gigabit MIMO-OFDM systems, In: IEEE Transactions on Circuits and Systems—I: Regular Papers, **58**(10) (2011)
13. TMS320C6678, Multicore fixed and floating-point digital signal processor, Data Manual, Texas Instruments, SPRS691E-November 2010-Accessed March 2014, http://www.ti.com.cn/cn/lit/ds/symlink/tms320c6678.pdf

Enhanced Color Image Method for Watermarking in RGB Space Using LSB Substitution

Mouna Bouchane, Mohamed Tarhda and Laamari Hlou

Abstract Nowadays, with the rapid advancement of internet and the development of digital contents the enormous use of digital data such as multimedia services on the web become a fact. As still images, videos and other works are available in digital form, the ease with each perfect copy can be made may lead large scale unauthorized copying. In this paper a new approach for still image watermarking using least significant bits LSB technique is presented. The proposed algorithm satisfies the invisibility of the embedded colored JPEG image into the cover data under normal viewing conditions. The experimental results based on Matlab simulation by testing Mean square error (MSE), Peak signal to noise Ratio (PSNR) and Normalized Correlation (NC) show that the quality of the watermarked image of our system is higher.

Keyword Digital watermarking · LSB technique · RGB image · PSNR · NC

1 Introduction

The information society, with its attendant technological facilities to broadcast and duplicate any type of digital media represents new challenges to intellectual property rights holder [1].

The original version of the book was revised: The spelling of the author's name was corrected. The erratum to this chapter is available at 10.1007/978-3-319-30301-7_62

M. Bouchane (✉) · M. Tarhda · L. Hlou
Electrical Engineering and Energy Systems Laboratory, Ibn Tofail University, BP. 133, Kenitra, Morocco
e-mail: mouna.bouchane@gmail.com

M. Tarhda
e-mail: tarhdamo@yahoo.fr

L. Hlou
e-mail: hloul@yahoo.com

© Springer International Publishing Switzerland 2016
A. El Oualkadi et al. (eds.), *Proceedings of the Mediterranean Conference on Information & Communication Technologies 2015*, Lecture Notes in Electrical Engineering 380, DOI 10.1007/978-3-319-30301-7_17

157

Digital watermarking based on image processing is the technology of embedding imperceptible and undetectable information such as personal information or logo into a cover data in order to proof the paternity of a digital work. It provides a possible solution to the problem of piracy and duplication of images, since it enables us to protect intellectual property rights by embedding secret information in it. This secret data called "Watermark" can be more or less visible digital signature as a text, image or photo. The most important challenge is to find the good representation of the compromise between robustness, invisibility and capacity of hiding information.

Depending on the application, digital watermarking techniques include the two main domains: Transform and Spatial one. The former is a space where the image is considered as a sum of the frequencies of different amplitudes, the process of embedding showed better robustness against compression, pixel removal, rotation and shearing. Nevertheless, it required the transition to the transform domain [2]. The latter is used to embed the watermark by directly modifying the pixel values of the original image. LSB technique is one of the most commonly algorithm employed in spatial domain because of its insertion progress. It's based on directly modifying the Least Significant Bits (LSB) of the host data. The invisibility of the watermark is achieved on the assumption that the LSB data are visually insignificant [3].

In this paper, we present a new approach of color image watermarking scheme in RGB space using LSB substitution. Most of researches have proposed the first LSB, the third LSB and even the forth LSB for hiding data because of the lack of embedding bits. The proposed method ensures watermark transparency by substituting the least significant bit on each RGB component of the cover data into the watermark. The capacity of embedding is achieved by applying a scaling to the cover image using a pseudo random factor, that obviously increase redundant blocks allowing us to embed more information.

The remaining paper is organized as follows:

In the next section, we describe some related work. In Sect. 3 we present the proposed algorithm. In Sect. 4 we provide experimental results that demonstrate the performance of our approach. Final remarks are concluding our paper in the last section.

2   Related Work

According to literature, several works are going on to make watermarking techniques immune towards attacks and to retain the originality of watermark by providing successful extraction of watermark with low error probabilities. In [4] a scramble is applied to the secret image before embedding process. The extraction algorithm is carried out starting by descrambling the watermarked image. This method also used the third and fourth LSB for hiding the data represented by an RGB watermark image.

In [5] a digital watermarking is presented using LSB and second LSB bit. The binary value of the watermark text is inserted in least significant bit place and the inverse of its correspond LSB bit in the place of the second LSB. In [6], the insertion method improves the stego-image quality by finding an optimal pixel after performing an adjustment process. Three candidates are picked out for the pixel value and compared to see which one has the closest value to the original pixel value with the secret data embedding. This however, makes the hiding capacity of the watermarked image very low.

To overcome the drawback of existing techniques, we would like to introduce a new alternative method by inserting RGB watermark in RGB host image using our watermarking approach.

3 Proposed Scheme for Watermarking

Based on LSB technique, we propose a new approach watermarking algorithm. First, we pre-scale the cover RGB image using a pseudo random factor (α). After that, we embed the MSBs of the watermark image into the LSB bit of the scaled cover. The aim of this operation is to increase the capacity of embedding and grow the robustness of our algorithm that hides the color image into the host one assuming that image is 24-bit. Thus, for every scaled pixel of 3bytes x α, in which the intensity of that color can be specified on scale of 0–255, we can embed all bits of the watermark whatever it size. The Fig. 1 shows the framework of the proposed method. Figure 2 and 3 present respectively the pixels sequences of the scaled cover

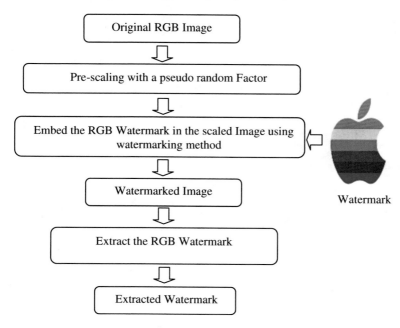

Fig. 1 The framework of the proposed method

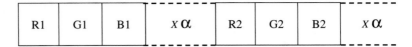

Fig. 2 Pixels of the cover image after scaling and storing in sequence

Fig. 3 Pixels of the watermark color image after storing in sequence

---

Algorithm: Watermark Embedding

   Input: RGB color (Cover) image (C) of size *m* x *n* and RGB color image Watermark (W)
                         of size *p* x *q*.
   Output: Watermarked image of size *m* x *n*

   Begin

   1. Read the Cover image C.
   2. Choose a pseudo random factor scaling.
   3. Store the components of each pixel of (C) in sequence.
   4. Read the watermark color image.
   5. Store (W) in sequence.
   6. Convert the watermark into binary form. Similarly convert the cover image into binary format.
   7. **Do loop**: Bitset each bit of the watermark (W) into the LSB position of cover image scaled [Red] byte, Bitset the next bit from (W) in the LSB position of C scaled [Green] byte, Bitset the next bit from W in the LSB of C Scaled [Blue] byte.

   End

Fig. 4 The watermark embedding algorithm

image and the watermark. Figure 4 shows the embedding algorithm and Fig. 5 shows the extracting method.

3.1   *Watermark Embedding*

Conventionally, digital color image is composed of elemental color pixels. Each pixel has three digital color components having the same number of bits, each

---

Algorithm: Watermark Extraction

Input: Watermarked color image (WC) of size (*m x n x* α).
Output: Watermark color image of size (*p x q*)

Begin

1. Read the watermarked Image.
2. Store the watermarked matrix in sequence.
3. Convert the sequence into binary form.
4. **Do loop**: Bitget each bit of the watermark image from the LSB of the watermarked data
   scaled [Red] byte, Bitget the next bit from the LSB position of WC scaled [Green] byte,
   Bitget the next bit from the LSB of WC scaled [Blue] byte.
5. Convert the sequence of bits extracted in decimal form.
6. Rebuild the matrix of the extracted watermark.

End

Fig. 5 The watermark extracting algorithm

component corresponding to a primary color. Usually red, green or blue, still referred to as RGB components word.

In this section, we describe the embedding process of our algorithm. We select RGB host image, then we choose a pseudo random factor α that we consider as private Key known only by the authorized person. We Store the 3 components of the scaled matrix and the watermark image in sequences as shown bellow. Finally, we embed all the binary bit of each component of the watermark in the LSB bit of the scaled image. Figure 4 depicts the embedding algorithm.

3.2 Watermark Extracting

In order to get the hidden image after receiving the watermarked one, the process of extraction is executed. We start by saving in sequences (R1*, G1*, B1*....) the decimal values of RGB components of the watermarked image, afterward we convert it into binary bits. Subsequently we extract the most significant bits of the watermark from the LSBs of each component of the watermarked binary sequences. We then obtain the watermark image by converting the extracted sequences bits into decimal and fill their values in the extracted matrix image.

4   Experimental Results and Discussion

4.1   Invisibility Test

In order to evaluate the effectiveness of our proposed method, subjective tests on the resultant watermarked images are performed without any attacks. Using four popularly 24 bit 256×256 color images namely; Lena, Peppers, House, Fruits as cover images, we embed and extract 256×256 RGB watermarks. Figure 6 shows the results of the embedding and extracting process. As can be seen, the embedded watermarks are visually imperceptible by human beings. Moreover, the watermarked images present less values of MSE when the high average PSNR values are 47.37, 51.3, 51.11 and 46.74 respectively, which further indicates that the hiding of complete watermark into all pixels of three channels of the host image does not affect the quality of the reconstructed image.

4.2   Robustness Test

Figure 7 shows the effect of applying several image processing attacks on the watermarked images (Lena and Peppers) and the extracted RGB Apple watermarks. We analyze the results in terms of the Normalized Correlation (NC) Values.

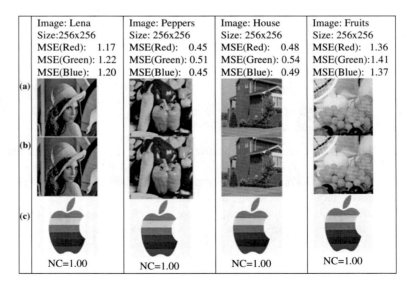

Fig. 6 The original image (**a**) The watermarked image (**b**) The extracted watermark (**c**)

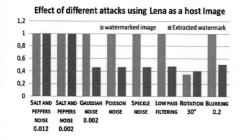
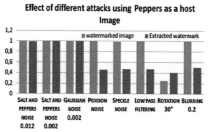

Fig. 7 Effect of salt and peppers noise, gaussian noise, poisson noise, speckle noise, low pass filtering, rotation and blurring attacks on the watermarked images and extracted watermarks measured by NC

Table 1 The comparisons between our approach and Golea et al. [7], Su et al. [8] and Chou and Wu [9]

Properties	Golea et al. [7]	Su et al. [8]	Chou and Wu [9]	Proposed method
Processing domain	Frequency domain	Frequency domain	Spatial domain	Spatial domain
NC of the watermarked image	0.92	0.97	–	0.99
NC of the extracted watermark	0.98	0.99	–	1.00
PSNR (dB)	34.45	40.38	33.35	43.68
Watermark length (bit)	32 × 32 × 24	64 × 64 × 24	128 × 128 × 24	256 × 256 × 24

4.3 Comparison

Under the same condition, we compare the proposed approach with the existing methods in spatial and frequency domain. We use the same original Peppers host image and the 8-color watermark employed in Golea et al. [7], Su et al. [8] and Chou and Wu [9]. The performance comparisons between different algorithms and their properties are illustrated in Table 1.

Simulation results validate that our method compared to the three existing methods maintain the high quality measures including NC equal to 1 and the PSNR values of the watermarked image and the extracted watermark. It is also confirm that the capacity of embedding of our approach is better.

5 Conclusion

The image watermarking is an important field that most of watermarking researches are focused on. Currently, there is a large demand for data hiding and data security especially in defense and civil applications.

This paper proposes a novel blind image watermarking scheme to embed color watermark into color cover image. With few steps for insertion, full watermark extraction without the requirement of the original host image or the original watermark and the pseudo random factor variation, the proposed watermarking leads to a great probability of imperceptible changes in least significant bit as human eye cannot suspect negligible changes in color of a pixel among so many others. Moreover the high capacity of hiding allows us to embed one hundred percent of the RGB watermark. Thus, we assume that our approach provides good watermark invisibility, an excellent embedding capacity and less computational complexity.

More deep works related to this research, should be centre on improving the robustness of our technique against geometrical attacks.

References

1. Smith, S.W.: The scientist and engineer's guide to digital signal processing (2011)
2. Manoharan, J., Vijila, C., Sathesh, A.: Performance analysis of spatial and frequency domain multiple data embedding techniques towards geometric attacks. I. J. secur. **4**(3), 28–37 (2010)
3. Tyagi, V., Kumar, A., Patel, R., Tyagi, S., Gangwar, S.S.: Image steganography using least significant bit with cryptography. J. Glob. Res. Comput. Sci. **3**(3) (2012)
4. Verma, R., Tiwari, A.: Copyright protection for RGB watermark image using in LSB. Int. J. Innovative Res. Stud. **3**(1) (2014)
5. Singh, A., Jain, S., Jain, A.: Digital watermarking method using replacement of second least significant bit (LSB) with inverse of LSB. I. J. Emerg. Technol. Adv. Eng. **3**(1) (2013)
6. Chan, C.K., Cheng, L.M.: Hiding data in images by simple LSB substitution computer. J. Pattern Recogn. Lett. **37**(3), 469–474 (2004)
7. Chou, C.H., Wu, T.L.: Embedding color image watermarks in color images. EURASIP J. Adv. Sig. Process. **1**, 32–40 (2003)
8. Su, Q., Wang, G., Jia, S., Zhang, X., Liu, Q., Liu, X.: Embedding color image watermark in color image based on two-level DC. Signal Image and Video Processing, Springer (2013)
9. Golea, N.E.H., Seghir, R., Benzid, R.: A blind RGB color image watermarking based on singular value decomposition. In: IEEE/ACS International Conference on Computer Systems and Applications (AICCSA), pp. 1–5. Hammamet (2010)

Eye Detection Based on Neural Networks, and Skin Color

Samir El Kaddouhi, Abderrahim Saaidi and Mustapha Abarkan

Abstract This paper proposes an eye detection method based on neural networks, skin color and eye template. First a multilayer perceptron neural network (MLP) using the retro propagation function with Gabor filter feature is used to detect faces in images. Next a method based on skin color and eyes template used to locate the eyes in detected faces. The results obtained are satisfactory in terms of quality and speed of detection.

Keywords Eye detection · Face detection · Gabor filter · MLP · Skin color · Eye template

1 Introduction

The eyes detection is often the first step in such applications such as face recognition, analysis of emotions and the driver fatigue control systems [1].

However, eyes detection is a difficult task because the computation time, the lighting conditions and the presence or absence of structural components such as glasses. To overcome these problems, various techniques have been developed over recent years and are classified into three methods: Template matching method [2, 3],

S.E. Kaddouhi (✉) · A. Saaidi · M. Abarkan
LSI Department of Mathematics Physics and Computer Science, Polydisciplinary
Faculty of Taza, Sidi Mohamed Ben Abdellah University, B.P. 1223,
Taza, Morocco
e-mail: samir.elkaddouhi@usmba.ac.ma

A. Saaidi
e-mail: abderrahim.saaidi@usmba.ac.ma

M. Abarkan
e-mail: mustapha.abarkan@usmba.ac.ma

A. Saaidi
LIIAN Department of Mathematics and Computer Science, Faculty of Science
Dhar El Mahraz, Sidi Mohamed Ben Abdellah University, B.P. 1223, Fez, Morocco

© Springer International Publishing Switzerland 2016
A. El Oualkadi et al. (eds.), *Proceedings of the Mediterranean Conference
on Information & Communication Technologies 2015*, Lecture Notes
in Electrical Engineering 380, DOI 10.1007/978-3-319-30301-7_18

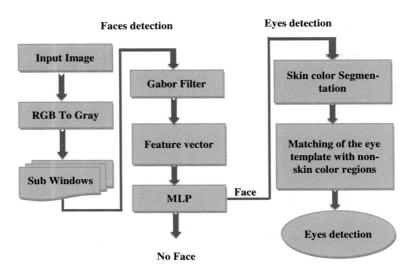

Fig. 1 Diagram of eyes detection by our method

geometry-based methods [4, 5], and Appearance-based methods [6, 7]. In template matching method, an eye template is prepared and then it is compared with the different regions of the image to locate the eyes. The geometry-based methods are focused on edge detection, or important regions of the eyes like the edges of the eye, the pupil, and the corners of the eyes. In appearance based methods, images are represented by a vector using a linear transformation and statistical methods such as principal component analysis (PCA) and independent component analysis (ICA).

In this paper, we propose a new method of eye detection. It is based on neural networks, and the skin color. Our method is processed into two steps: In the first step, face are detected by a multilayer perceptron neural network (MLP) using the retro propagation function trained with Gabor filter feature. In The second step, eyes are detected by the matching of the eye template with small regions determined after elimination of the skin regions.

The different steps of our method of eyes detection are presented in Fig. 1.

2 Related Work

The eyes detection is a current subject which has been studied by many researchers in recent years. Many eye detection methods are proposed, namely:

The eye detection method employed by Bansode et al. [2] uses an eye template and the normalized cross-correlation function for the detection of eyes.

The paper [3] presents a method for detecting eyes in face images that integrates cross-correlation function for different eye template. This method involves three steps: The first is to find the eye candidates from face images. The second calculates the cross-correlation between an eye template and the different candidates regions to determine the area that contains the maximum correlation value which is the center of an eye. The third re-correct detection when the correlation function does not give high values for the eyes regions, by multiplying correlation matrices of three eye template. The advantage of this method is that it can increase the detection rate compared to the use of the simple correlation.

The paper [4] provides an algorithm for eye detection and localization using a salient map. This algorithm uses an improved salient-point detector to extract salient points in an image (the visually meaningful points), and Then, the improved salient-point detector is extended for eye detection and localization (eyes are always the most varied blocks in a face image).

The eyes detection method proposed in [6], is based on new features rectangular. These features used to form a cascaded AdaBoost classifier for the detection of the eyes. Then the geometric characteristics of the symmetry of the eyes are used for correction of the detection of eyes.

Goutam Majumder et al. [8] propose a method for detecting eye using a feature detection technique, called fast corner detector. The method applies the fast corner detector on the grayscale images after a lighting stretching step which reduces the illumination effect.

Vijayalaxmi et al. [7] performing the detection of the eye by using a neural network and a Gabor filter. The proposed algorithm is divided in two stages: training and detection stage. First a set of images of the eyes and the non-eyes used to train the neural network, and the eyes are detected using the neural network trained.

3   Our Method

Our suggested method for eye detection contains two algorithms. The first algorithm involve use a simple and effective method for face detection which is based on a multilayer perceptron neural network (MLP) using the retro propagation function trained with Gabor filter feature. The second algorithm consists of locating eyes in the face detected by eliminating the skin color regions in the face then the remainder regions are compared with a model eye.

3.1   Faces Detection

Face detection has achieved enormous success in recent years. It is the basis of several works [9–14]. In this paper we used a multilayer perceptron neural network with Gabor filters. This system begins with the convolution of an image with a series of Gabor filters at different scales and orientations. Then these features are used in the input of the classifier, which is a multilayer perceptron neural network (MLP) using the retro propagation function.

3.1.1 Gabor Filters

In image processing a Gabor filter is a linear filter used for the face detection. Frequency and direction of the representations of Gabor filters are similar to those of the human visual system. In space, a 2D Gabor filter is a Gaussian kernel function modulated by a sinusoidal plane wave. Gabor filter extracts the image information preprocessed using 8 orientations and 5 frequencies [9].

The Gabor representation of an image is given by the convolution of the image with different kernels Gabor. Figure 2 shows a face image, a bank of 40 Gabor filters with different scales and orientations, and correspondence of the image filtered by each Gabor filter.

3.1.2 Multilayer Perceptron Neural Network

A neural network is composed of a set of interconnected neural giving birth to networks with varying structures. In our article, we used the multi-layer perceptron structure. Such structure distributes the information of the input layer, made by the neural receiving primitive information, to the output layer, which contains the final neurons transmitting output information processed by the entire network, crossing one or more intermediate layers, called hidden layers [10].

The developed neural network uses a retro propagation function to optimize learning phase. It is organized in a multilayer structure (MLP) with a single hidden layer. For this layer, we selected a number of 10 neurons to give slightly better results than the layers formed by 5, 15, 20 or 25 formal neurons. On the other hand, the input layer is formed by a number of neurons, which corresponds to the size of the feature vectors. The output layer is formed of a single neuron.

Face detection using this network proceeds in two stages: training and classification. In the training stage a set of 69 faces images and 59 non-faces images filtered with a bank of 40 Gabor filters are used to train the network. In the classification stage, any image under test is divided into sub-windows that are presented to the network trained for classification. Figure 3 shows the detection of

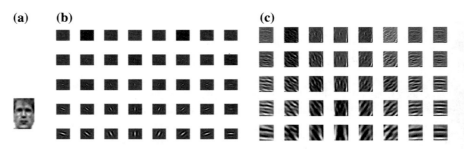

Fig. 2 Application of gabor filters on an image. **a** Face image. **b** Gabor filter kernel with different scales, orientation. **c** Representation of the image with different kernels gabor

Fig. 3 Faces detection

faces in an image by the neural network. Figure 3 present the detection of faces in an image by the neural network.

3.2 Eyes Detection

The objective of this step is to locate the eyes in each face detected based on the skin color and the eye template. In general, the human face contains skin regions and non-skin regions (eyes, mouth, beard, eyebrows). To detect the eyes, the non-skin regions are compared with the eye template.

3.2.1 Elimination of Skin Regions

In the human face the eye color is different from the skin color. For this skin region are segmented in three color spaces (RGB standard, YCbCr and HSV) using the rules defined in Table 1 [11]. Then those regions are removed to keep only the non-skin regions that may be the eye.

Table 1 Rules defining the skin regions [11]

Space	Rules	
Normalized RGB	A	$0.45 \leq r \leq 0.7$
	B	$0.23 \leq g \leq 0.55$
YCbCr	C	$77 \leq Cr \leq 127$
	D	$133 \leq Cb \leq 173$
HSV	E	$0 \leq H \leq 0.2$
	F	$0.2 \leq S \leq 0.7$

Fig. 4 Eyes detection

3.2.2 Matching of the Template Eye with Non-Skin Regions

This step is to create an eye template, with the average of eyes images for many people, and then calculate the correlation between this template and the different non-skin regions using the normalized cross-correlation function (NCC) defined by the formula (1) [2].

$$NCC(x, y) = \frac{\sum\limits_{x,y} \delta_{I(x,y)} \delta_{T(x-u,y-v)}}{\left\{ \sum\limits_{x,y} \delta_{I(x,y)}^{2} \sum\limits_{x,y} \delta_{T(x-u,y-v)}^{2} \right\}^{0.5}} \tag{1}$$

With:

$$u \in 1, 2, 3, \ldots, p, \quad v \in \{1, 2, 3, \ldots, q\}, \quad x \in 1, 2, 3, \ldots, \quad m - p + 1, \quad y \in 1, 2, 3, \ldots, \quad q - n + 1$$
$$\delta_{I(x,y)} = I(x, y) - \overline{I_{u,v}}, \quad \delta_{T(x-u,y-v)} = T(x - u, y - v) - \overline{T}$$
$$\overline{I_{u,v}} = \frac{1}{pq} \sum_{uv} I(x, y), \quad \overline{T} = \frac{1}{pq} \sum_{uv} T(x - u, y - v)$$

I is the input image, T the template, $\overline{T}$ is the average of template, et $\overline{I}$ is the average of the image in the template region.

Figure 4 shows the eyes detection in the face image.

4   Experimental Results

To evaluate our method, we defined two indicators: the false detection rate (the number of false detected eyes in relation to the number of detections) and correct detection rate (the number of eyes detected on the actual number of faces in the image).

Our method is tested on 50 images that contain 282 eyes. We obtained important results (Table 2). We detected 266 eyes and 18 false detections. We find that the correct rate exceeds 94 %, compared to the false rate detection (6.33 %).

Table 2 Results obtained

	The false detection rate	The correct detection rate
Our method	**6.33 %**	**94.32 %**
Wenkai, Xu et al. [5]	9.05 %	92.55 %
Vijayalaxmi et al. [7]	7.85 %	91.48 %

Table 3 Steps of eye detection by our method

Input image	Faces detection	Eyes detection

To demonstrate the effectiveness of our method, our results are compared with the results obtained by the methods of detecting eyes presented in [5] and [7] (Table 2).

It is shown in Table 2 that our method gives satisfactory results compared to other approaches. It gives the highest correct detection rate and the lowest false detection rates.

The robustness of our method resides in the quality of detection of eyes achieved a very high correct detection rate (94.32 %) and a very low false detection rate (6.33 %).

These results are obtained by the application of a neural network for the face detection, which has the ability to detect faces in images with complex backgrounds, and in noisy images. And to detect eyes our method uses a simple and robust method that is based on the elimination of the regions of the skin color, then matching a template eye with the remaining non-skin regions.

To illustrate the different stages of our method, we have shown in Table 3, the results for three images selected from 50 images used to test our method and that are characterized by complex background.

5 Conclusion

In this paper, we have presented an eye detection method based on neural networks, the skin color, and the application of an eye template. The hybridization of these three methods help increase the quality and speed of the method which gives very satisfactory results. We obtained 94.32 % true detection, against 6.33 % of false detections.

References

1. Al-Rahayfeh, A., Faezipour, M.: Eye tracking and head movement detection: A state-of-art survey. IEEE J. Transl. Eng. Health Med. **1** (2013)
2. Bansode, N.K., Sinha, P.K.: Efficient eyes detection using fast normalised cross-correlation. Sig. Image Process. Int. J. (SIPIJ). **3**, 53–59 (2012)
3. Kutiba, N.: Eye detection using composite cross-correlation. Am. J. Appl. Sci. **10**, 1448–1456 (2013)
4. Muwei, J., Kin-Man, L.: Fast eye detection and localization using a salient map. The Era of Interactive Media, Springer, pp. 89–99 (2013)
5. Jiatao, S., Zheru, C., Jilin, L.: A robust eye detection method using combined binary edge and intensity information. Pattern Recognit. **39**, 1110–1125 (2006)
6. Xu, W., Lee, E.J.: Eye detection and tracking using rectangle features and integrated eye tracker by web camera. Int. J. Multimedia Ubiquitous Eng. **8**, 25–34 (2013)
7. Vijayalaxmi: Neural network approach for eye detection. Comput. Sci. Inf. Technol. **2**, 269–281 (2012)

8. Goutam, M., Mrinal, K.B., Debotosh, B.: Automatic eye detection using fast corner detector of north east indian (NEI) face images. In: International Conference on Computational Intelligence: Modeling Techniques and Applications (CIMTA), vol. 10, Elsevier, pp. 646–653 (2013)

9. Puneet, K.G., Mradul, J.: A review on robust face detection method using gabor wavelets. In: International Conference on Computational Intelligence: International Journal of Innovative Research in Science, Engineering and Technology, vol. 2, pp. 604–611 (2013)

10. Lalita, G., Nidhish, T.: Face detection using feed forward neural network in matlab. Int. J. Appl. Innovation Eng. Manag. 3, 208–210 (2014)

11. El Kaddouhi, S., Saaidi, A., Abarkan, M.: A New robust face detection method based on corner points. Int. J. Softw. Eng. Appl. 8, 25–40 (2014)

12. Deepak, G., Joonwhoan, L.: A robust face detection method based on skin color and edges. J. Inf. Process. Syst. 9, 141–156 (2013)

13. Viola, P., Jones, M.: Robust real-time face detection. Int. J. Comput. Vision 57, 137–154 (2004)

14. Chuan, L.: Face detection algorithm based on multi-orientation gabor filters and feature fusion. TELKOMNIKA 11, 5986–5994 (2013)

Simulated Annealing Decoding
of Linear Block Codes

Bouchaib Aylaj and Mostafa Belkasmi

Abstract In this paper, we present a hard-decision decoding algorithm using Simulated Annealing (SA) technique. The main idea is to find the optimal solution of the transmitted codeword by a process hopping between two different tasks. The simulations, applied on some binary linear block codes over the AWGN channel, show that the Simulated Annealing decoder has the same performance as the Berlekamp-Massey Algorithm (BM). Furthermore SA Decoder is more efficient compared to other Decoder based on Genetic algorithms in terms of performance and run time.

Keywords Simulated annealing · Hard-decision decoding · Errors correcting codes

1 Introduction

Error-correction coding is essentially a signal processing technique that is used to improve the reliability of data transmitted over communication channels (transmission media), such as a telephone line, microwave link, or optical fiber, where the signals are often affected by noise. The encoder generates code words by addition the redundancy symbols at data source. The decoder attempts to search the most probable transmitted code word by various techniques. The decoding techniques can be categorized into two: hard decision and soft decision decoding. Hard decision decoding uses binary inputs resulting from thresholding of the channel output and usually it uses the Hamming distance as a measure [1]. However, soft

B. Aylaj (✉)
Department of Maths, MMID, Faculty of Sciences, Chouaib Doukkali University,
El Jadida, Morocco
e-mail: bouchaib_aylaj@yahoo.fr

M. Belkasmi
SIME Labo, ENSIAS, Mohammed V University, Rabat, Morocco
e-mail: m.belkasmi@um5s.net.ma

© Springer International Publishing Switzerland 2016
A. El Oualkadi et al. (eds.), *Proceedings of the Mediterranean Conference on Information & Communication Technologies 2015*, Lecture Notes in Electrical Engineering 380, DOI 10.1007/978-3-319-30301-7_19

175

decision decoding uses real inputs, i.e. The informations in the channel output are not thresholded and the measure used is the Euclidean distance [1]. Decoding operation is an NP-hard problem and was approached in different ways. The optimum decoder is Maximum Likelihood [2], but its high complexity gives rise to other suboptimal techniques with acceptable performance such as Turbo codes and LDPC. Recently this problem has been tackled with heuristic methods. These latest show very good solutions. Among related works, one idea used A* algorithm to decode linear block codes [3], another one uses genetic algorithms for decoding linear block codes [4] and a third one uses neural networks to decode of Convolutional codes [5]. The Simulated Annealing (SA) method has been introduced by El Gamal et al. [6] in coding theory. Mathematically this method has been proved that it is possible to converge to globally solution [7]. In [6] the authors have used SA to construct good source codes, constant weight error-correcting codes, and spherical codes. Other works were lead for the use of the SA in computing the minimum distance of linear block codes [8, 9]. However, the authors of [10] have applied SA to construct good LDPC codes. In the decoding of error correcting codes, in our knowledge SA is not used as a principal method for decoding. In this paper we introduce a hard decision decoding algorithm of linear block codes based on a simulated annealing method. The decoder goal is find rapidly the optimal solution of the transmitted codeword (global state of minimum energy) by jumping between two tasks. The rest of this paper is organized as follows. On the next section, we give an introduction on the errors correcting codes. In Sect. 3 we present the proposed decoder of binary block codes based on SA technique. In Sect. 4 we present the simulation results. Finally, a conclusion and perspectives of this work are outlined in Sect. 5.

2 Error Correcting Codes

Essentially there are two large families of error correcting codes: block codes and convolutional codes. The family of block codes contains two subfamilies: nonlinear codes and linear codes. Here, our work is designed to linear block codes, for this we give the principal features for this class of codes.

Let $F(2) = \{0, 1\}$ be the binary alphabet and $C(n, k, d)$ a block code where n, k and d represent the length, dimension and minimum distance of the code respectively, with the code rate is the ratio k/n and t is the error correcting capability of C. Each element $V \in C$ is called a codeword, there are a total of 2^k codewords. The code C is a k-dimensional subspace of the vector space $F(2)^n$. The code C is called linear if the modulo-2 sum of two codewords is also a codeword. The Hamming weight $W_H(V)$ is the number of nonzero components of V and the Hamming distance noted $d_H(V_1, V_2)$ between two words V_1 and V_2 is the number of places in which V_1 and V_2 differ. The minimum Hamming distance d (or the minimum distance) of the code C is the Hamming distance between the pair of codewords with smallest Hamming distance. Let

$$d_H = d_H(V_i, V_j) \tag{1}$$

It can easily be proved that the Hamming distance of the linear block code C between two codewords is also equal to the Hamming weight of the modulo-2 sum of two codewords is written as follows

$$d_H = W_H(V_i \oplus V_j) \quad \forall \quad V_i, V_j \in C \tag{2}$$

Let $G = (g_{ij})_{k \times n}$ be the generator matrix of the code C, whose rows constitute a set of basis vectors for subspace $F(2)^k$. Then every codeword $V = (v_1, v_2, \ldots, v_n)$ can be uniquely represented as a linear combination of the rows of G.

$$\forall 1 \leq j \leq n \quad v_j = \oplus_{i=1}^{k} u_i g_{ij} \tag{3}$$

where $\oplus$ denotes modulo-2 sum, $U = (u_1, u_2, \ldots, u_k) \in (0, 1)^k$ information vector. The code $C^{\perp}(n, n-k)$ defined by Eq. (4) where $\langle,\rangle$ denotes the scalar product.

$$C^{\perp} = \left\{ V' \in F(2)^n : \quad \forall \quad V \in C, \quad \left\langle V, V' \right\rangle = 0 \right\} \tag{4}$$

is also linear and called the dual code of C, its generator matrix is $H = (h_{ij})_{n-k \times n}$ also called parity-check matrix of code C. The resultant vector $S(r) = Hr$ called the syndrome. If the received word r contains no error then the syndrome $S(r) = 0$. Examples of linear block codes [1]: Quadratic-Residue (QR) codes and Bose, Ray-Chaudhuri and Hocquenghem (BCH) codes.

3 The Proposed Simulated Annealing Decoder (SA Decoder)

3.1 Description of SA Decoder

If we conceptualize hamming distances between the hard version of the received word and a codeword as an atom states of energy, then the analogy between the lowest energy of the atom and the minimum of the Hamming distance is apparent. The correspondence between Physical Annealing and our decoding problem is in Table 1.

The SA Decoder used for estimating the decoded word is based on two tasks of treatment, task1 and task2 which are controlled by the SA method. Having these two tasks allows us to search the good solution locally in the case of task1 ensured by the perturbation 1 and to perform an external research in the case of task2

Table 1 Analogy between SA and our decoding problem

Physical annealing our decoding problem	
State	Feasible solution of the received word
Energy	objective function = hamming distance
Change of State	Neighboring Solutions (Perturbation 1 or 2)
Temperature	Control parameter (number of iterations)
crystalline solid state	Optimal solution = decoded word

ensured by the perturbation 2. So, it's possible to exploit the search space and the movements between its different regions. The research is strengthened by accepting all movements improvers and it's varied by accepting all movements non improvers. The variety decreases with the decrease in temperature throughout of the algorithm. The algorithm 1 presents SA Decoder.

Algorithm 1 The basic algorithm of SA Decoder

Input(s): r, t capability of code, N Iterations number of T_i, task $i = 1$, T_i, T_f
Output(s): Decoded word

Calculate h hard version of the received word r
If$(S(h) == 0)$ **Then** decoded word $\leftarrow h$
Else
Calculate information vector U of h
Identification of the least reliable positions in r
While $(T_i > T_f)$**do:**
    While $(iterations < N)$**do:**
        Get neighborhood state U^* from task i
        Evaluate $E(h^*)$
        If $((E(h^*) \leq t+1)$ or $(rand(0,1) \leq \exp(-E(h^*)/T_i))$ **Then**
        $U \leftarrow U^*, h \leftarrow h^*$, $Nbr\_accpt + +$
        End If
    End While
    If (the criterion of transion is satisfied?) **Then** change task i to task $i \pm 1$
    End If
    $T_i \leftarrow cool(T_i)$
End While
Decoded word $\leftarrow h$
End If

The tasks 1 and 2 are assured respectively by the following perturbations:

- **Perturbation 1**: Generate an error pattern $\beta \in \{0,1\}^k$ where $1 \leq W_H(\beta) \leq t$ applied over the least reliable positions of r
- **Perturbation 2**: Generate an error pattern $\alpha \in \{0,1\}^k$ where $1 \leq W_H(\alpha) \leq t$ applied over positions of U.

Simulated Annealing Decoding of Linear Block Codes

The criterion of transition from task1 to task2 is: Nbr_accpt = 0, Where Nbr_accpt is the number of the new solutions accepted. The criterions of transition from task2 to task1 are: Nbr_accpt ≠ 0 or $rand(0, 1) \leq exp(-F/T)$?,

The variable F is the average of $E(h^*)$ i.e. $(E(h^*)accepted/number\ of\ iterations)$

Cost function To evaluate the feasible solutions a function called a cost function or objective function denoted E can be developed as follows:

Let $h = (h_1, h_2, \ldots, h_n) \in [0, 1]^n$ is the hard version of the received word and $U = (u_1, u_2, \ldots, u_k) \in [0, 1]^k$ is the information vector of h. For $h^* = (h_1^*, h_2^*, \ldots, h_n^*) \in C$ the information vector of h^* is $U^* = (u_1^*, u_2^*, \ldots, u_k^*) \in [0, 1]^k$, with Eq. (3) we have

$$\forall\ 1 \leq j \leq n \quad h_j^* = \oplus_{i=1}^{k} u_i^* g_{ij} \tag{5}$$

In Eq. (2) we have $d_H = W_H(h \oplus h^*)\ \forall\ h^* \in C$, noting that $W_h(V) = \sum_{j=1}^{n} v_j$ then

$$d_H = W_h(V) = \sum_{i=1}^{n} h_i \oplus h_i^* \quad \forall\ h^* \in C \tag{6}$$

Replacing Eq. (5) in Eq. (6) we have the cost function defined as follows:

$$E(h^*) = \sum_{i=1}^{n} \left[h_i \oplus \left(\oplus_{j=1}^{k} u_j^* g_{ij} \right) \right] \quad \forall\ U^* \in [0, 1]^k \tag{7}$$

The modified SA is brought to search the information vector U^* associated to the codeword h^*, this information vector converges the cost function $E(h^*)$ towards an value less than or equal to $(t + 1)$ of linear block code C.

3.2   Control Parameters of SA Decoder

- **Initial and final temperature**: The initial and final temperature respectively T_i and T_f are determined by the acceptance probability P as follows:

$P_H = exp(-E_{max}(h^*)/T_i)$ where $E_{max}(h^*) \leq k$, P_H should be high probability then

$$T_i \leq \frac{-k}{lnP_H} \tag{8}$$

$P_l = exp(-E_{min}(h^*)/T_f)$ where $E_{min}(h^*) \geq 1$, P_l should be small probability then

$$T_f \geq \frac{-1}{lnP_l} \qquad (9)$$

T_f is used as stop criterion for the SA decoder during the simulation

- **Reducing rate**: We used a geometric rule for decreasing temperature:

$$T_{i+1} = \theta T_i \quad \theta \in [0.7, 0.9] \qquad (10)$$

- **Number of iterations** (N_i): Let N be the number of iterations at each temperature T which is constant during the simulation. The number of iterations (number of perturbations) is:

$$N_i = N * (number\ the\ variations\ of\ temperature) \quad N \in [10, 300] \qquad (11)$$

4 Simulation Results

In this work, the simulation program developed in language C and a computer with Intel Core(TM) 2 Duo CPU T5750 @ 2.00 GHz processor and 2 GB RAM. All simulations are obtained by using the parameters given in Table 2. We note that the SA Decoder parameters are obtained by a preliminary optimization.

As a first step, in order to evaluate the error correcting performance of the SA Decoder: the averaged evolution of the cost function $E(h^*)$ is performed, each of the averaged evolution curve is obtained by 500 transmitted blocks of same errors weight, during an iterations number for BCH(31,16,7) code. We plot the function: $E(moy) = 1/(1 + E_{moy}(h^*))$. Figure 1a show that all $E(moy)$ curves have converged before complete iterations and we note that the number of iterations and the errors weight are two main factors which affect convergence $E(h^*)$ of SA Decoder. After that, as a second step, to show the impact of the iterations number on the error correcting performance of the SA Decoder, we give in Fig. 1b the correction

Table 2 Simulation parameters

Simulation parameter	Value	SA Decoder parameter	Value
Channel	AWGN	Reducing rate	$\theta = 0.9$
Modulation	BPSK	Starting temperature	$T_s = 0.2$
Min. number of residual bit in errors	200	Final temperature	$T_f = 0.001$
Min. number of transmitted blocks	1500		

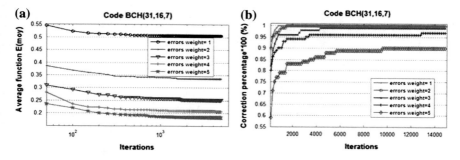

Fig. 1 a Averaged of E(moy) during 5000 iterations. **b** The correction percentage

percentage results of each errors weight for BCH (31,16,7) codes during iterations. The SA Decoder allows the correction of about 100 % of errors weight 1 in 750 iterations, 100 % of errors weight 2 in 1000 iterations, 99 % of errors weight 3 in 4300 iterations, 97 % of errors weight 4 in 10,000 iterations, 90 % of errors weight 5 in 9500 iterations for BCH(31,16,7) of error correcting capability equal to 3. The curves are obtained by 500 transmitted blocks of same errors weight.

Figure 2 compares the performance of SA decoder and Berlekamp-Massey (BM) algorithm [11] for the codes BCH (15,7,5) and BCH (31,16,7). We note that the SA Decoder outperformed the performance of BM Decoder.

Figure 3 compares the performance of SA decoder and OSD-0 Decoder [12] for the codes QR (41,21,9) and QR (71,36,11). We note that the SA Decoder allows to win about 0.7 dB in 20^4 iterations compared to the OSD-0 at BER = 10^{-5} for QR (71,36,11) code.

Figure 4 compares the performance of SA decoder compared to AG Cardoso [4] Decoder. We note that the SA Decoder outperformed the performance of AG Cardoso Decoder for BCH (31,16,7) code and they have the same performance for BCH(63,30,13). As shown in Table 3, the required computation time (time

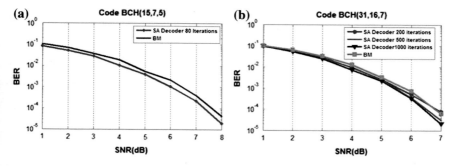

Fig. 2 Performance SA Decoder vs. BM: **a** for BCH(15,7,5) and **b** for BCH(31,16,7)

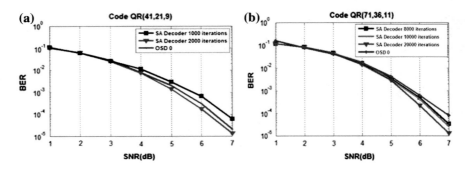

Fig. 3 Performance SA Decoder vs. OSD-0 decoder: **a** for QR (41,21,9) and **b** for QR (71,36,11)

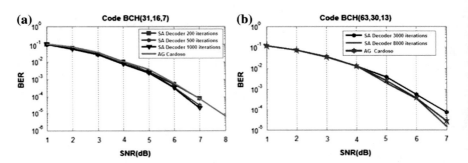

Fig. 4 Performance SA Decoder vs. AG Cardoso decoder: **a** for BCH(31,16,7) and **b** for BCH (63,30,13)

Table 3 SA Decoder and AG Cardoso Decoder in term of time complexity

Code	SA Decoder time complexity (s)	AG Cardoso Decoder time complexity (s)
BCH (31, 16, 7)	29	436
BCH (63, 30, 11)	75	1819

complexity) is very small with SA Decoder compared to AG Cardoso. The time complexity corresponds to the necessary time for decoding 1000 blocks, for BCH (31, 16, 7) and BCH (63, 30, 11) codes.

5 Conclusion

In this paper, we have presented a Simulated Annealing Decoder which can be applied to any linear block codes. The fact that we have two tasks into the SA technique speeded up the optimization process. The simulations show that it is less complex in term of run time than the decoder based on Genetic Algorithm of Cardoso and it has the same performance as the algebraic decoder BM. AS a future work we plan to apply this decoder to other codes and other channels.

References

1. Clarck, G.C., Cain, J.B.: Error-Correction Coding for Digital Communications. Plenum, New York (1981)
2. Roman, S.: Introduction to Coding and Information Theory. Springer, New York (1996)
3. Han, Y.S., Hartmann, C.R.P., Chen, C.C.: Efficient maximum-likelihood soft-decision decoding of linear block codes using algorithm A*. In: Technical Report SUCIS-91-42, Syracuse University, Syracuse, NY, 13244 (December, 1991)
4. Cardoso, A.C.M., Arantes, D.S.: Genetic Decoding of Linear Block Codes. Congress on Evolutionary Computation, Washington, DC (1999)
5. Sujan, R. et al.: Adaptive soft sliding block decoding of convolutional code using the artificial neural net-work. Transactions on Emerging Telecommunications Technologies (2012)
6. El Gamal, A., Hemachandra, A., Shperling, I., Wei, V.K.: Using simulated annealing to design good codes. IEEE Trans. Inf. Theory **33**, 116–123 (1987)
7. Gidas, B.: Non-stationary Markov chains and convergence of the annealing algorithm. J. Stat. Phys. **39**, 73–131 (1985)
8. Zhang, M., Ma, F.: Simulated annealing approach to the minimum distance of error-correcting codes. Int. J. Electron. **76**(3), 377–384 (1994)
9. Aylaj, B., Belkasmi, M.: New simulated annealing algorithm for computing the minimum distance of linear block codes. Adv. Comput. Res. **6**(1), 153–158. ISSN: 0975-3273. E-ISSN: 0975-9085 (2014)
10. Yu, H., et al.: Systematic construction, verification and implementation methodology for LDPC codes. EURASIP J. Wireless Commun. Networking **2012**, 84 (2012)
11. Berlekamp, E.R.: Algebraic Coding Theory. Aegean Park Press (1984)
12. Fossorier, M.P.C., Lin, S.: Soft decision decoding of linear block codes based on ordered statistics. IEEE Trans. Inf. Theory **41**, 1379–1396 (1995)

Part IV
Wireless and Optical Technologies and Techniques

Interaction Between Human Body and RF-Dosimeter: Impact on the Downlink E-field Assessment

Rodrigues Kwate Kwate, Bachir Elmagroud, Chakib Taybi, Mohammed Anisse Moutaouekkil, Dominique Picard and Abdelhak Ziyyat

Abstract The personal radiofrequency dosimeters are increasingly used for the workers exposure estimation to the electromagnetic radiations in terms of reference level of the fields. The presence of the human body at the vicinity of dosimeter compromises its measurement accuracy. In this paper, the study of interaction between personal radiofrequency dosimeter and human body is presented. First, we deal with the error introduced by the presence of the body and an approach to reduce this impact by adding elements in the dosimeter structure. Indeed, by introducing a metal ground plane and optimizing the height of antenna-ground plane, we can estimate at 10 % close the level of incidental field, that is to say a considerable reduction impact of the body on the dosimeter. We also study the stability of the intrinsic characteristics of the dosimeter (isotropy and front to back ratio). Different simulations were performed in the GSM 900 band with the finite integration technique. In the context of the design of worn-body radiofrequency dosimeters it's very useful to master these interaction results.

Keywords Personal radiofrequency dosimeter · Human body · Metal plate · Dipole antenna · FIT · Specific absorption rate (SAR)

R.K. Kwate (✉) · B. Elmagroud · C. Taybi · A. Ziyyat
Electronic and Systems Laboratory, Faculty of Sciences,
Mohammed Premier University, Oujda, Morocco
e-mail: rkwate2009@gmail.com

D. Picard
DRE, Signals and Systems Laboratory, Centrale Supelec,
University of Paris-Sud 11, Orsay, France

M.A. Moutaouekkil
Information Technology Laboratory, National School of Applied Sciences,
Chouaïb Doukkali University, El Jadida, Morocco

© Springer International Publishing Switzerland 2016
A. El Oualkadi et al. (eds.), *Proceedings of the Mediterranean Conference on Information & Communication Technologies 2015*, Lecture Notes in Electrical Engineering 380, DOI 10.1007/978-3-319-30301-7_20

1   Introduction

Since recent years, the assessment of electromagnetic field exposure requires more attention. Epidemiological studies are used to qualify the potential risk of that kind of human exposure. There are now carried out largely by using a personal radiofrequency dosimeter [1]. The dosimeter compares the levels of measured fields with international reference value in the ICNIRP guidelines [2] and gives us alert when the difference is important. However, commonly used of that device is highly affected by the presence of the human body. Its face some uncertainties due to several factors like [1, 3, 4]: the position on the body (torso, limb, etc.), the electromagnetic environment the incident field and body interaction, etc. In [5], it has even been proposed a parametric model to predict the SAR function of the angle of incidence field, the frequency, the polarization and the morphology of the human body. Among these factors, the main challenge of a personal dosimeter is to overcome the interaction with the incident field and human body. It's in this context that recent studies have been conducted.

In order to develop a better technique to measure the exposure level based on a personal dosimeter, although recent studies have focused on the interaction between the incident field and human body. It has produced several results such as multiple and distributed nodes acquisitions connected by electronic embedded systems to create an omnidirectional measurement [6]. This approach is very effective, but the manufacturing complexity and convenience of all raise many other questions. For example, it cannot be easily used on a large scale of population, but restricted to professionals. Other things, by focusing mainly on the interaction between the incident field and the human body, we neglect the possible effects of the presence of man on the reliability degree of the personal dosimeter. Indeed, trying to correct the problem would be greatly important to characterize the interaction between the human body and the dosimeter. However, very few studies have focused on the interaction between the human body and the dosimeter for measuring fields in close proximity to the body. In this paper, we are interested on how metal plate can interact with the dosimeter at the vicinity of human body. In order to try to reduce the margins of errors made by the dosimeter when it measures incident field.

The antenna and metal plate interaction has already been discussed [7, 8]. Results show us how the parameters of antenna are changed under influence of metal plate. For example, for a dipole antenna, when the distance between antenna and metal plate is increasing, we can see many radiation pattern lobes and many of antenna parameters are changing periodically.

2 Methods

2.1 Numerical Method of Analysis

In this work, numerical simulations are performed by using CST MWS software (Microwave Studio, Computer Simulation Technology) based on finite integration technique. To have an acceptable level of accuracy, we applied a mesh of about 25 million cells with 16 points per wavelength. To minimize reflection wave on the border, we applied 9 perfectly matched layers for absorbing waves at these locations. We used a machine with 4 GB RAM memory and dual core processor with 2.6 GHz frequency for each. Overall, it took us for one simulation about 10 h.

2.2 Personal Dosimeter and Metal Plate

For the isolated radiofrequency dosimeter, we use the half-wave dipole antenna with following characteristics (Fig. 1): resonant frequency f = 942 MHz (GSM), bandwidth at −10 dB is 150 MHz, length, L = 14.645 cm, this length is calculated as in Eq. (1). Z_{in} = 73.03 Ω (Input impedance), G_{max} = 2.2 dBi (Max Gain), opening angle at −3 dB in the E-plane is 78.0°. In Fig. 1a we can also see how the presence of human body modifies the frequency and bandwidth. The frequency offset is 13 MHz, bandwidth is around 175 MHz.

$$L = 0.5 \times \frac{c}{f\sqrt{\varepsilon_r}} \tag{1}$$

C: Speed of light in vacuum and ε_r: Dielectric constant of material.

Figure 2 present the finite-dimensional square metal plate with side L_{mp} = 18.31 cm (1.25 * L). This plate is positioned vertically at a distance d between the torso and dosimeter. Note that it has also been shown in [8] that the

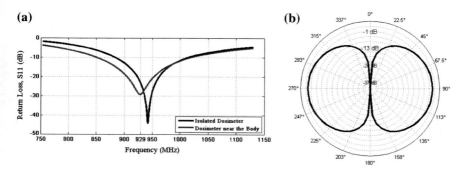

Fig. 1 Dipole antenna parameters, **a** return loss, **b** radiation pattern

Fig. 2 Position of metal plate between antenna and body

surface has a very small impact on the characteristics of the antenna, since the frequency offset and the ratio of bandwidths are almost insignificant, what matters most is the separation distance. The distance between the antenna and the body is constant (4 cm) because it is not the aim of our study and is the common distance in practice between body and dosimeter.

2.3 Human Body Model

We use the Zubal phantom. This phantom was segmented from CT scans (CT and MRI) of a 35 year old male, 178 cm height 70 kg. This phantom contains over 100 tissue and organs whose dielectric properties and density were determined in accordance with [9] following the model of Cole-Cole. For reasons of computation time we use homogeneous phantom whose indices are those of the skin for a frequency of 942 MHz.

The incident field is a plane wave with uniform density of 1 W/m^2 in the direction of incidence. We use different incidence field: front (0°), left (90°) and back (180°); five distances of separation 1 cm ($\approx$L/15), 1.25 cm ($\approx$L/12), 1.5 cm ($\approx$L/10), 1.75 cm ($\approx$L/8) and 2 cm ($\approx$L/7) and two cases (with or without metal plate). We performed around 30 simulations.

3 Simulations and Results

3.1 Incident Field and Human Body Interaction

Exposure wave in the far field is considered in this section. Sources of exposure are indicated by a monochromatic plane wave and vertically polarized in all simulations. In Fig. 3, for the distribution of fields E, H, and the power density **S** on the plane Z = 40 cm (the torso area of the phantom).

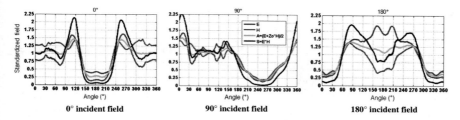

Fig. 3 Electromagnetic field around de body, (*Line 1*) E-Field, (*Line 2*) H-Field, (*Line 3*) standardized electromagnetic fields at 3 cm distance around the phantom in the plan Z = 40 cm (torso)

$$S = E * H \tag{2}$$

We also calculate the average A of the electric field E and the magnetic field H for each angle value around the phantom following the formula:

$$A = \frac{E + Z_0 * H}{2}. \tag{3}$$

In this formula, $Z_0(377 \ \Omega)$ is the free space impedance wave. The objective is to check if the variation of the two fields can compensate each other in our specific case. We can see three major areas, for 0° incident field:

- First, between [0°, 110°] and [240°, 360°], the side of the incident wave (torso and upper limbs), one of the fields is high when the other is low. Power density S follows the shape of the field E: is the reflection area of the incident plane wave. In this area the average **A** varies much less than each of the other fields and its value remains around 1.
- Then, between [150°, 200°], each field is at its lowest value. This area is the main problem of the personal assessment of exposure. This is the shadow area. The average A does not make improvement.
- Finally, we have an intermediate zone between reflection and shadow areas. This is the diffraction area.

Note that all the different areas depend on the incident field as can be seen in 90° and 180° incident field. The findings remain the same for the different fields (E, H, S and A)

With these observations, we understand the effect of interaction between an incident field and the human body. This interaction inevitably occurs on the performance of a radio frequency dosimeter near the body.

3.2   *RF-Dosimeter and Human Body Interaction*

We can observe in Fig. 4 a loss of omnidirectional nature of the reference dosimeter
resulting in an asymmetrical directivity diagram. To assess the evolution of pattern,
the maximum directivity and the front to rear ratio are going to be investigated.
Emphasis will be placed on the latter, because it reflects the imbalance introduced
by the disturbance in the symmetric diagram of isolated dosimeter in free space.
The ultimate goal is to highlight possible correlations between diagram and distance. In Fig. 4a, we see how the human body has greatly played the role of
reflector for this dosimeter, increasing the aperture (78° to almost 150°). In Fig. 4b,
we find that the small distance variation (1–2 cm) does not have a great effect on the
radiation pattern of dosimeter incorporating metal plate. But the change of pattern is
still significant compared to the absence of the metal plate, especially by the
appearance of an additional lobe parasite compared to the case of Fig. 4a. In Fig. 4c,
we compared all diagrams in the same graph; marking observation is the reduction
of the aperture with the metal plate, but increasing the gain.

As we can see in Table 1, directivity is almost tripled (from 2.23 to 7.33 dBi) with
the human body and it is multiplied by 4 (from 2.23 dBi to approximately 8 dBi)
with the additional presence of the metal plate. The front to rear ratio is almost tripled
(from 7.84 dB to approximately 23 dB) with the presence of the metal plate. This
interaction is approximately similar in [7] without the human body.

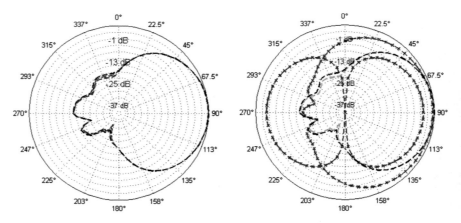

Fig. 4 Diagram Pattern, -x-x (*blue color*)—isolated dosimeter; -x-x (*red color*)—dosimeter and
Human body without metal plate; -x-x- (*black color*), -x-x (*green color*)—dosimeter and human
body and metal plate

Table 1 Directivity and front to rear ratio

		Isolated Dosimeter	Dosimeter and HB	Dosimeter with metal plate near de human body, with d (cm)				
				1	1.25	1.5	1.75	2
Max directivity (dBi)	Front	2.23	7.33	8.09	8.05	8.02	7.98	7.94
	Rear	2.23	−0.51	−15.68	−15.58	−15.49	−15.37	−15.26
Front/rear (dB)		0	7.84	23.77	23.64	23.52	23.36	22.20

3.3   Incident Field Level Measured

We collect the fields measured by the dosimeter. On the outskirts we do it for a plane wave oriented vertically and placed in front of the body torso. We note that, when the distance increases, the measured voltage is approaching the one measured by the isolated dosimeter. Particularly, when d is between 1.5 cm ($\approx$L/10) and 1.75 cm ($\approx$L/8), the gap is almost lower than −1 dB. Unlike, without the metal plate, for the distances 7 cm ($\approx$L/2) and 4 cm ($\approx$L/4) of the human body, we see the effects of reflection that bring the difference to 4 dB on the one hand and −6 dB of the other hand (Fig. 5). Another prominent finding is when approaching the high frequencies, the gap is narrowing ($<$−2 dB) between the measurement made by isolated dosimeter and that done by the dosimeter in the presence of the metal plate and human body.

For other incident fields, when dosimeter is at 4 cm of torso, we selected two positions of the metal plate, 1.5 and 1.75 cm; this is done according to the good results obtained for these positions in the event of a frontal impact of field. We can

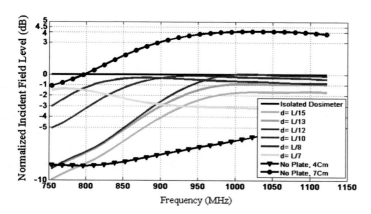

Fig. 5 Normalized incident field level measure by dosimeter for different position of metal plane and vertical plane wave

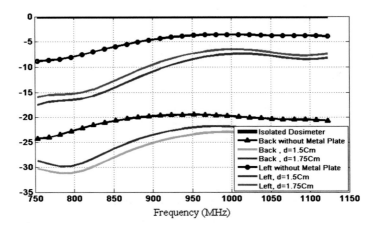

Fig. 6 Normalized incident field level measure by dosimeter for other incident field

observe in Fig. 6, the masking effect is more important than the effect of diffraction or reflection. Indeed all the fields level measured by dosimeter are low compared to the baseline measurement, is a shift of approximately −25 to −20 dB against −10 to −3 dB for a side impact of field.

We realize that the interaction between the human body and the dosimeter is transformed into interaction between the isolated dosimeter and metal plate. This reflects the fact that a thorough study of the structure and shape of the metal plate used to inform us about the possibility of a compromise on the different angles of incidence.

4 Discussion

In the majority of work, when it comes to study the potential E-field assessment errors associated with personal radiofrequency dosimeter in electromagnetic exposure, we mainly focus on angles of arrivals, position of dosimeter on body, environment, etc. We wanted to present in this parametric study the fact that the uncertainties evaluation exposure also greatly depends on the human body and dosimeter coupling. In this document, the initial goal was double.

First, we present the interaction between human body and incident field. The electromagnetic fields in the presence of the phantom differ significantly from the incident field. Our results are in agreement with those of [10]. This underestimation of the electromagnetic fields must be taken into account during the use of the personal dosimeters. The use of the average of the electric field and the magnetic field in the presence of the phantom makes it possible to obtain values closer to the incident fields around the direction of incidence (reflection area). But this solution does not give good performances in the shadow area. It will be interesting to study that kind of possibility (average, regression or correction law) to obtain values closer to the incident field, thus, it will be able to have a Smart RF-Dosimeter.

After, we studied the effect of the metal plate on the body and dosimeter coupling. With the human body exposed by a plane waves, we target the reflection area (torso) to position our dosimeter and metal plate, we found that, for a good distance between the dosimeter and the metal plate it's possible to reduce the reflection effect of the human body and by the same occasion reduce the gap between the measurement of the wave made by the isolated dosimeter and near the human body. For example, we can achieve a reduction of gap of around -1 dB (90 %) in a wide frequency range (800–1100 MHz) and for a distance of 1.75 cm (wavelength/18). This result is already a good indicator that confirms our idea of proceeding to a correction algorithm based on intermediate elements as a metal plate. We have also studied other performances of dosimeter at the vicinity of human body. For this, the results show that the radiation pattern is significantly influenced by the body. But the metal plate has allowed us to increase the gain and front to rear ratio in the direction of reception. Correction protocol that we will develop will also consider this feature to prevent mismatching. In this context, one of our future works is to highlight the impact of key criteria antenna design for personal dosimeters, namely the type of antenna and feeding methods. We also study the stability of the intrinsic characteristics (isotropy, bandwidth and return loss) of the different antennas family's close to the body.

5 Conclusion

Understanding and predicting in which way the presence of the human body alters the performance of personal dosimeters, is another great part of fundamental aspect when the dosimeter is used for the assessment of personal exposure level. Try to correct this alteration is even more important. It is clear that the body greatly influences the dosimeter performance particularly in terms of the measured field, radiation pattern, return loss, frequency and bandwidth. What really caught our attention is the influence of the human body in the ability of dosimeters to measure exactly incident field level. Underestimations at this level were very remarkable. In this correction worries, we introduced a metal plate in the structure of dosimeter. Its effect was good in terms of reductions of that gap.

References

1. Blas, J., Lago, F.A., Fernández, P., Lorenzo, R.M., Abril, E.J.: Potential exposure assessment errors associated with body-worn RF dosimeters. Bioelectromagnetics **28**, 573–576 (2007)
2. ICNIRP: Guidelines for limiting exposure to time-varying electric, magnetic and electromagnetic fields (up to 300 GHz). Health Phys. **44**, 1630–1639 (1998)

3. Bahillo, A., Blas, J., Fernández, P., Lorenzo, R.M., Mazuelas, S., Abril, E.J.: E-field assessments errors associated with RF-Dosimeters for different angles of arrival. Radiation Protection Dosimetry, October 16, 2008 (2008)
4. Iskra, S., McKenzie, R., Cosic, I.: Factors influencing uncertainty in measurement of electric fields close to the body in personal RF dosimetry. Radiat. Prot. Dosim. **140**, 25–33 (2010)
5. Conil, E., Hadjem, A., El Habachi, A., Wiart, J.: Whole body exposure at 2100 MHz induced by plane wave of random incidences in a population. C.R. Phys. **11**, 531–540 (2010)
6. Thielens, A., De Clercq, H., Agneessens, S., Lecoutere, J., Verloock, L., Declercq, F., et al.: Personal distributed exposimeter for radio frequency exposure assessment in real environments. Bioelectromagnetics **34**, 563–567 (2013)
7. Raumonen, P., Sydanheimo, L., Ukkonen, L., Keskilammi, M., Kivikoski, M.: Folded dipole antenna near metal plate. In: IEEE, Antennas and Propagation Society International Symposium, vol. 1, pp. 848–851. (2003)
8. Ghannay, N., Ben Salah, M.B., Romdhani, F., Samet, A.: Effects of metal plate to RFID tag antenna parameters. IEEE Conference on Mediterranean Microwave Symposium (MMS), pp. 1–3 (2009)
9. Gabriel, S., Lau, R.W., Gabriel, C.: The dielectric properties of biological tissues: III. Parametric models for the dielectric spectrum of tissues. Phys. Med. Biol. **41**, 2271 (1996)
10. Picard, D., Berrahma, K.: Radiofrequency electromagnetic fields in the vicinity of the body. In: 6th International Workshop on Biological Effects of Electromagnetic Fields (2010)

Effect of Duty Cycle of Bragg Filters on the Reflection Coefficient and Bandwidth

Abdelaziz Ouariach, Kamal Ghoumid, Rachid Malek, Ali El Moussati, Abdelkrim Nougaoui and Tijani Gharbi

Abstract In this paper, we investigate through a theoretical model the effect of duty cycle on reflectivity and bandwidth of optical filters such as Bragg filters. Our work is dedicated to the effect of duty cycle (r) and form factor (Z) on reflectivity, bandwidth and coupling coefficient evolution for different values of the corrugation depth l. The theoretical results are verified by experimental measurements realized on $LiNbO_3$:Ti.

Keywords Bragg filter · Bandwidth · Reflectivity · Duty cycle

1 Introduction

In parallel with the huge progress made in materials sciences, several accomplishments have been reached in nano-materials field which enabled us to achieve important tasks in the nanotechnolgy domain within the recent decades. For instance, several works demonstrated that electronic, magnetic, optical and acoustic properties of materials could be controlled and manipulated at the nanoscale-area. Concerning the optical properties, first note the appearance of small size structures

A. Ouariach (✉) · K. Ghoumid · A.E. Moussati
Department of Electronics, Informatics and Telecommunication,
Ecole National des Sciences Appliquées D'Oujda (ENSAO),
University of Mohammed Premier, Oujda, Morocco
e-mail: Ouariach.abdelaziz@gmail.com

A. Ouariach · A. Nougaoui
Laboratoire LDOM, Department of Physics,
University of Mohamed Premier, 60000 Oujda, Morocco

R. Malek
CEEP/LETAS, UMP Oujda—ENSA, Oujda, Morocco

T. Gharbi
Nanomedcine Lab, CHU Jean MINJOZ, University of Franche-Comté,
25030 Besançon, France

© Springer International Publishing Switzerland 2016
A. El Oualkadi et al. (eds.), *Proceedings of the Mediterranean Conference on Information & Communication Technologies 2015*, Lecture Notes in Electrical Engineering 380, DOI 10.1007/978-3-319-30301-7_21

197

that can serve the field of integrated optics, optomechanical systems (MOMS), photonic crystal with complete band gap, quantum wells (in electronics) etc. The reduced size of such systems enables the realization of nanomaterials combining mixed interesting physical properties, which gave birth to optoelectronics, acousto-optics, magneto-optics etc. Lithium Niobate (LiNbO$_3$) is a material that has been exploited in such domains due to its favorable optical, mechanical, ferro-electric and acousto-vibrational properties. For example, it's very much employed to fabricate photonic devices [1, 2].

Lithium Niobate has also played an important role in transduction, acoustics and signals recovering. Indeed, in recent decades it has attracted the attention of researchers for its electro-optical couplings properties and acousto-optic with a certainly impact in non-linear optics. In this same direction, the fabrication and development of Bragg gratings lithium niobate has acquired particular interest especially for specific functions as the realization of filters, mirrors or couplers narrow spectra [3–5].

The Bragg grating structures are consisting by a periodic perturbation of the effective refractive index in the waveguides optics. Especially, the filtering function that is targeted in this contribution, and to meet the high demand received by the industry. Indeed, the design of the optical filters is complicated; consequently, it must be controlled to achieve the desired performance in terms of reflectivity, bandwidth [Width Band Mid-maximum (FWHM)] and Central wavelength typically used to mark the peak the transmission function of the filters optical. Several techniques have been used for the fabrication of Bragg grating, we can find the following methods such as the photo-refractive [6, 7], the protonic exchange [8, 9], laser ablation [10, 11], the Reactive Ion Etching (RIE) and the Depth Reactive Ion Etching (DRIE) [12, 13]. In terms of micro/nano-systems, recently band pass filters photonic crystals have attracted great attention due to their applications in the processing of information at very broadband.

In this paper, we have proposed a theoretical model that we have validated with experimental results. This model focused on the effect of physical and geometrical parameters on the coupling coefficient, reflectivity and bandwidth. These coefficients in concerning question are in particular the duty cycle r, the order of Bragg m and the form factor Z.

2   Theory and Simulation

2.1   Model Description

A Bragg grating (BG) is generally an optical multilayer structure composed of two supports alternating with different refractive index. Figure 1 shows a representation of the studied model where an index corrugation is etched inside a waveguide whose index is n$_2$ and width is d.

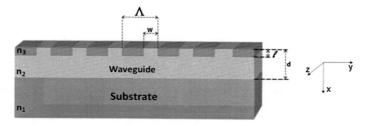

Fig. 1 Basic schema of Bragg grating etched in a Ti:LiNbO₃

Index modulation is characterized by a depth l, a period Λ and a cyclic ratio $r = w/\Lambda$. The total number of periods is N and the length of corrugation is: $L = N*\Lambda$.

The BG represented in Fig. 1 reflects a frequency band around a wavelength so-called Bragg's wavelength according to the following equation:

$$2n_{eff} * \Lambda = m * \lambda_B. \tag{1}$$

n_{eff} is the effective index, m is the Bragg's order, Λ the period of the BG and λ_B the wavelength Bragg.

This model is generally used to obtain a very high reflections of the light around a given central wave length.

2.2   Theory and Simulation Results

In this structure, we are interested in the determination of the coupling coefficient K cited in Ref. [4] that allows obtaining the values of the reflectivity and the bandwidth based on the resolution of Maxwell's equation associated to the transfer matrix approach. The expression of the coupling coefficient K is:

$$
\begin{aligned}
k = \frac{K_0^2(n_2^2 - n_3^2)}{2\pi m \beta N^2} &\sin\left(m\frac{\pi w}{\Lambda}\right)\{l_2 + \frac{\sin(2hl_2)}{2h} + \frac{q}{h^2}[1 - \cos(2l_2 h)] \\
&+ \frac{q^2}{h^2}[l_2 + \frac{\sin(2hl_2)}{2h}] + \frac{1}{q}[1 - e^{-2ql_1}]\}
\end{aligned}
\tag{2}
$$

with:

$$l_1 = \frac{w}{\Lambda}l \quad \text{and} \quad l_2 = (1 - \frac{w}{\Lambda})l$$

It is obvious from Eq. (2) that the coupling coefficient depends on several opto-geometric parameters such as the Bragg order m, the depth of corrugation l,

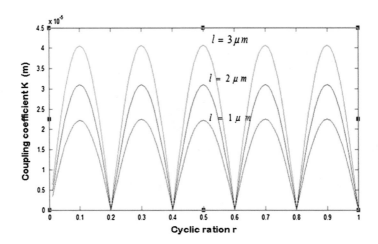

Fig. 2 The coupling coefficient K versus the cyclic ratio r for different values of etched depth l (simulation data: $n_1 = 2, 2112, n_2 = 2, 2174, n_3 = 1, d = 5$ μm and $m = 5$)

the index variation ($n_2 - n_3$) and the duty cycle. $r = w/\Lambda$. This important equation has a very important term which is the form factor:

$$Z = \sin(m\pi w/\Lambda) \quad \text{with} \quad r = W/\Lambda$$

We can optimize the reflectivity R and the coupling coefficient K by the simultaneous control of the duty cycle r and the Bragg order m. Indeed, as it's shown in Fig. 2 which represents the curve of the coupling coefficient versus the cyclic ratio for different values of engraved depth l. For $m = 5$ and $r = 0.1; 0.3; 0.5; 0.7...$ the form factor $Z = \pm 1$, consequently $|K|$ is maximum. And for $m = 5$ and $r = 0.2; 0.4; 0.6...$ the form factor $Z = 0$, which corresponds to $|K| = 0$.

However, we can deduce that for a duty cycle $r = 0.5$, the coupling coefficient K is equal to zero for a same order $m = 2, 4, 6, 8....$ And for a duty cycle $r = 0.33$, the coupling coefficient K is equal to zero for orders which are multiples of 3 i.e. $m = 3, 6, 9, 12...$

From Eq. (2) and Fig. 2 it appears that the effect of the duty cycle is periodic with a period $T = 1/m$ and that its maximum value is obtained whenever the duty cycle r_j satisfies Eq. (3) with $0 < j < m - 1$:

$$r_j = \frac{1}{2m} + \frac{j}{m} \tag{3}$$

As for the effect of duty cycle on the reflectivity R versus wavelength λ, it is clear from the results shown in Fig. 3 that increasing the reflectivity is done to the detriment of the mi-height bandwidth which increases to very specific value of r. This means that the form factor Z plays a very important role on the reflectivity and the bandwidth.

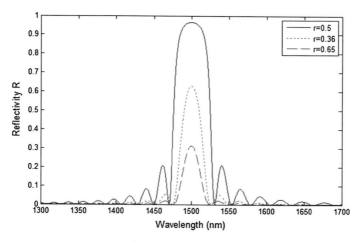

Fig. 3 Reflectivity R versus wavelength λ for different values of the cyclic ration r. (simulation data: $n_1 = 2, 2112$; $n_2 = 2, 2174$; $n_3 = 1$; $d = 5$ μm; $m = 5$; $l = 2.7$ μm; $\Lambda = 1.05$ μm)

2.3 Effect of Inclination

Lithium niobate is one of the hardest materials to etch. Indeed, etching of this material brings up numerous problems. One of these is the fact that the obtained etched furrows get inclined inwards instead of being vertical ($\alpha = 90°$). This inclination is explained by the redeposition of the material during its etching. A large amount of the removed material is redeposited on the etched furrows during the process. The resulting furrow has a trapezoidal shape as illustrated in Fig. 4.

The most important consequence of this effect relates to the duty cycle which is no longer constant. The value of the former depends then on the inclination slope and does increase with the depth of the etching as illustrated in Fig. 5 for several angles of inclinations. One notice that initially the so-called surface duty cycle (r) is

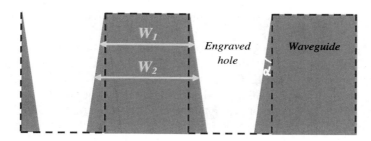

Fig. 4 Effect of the inclination due to the etching of a Bragg grating of period Λ (right flanks representing the ideal case are in *dotted lines*). The duty cycle increases according to the etched depth

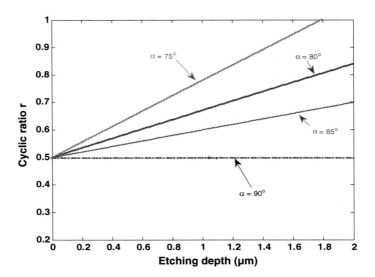

Fig. 5 Influence of furrows inclination on duty ratio-increasing with etched depth (y). At first, one has r = 0.5 and then the former gets higher with y. With an angle of inclination set to α = 75°. The maximum depth engraved is y_{max} = 1.8 μm

equal to 0.5, and then increases with the etched depth. The more the latter is important the more the duty cycle gets higher $r_2 > r_1$. This increasing is bigger if the angle of inclination α is important.

3   Experimental Study

3.1   Realization

Concerning the experimental part of this work, we did set up a method of Bragg grating nanostructuring based on focused ion beam engraving technique (FIB). This technique offers the possibility of designing Bragg gratings whose dimensions can be lower than 100 nm.

The guiding structure on which Bragg grating (BG$_s$) have been engraved is single-mode around λ = 1.50 μm. This waveguide is fabricated by standard diffusion at a temperature of T = 1080 °C during 10 h. A titanium ribbon having a width of 7 μm and a thickness of 80 nm is deposited by sputtering on a lithium niobate sample (LiNbO$_3$) of type X-cut, Z propagation, TE polarization [13, 14]. Once the waveguide is elaborated, the next step consists of engraving the BG. The latter is realized by the mean of FIB technique as mentioned above [12, 13]. Figure 6 illustrates an upper view of Bragg grating engraved by (FIB), of period Λ = 1.05 μm, a duty cycle r = 0.5 (Fig. 6a) and r = 0.65 (Fig. 6b).

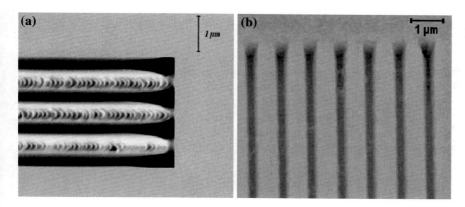

Fig. 6 Picture of Bragg grating (*upper view*) etched by FIB, of period Λ = 1.05 μm and a cyclic ratio r = 0.5 (Fig. 6a) and r = 0.65 (Fig. 6b)

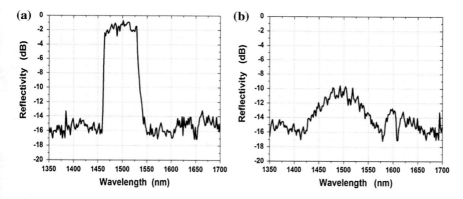

Fig. 7 Experimental reflectivity versus wavelength for Bragg grating (BG). (l = 2.7 μm; N = 60; m = 5; and Λ = 1.05 μm)

3.2 Results and Discussion

We have performed experimental measurements on Ti-diffused LiNbO3 based waveguides which have been made using FIB-process. Figure 7 represents the reflectivity curve versus the wavelength λ for cyclic ratios, r = 0.5 (Fig. 7a) and 0.65 (Fig. 7b).

We observe an important reflectivity that reaches 96 % within a whole wavelengths range centered on 1500 nm for r = 0.5. For r = 0.65 (within the same experimental conditions) one notes a significant falling down of reflectivity of almost 30 % in Fig. 5b. Comparison between experimental data presented in this paragraph and theoretical results (given in previous section) indicates that a good agreement is achieved overall.

4   Conclusion

In this work, we have presented both theoretical and experimental results relative to the effect of physical and geometrical parameters on the reflectivity, bandwidth and coupling coefficient (K) evolution (for different values of corrugation depth l) of Bragg gratings. Specifically, we have investigated the influence of duty cycle r and Bragg order m. Theoretical and experimental data show a good agreement in general. Our output data demonstrate the possibility of designing highly reflective optical components (optical filters) by properly adjusting the parameters in question (r, m).

References

1. Rabiei, P., et al.: Lithium niobate ridge waveguides and modulators fabricated using smart guide. Appl. Phys. Lett. **86** (2005) art no. 161115
2. Ulliac, G., Courjal, N., Chong, H.M.H., De La Rue, R.M.: Batch process for the fabrication of LiNbO$_3$ photonic crystals using proton exchange followed by CHF$_3$ reactive ion etching. Opt. Mater. **31**, 196–200 (2008)
3. Ghoumid, K., Benkalfat, B., Ferrière, R., Ulliac, G., Gharbi, T.: Wavelength-Selective Ti: LiNbO$_3$ multiple Y-Branch coupler based on Focused Ion Beam milled Bragg reflectors. IEEE J. Lightw. Technol. **29**, 3536–3541 (2011)
4. Ghoumid, K., Mekaoui, S., Ouariach, A., Cheikh, R., Nougaoui, A., Gharbi, T.: Tunable filter based on cavity electro-optic modulation for DWDM applications. Opt. Commun. **334**, 332–335 (2015)
5. Schiek, R., et al.: Nonlinear directional coupler in periodically poled lithium niobate. Opt. Lett. **24**(22), 1617–1619 (1999)
6. Becker, Ch., Greiner, A., Oesselke, Th., Pape, A., Sohler, W., Suche, H.: Integrated optical Ti: Er:LiNbO$_3$ distributed Bragg reflector laser with a fixed photorefractive grating. Opt. Let. **23**, 1194–1196 (1998)
7. Ghoumid, K., Ferriere, R., Benkelfat, B.-E., Mekaoui, S., Benmouhoub, C., Gharbi, T.: Technological implementation Fabry- Pérot cavity in Ti:LiNbO$_3$ waveguide by FIB. IEEE. Phot. Technol. Lett. **24**, 231–233 (2012)
8. Ghoumid, K., Ferriere, R., Benkelfat, B.-E., Ulliac, G., Salut, R.: Fabrication de miroirs de Bragg par faisceaux d'ions focalisés dans des guides Ti:LiNbO$_3$. Proc. J. EOS Ann. Meet. TOM **4**, 1–2 (2008)
9. Ghoumid, K., Ferriere, R., Benkelfat, B.-E., Ulliac, G., Salut, R., Rauch, J.-Y., Gharbi, T.: Effect of depth etching on Bragg reflectors realized by Focused Ion Beam in Ti:LiNbO$_3$ waveguide. Int Proc. SPIE. 7386, 738613-1 (2009)
10. Sidorin, Y., Cheng, A.: Integration of Bragg gratings on LiNbO$_3$ channel waveguides using laser ablation. Electr. Let. **37**, 312–314 (2001)
11. Cremer, C., Schienle, M.: RIE etching of deep Bragg grating filters in GaInAsP/InP. Electr. Let. **25**, 1177–1178 (1989)
12. Ghoumid, K., Elhechmi, I., Mekaoui, S., Pieralii, C., Gharbi, T.: Analysis of optical filtering in waveguides with a high index modulation using the extented coupled mode theory by hybridization of a matrix method. Opt. Commun. **289**, 85–91 (2013)
13. Ghoumid, K., Ferrière, R., Benkelfat, B.-E., Guizal, B., Gharbi, T.: Optical performance of Bragg gratings fabricated in Ti: LiNbO$_3$ waveguides by Focused Ion Beam milling. IEEE J. Lightw. Technol. **28**, 3488–3493 (2010)
14. Schmidth, R.V., Kaminow, R.I.: Metal-diffused optical waveguides in LiNbO$_3$. Appl. Phys. Lett. **25**, 458–460 (1974)

Brand Appeal Versus Value Added Services: When Does an MVNO Got a Share on the Mobile Market?

Hajar El Hammouti and Loubna Echabbi

Abstract In this paper, we investigate positioning on the mobile market of a Mobile Virtual Network operator (MVNO). We consider a market segmentation based on brand appeal, taking into account a segment of customers that are attached to the brand of the traditional operator and another segment where customers are less sensitive to the brand appeal of the operator. We consider three cases depending on the presence of the Mobile Network operator (MNO) on each segment and look for assumptions allowing a global Nash equilibrium.

Keywords Mobile network operator · Mobile virtual network opertor · Game theory · Nash equilibrium

1 Introduction

The telecommunication industry is one of the leading industries in the world. The last decades have seen an explosion of the use of mobile phones, especially with the access to the Internet. Such growth has led to a gradual evolution in the nature of mobile operators. Thus, Mobile Virtual Network Operators (MVNOs) have emerged. The MVNO refers to the organization who provides or resells mobile voice and data services to end users leasing the mobile spectrum of an existing MNO (Mobile Network Operator) by connecting to its wireless network but not obtaining the license for use of the radio spectrum [1]. These emerging operators are gaining popularity throughout the world especially in US, Europe and Asia as described in [2–4].

Almost all analysts agree that MNOs can take advantage of MVNOs to increase their revenue and generate economies of scale for their networks. Yet, the MVNO

H.E. Hammouti (✉) · L. Echabbi (✉)
Department Telecommunications Systems, Networks and Services,
STRS National Institute of Posts and Telecommunications, Rabat, Morocco
e-mail: elhammouti@inpt.ac.ma

L. Echabbi
e-mail: echabbi@inpt.ac.ma

© Springer International Publishing Switzerland 2016
A. El Oualkadi et al. (eds.), *Proceedings of the Mediterranean Conference on Information & Communication Technologies 2015*, Lecture Notes in Electrical Engineering 380, DOI 10.1007/978-3-319-30301-7_22

should design a competitive strategy, especially by looking for a service differentiation and reaching new customer niches. In this respect, authors in [5] present a Cournot approach with two stages to model the MNO-MVNO competition. They conclude that a successful MNO-MVNO business model relays on differentiation of services offered by the two operators.

Hence, Lillehagen et al. in [6] show that there are a large number of sustainable business models that are built on key success factors. One of these factors is the pricing model of the operators [7, 8]. The idea of a sustainable business model may be a bit complex once considering the uncertainty concept either when leasing spectrum [9, 10] or when predicting customers' or rivals' behavior.

In this work, we propose an MNO-MVNO competition model based on [11], taking into account two main parameters: brand appeal and value added services, and present the market structure according to the pricing strategies adopted. Unlike [11], we investigate situations where a global Nash equilibrium can be defined in order to present a global steady-state revenue for both operators, depending on the market structure.

Thus, this paper is structured as follows. First, we will define the main notations and describe the game modeling. Both MNO's and MVNO's demands and profits are formulated in a general way, and cases are positioned with respect to the operators' strategies. In Sect. 3, we will consider the global Nash equilibrium and specify its conditions. We will propose also numerical investigations that illustrate the global Nash conditions and exemplify a steady-state situation profitable to both operators.

2   Main Notations and Game Modeling

2.1   Main Notations

Let us consider a market where two operators are present: MNO and MVNO. The MNO, mostly the incumbent operator, has a strong reputation among customers. Thus, it bases its business model, firstly, on its brand awareness. The MVNO, newly introduced in the market, bases its business model on the value added services proposed to its customers. Besides, we consider a two-segmented market: the first segment S_1 where mobile users have a high perception of brand appeal, and the second, S_2, where costumers are less interested in the brand and more sensitive to the added value. In both segments, customers may refuse to subscribe with any operator if the prices do not match their expectations. We consider the following parameters:

- I_1 and I_2, the brand appeal of the MNO and MVNO. It is supposed that $I_1 \geq I_2$. The perception of the brand can be obtained for example by a survey where costumers evaluate the operator's brand by a score from 1 to 10.

- V_1 and V_2 are the added value of the MNO and MVNO. We suppose that $V_1 \leq V_2$ since MVNO should offer better added value. Again, these values may relate to a score obtained from a market study as for the brand appeal.
- P_1 and P_2 are the unitary prices charged by MNO and MVNO to their customers.
- ω is the unitary wholesale price charged by the MNO to the MVNO. Thus, the MVNO's retail price should be at least equal to this wholesale price.
- μ is the margin set by the MVNO. Thus $P_2 = \omega + \mu$.
- N designates the target population. With $0 \leq R \leq 1$ a proportion of S_1.
- α_1 and α_2: the customer sensitivity to the brand appeal according to the segment to which the customer belongs. α_1 for customers in S_1 and α_2 for those in S_2, with $\alpha_1 \leq \alpha_2$.
- β models the sensitivity to the added value. It is supposed, for the sake of simplicity, to be a uniform distribution on a bounded interval $[0; \beta_{max}]$.

Users choose to subscribe with MNO or MVNO depending on their utility $\alpha_i I_j + \beta V_j - P_j$, with $j = 1$ for the MNO and $j = 2$ for the MVNO, and $i \in \{1, 2\}$ according to the segment to which the customer belongs. Only three cases are possible (and detailed in our work in [11]). Case 1, where the MNO is present in both segments, in S_2, and perforce in S_1. Case 2, where the MNO leaves competition on S_2 and shares potential customers in S_1 with the MVNO. Case 3, where the MVNO is the only operator present in both segments.

Since the customer's utility depends on he/her sensitivity to the added value, we can set thresholds on β defining the market share of each operator as described in the Fig. 1 with, on the first segment:

$$\beta_{11} = \frac{P_1 - \alpha_1 I_1}{V_1}, \quad \beta_{21} = \frac{P_2 - \alpha_1 I_2}{V_2}, \quad \beta_{31} = \frac{(P_2 - P_1) - \alpha_1 (I_2 - I_1)}{V_2 - V_1}. \quad (1)$$

And, on the second segment:

$$\beta_{12} = \frac{P_1 - \alpha_2 I_1}{V_1}, \quad \beta_{22} = \frac{P_2 - \alpha_1 I_2}{V_2}, \quad \beta_{32} = \frac{(P_2 - P_1) - \alpha_2 (I_2 - I_1)}{V_2 - V_1}. \quad (2)$$

The order of these thresholds depends on the cases of the market as it is presented in Fig. 1 (please find more details in [11]).

2.2 The MNO's and MVNO's Demands

As described on Fig. 1, The MNO is present in the segment i only if $\beta_{1i} \leq \beta_{3i}$. Hence:

Fig. 1 The three possible cases of the mobile market structure

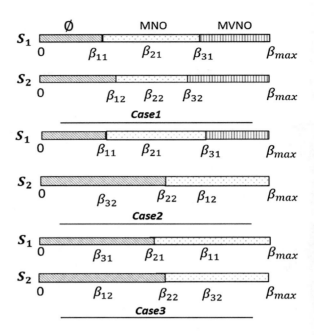

$$DMNO(\omega, \mu) = \max\left(0, \int_{\beta_{11}}^{\beta_{31}} \frac{NR}{\beta_{\max}} dx\right) + \max\left(0, \int_{\beta_{12}}^{\beta_{32}} \frac{NR}{\beta_{\max}} dx\right)$$

$$= \max\left(0, \frac{NR}{\beta_{\max}}(\beta_{31} - \beta_{11})\right) + \max\left(0, \frac{N(1-R)}{\beta_{\max}}(\beta_{32} - \beta_{12})\right). \tag{3}$$

Note that for μ fixed, this function is a piecewise linear function, with respect to ω, with thresholds that allow the transition from one case to another as follows:

$$\omega_1(\mu) = -(\mu - P_1 + \alpha_1 dI - \beta_{11} dV), \tag{4}$$

$$\omega_2(\mu) = -(\mu - P_1 + \alpha_1 dI - \beta_{12} dV). \tag{5}$$

In the same way we define the MVNO's demand:

$$DMVNO(\omega, \mu) = \min\left(\frac{NR}{\beta_{\max}}(\beta_{\max} - \beta_{31}), \frac{NR}{\beta_{\max}}(\beta_{\max} - \beta_{21})\right)$$

$$+ \min\left(\frac{N(1-R)}{\beta_{\max}}(\beta_{\max} - \beta_{32}), \frac{N(1-R)}{\beta_{\max}}(\beta_{\max} - \beta_{22})\right). \tag{6}$$

Like the MNO's demand, the MVNO's one is a continuous piecewise linear function with respect to μ, with thresholds that allow the transition from one case to another as follows:

$$\mu_1(\omega) = \frac{(P_1 - \alpha_1 dI)V_2 - \omega V_1 - \alpha_1 I_2 dV}{V_1}, \tag{7}$$

$$\mu_2(\omega) = \frac{(P_1 - \alpha_2 dI)V_2 - \omega V_1 - \alpha_2 I_2 dV}{V_1}. \tag{8}$$

2.3 The Proposed Pricing Model

We consider a sequential game where the MNO first fixes the wholesale price ω and the MVNO follows by deciding its margin μ. We define the MNO's and MVNO's profits with respect to ω and μ:

$$PMNO(\omega, \mu) = DMNO(\omega, \mu) * P_1 + DMVNO(\omega, \mu) * \omega, \tag{9}$$

$$PMVNO(\omega, \mu) = DMVNO(\omega, \mu) * \mu. \tag{10}$$

The objective of each operator is to maximize its profit while $\omega \leq P_1$. We suppose that regulation imposes this constraint for the sake of equity with regards to MVNO.

Once ω and μ are set, one of the 3 former defined cases arises and the profits can be calculated accordingly.

Using the thresholds on ω and μ, we can situate cases on a diagram as described in Fig. 2. Notice that $\omega_1(\mu)$ (resp. $\omega_2(\mu)$) is the inverse function of $\mu_1(\omega)$ (resp. $\omega_2(\mu)$) (We have, indeed, $\omega_1(\mu_1(\omega)) = \omega$).

Fig. 2 Cases' positioning with respect to ω and μ

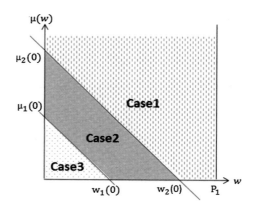

2.4 Local Nash Equilibrium

In our optimization problem, each operator is selfish and wants to maximize its own profit. In game theory, this problem has a well know solution concept called the *Nash equilibrium* (N.E). A Nash equilibrium is a situation where each player's strategy is a best response to its opponents. It is local when it is verified only for a specific situation, for example, when considering a specific case. In our game model, we consider that the optimization problem takes place only locally while considering values of ω an μ that keeps operators on a specific case. To find the local N.E for each case, since profits are concave functions, we derive MNO's profit with respect to ω and MVNO's profit with respect to μ, and solve the linear system obtained. Then, we make sure that the solution belongs to the intervals that define the case. This solution is the local N.E.

3 Numerical Illustrations and Results Analysis

3.1 Global Nash Equilibrium

The objective of this section is to find some realistic situations where the local N.E in case 2 becomes a global one. The choice of the equilibrium of case 2 as a global equilibrium is due to the fact that this situation (case 2) is more likely to be interesting for both operators. On one hand, it is more interesting for the MNO to be in case 2 than case 3 where he is absent from both segments. On the other hand, the MVNO would prefer to be in case 2 rather than case 1 because the MNO in case 2 does not share MVNO's potential costumers (costumers in S_2). Next, we will investigate the impact of the variation of the game parameters taken into consideration the Nash condition. Yet, due to pages limitation, we will present only comparison between case 2 and 1. Case 3 is of very little interest since in reality the MNO is unlikely to yield both market shares. As a leader, the MNO will set its wholesale price accordingly. Thus, for some specific metrics, we will check the following assumptions that define the global Nash equilibrium (with comparison to case 1), and discuss the results:

$$PMNO_1\left(\omega_{max/\mu_2^*}, \mu_2^*\right) \leq PMNO_2\left(\omega_2^*, \mu_2^*\right) \tag{11}$$

and

$$PMVNO_1\left(\omega_2^*, \mu_{max/\omega_2^*}\right) \leq PMVNO_2\left(\omega_2^*, \mu_2^*\right), \tag{12}$$

with $\omega_{max_{i/\mu_2^*}}$ maximizes $PMNO_i(\omega, \mu_2^*)$, $i \in 1, 3$, and $\mu_{max_{i/\omega_2^*}}$ maximizes $PMVNO_i$ (ω_2^*, μ), $i \in \{1, 3\}$. From these conditions, we can deduce new conditions on ω and

μ (which are not included in this version of the paper due to pages limitation) that we will be simulating in the next subsections. Each time, we simulate one condition on ω and another on μ. Note that these conditions are only sufficient conditions and are satisfied when the 3D-curve is negative.

3.2 Wide Difference Between Sensitivities to Brand Appeal

In Figs. 3 and 4, we consider a wide difference between the brand sensitivities ($\alpha_1 = 0.8$ and $\alpha_2 = 0, 2$). We find in this case that the Nash conditions are all the time satisfied. We conclude that when potential customers are either very attached to the brand appeal or weakly interested in it, the MNO needs the MVNO to acquire the last category of customers. The MNO can then take advantage of the MVNO to attain un-targeted segments.

3.3 Small Difference Between Sensitivities to Brand Appeal

Now, let us suppose $\alpha_1 = 0.6$ and $\alpha_2 = 0.4$ as illustrated in Figs. 5 and 6. In this case, the entire population has almost the same sensitivity to the brand appeal. The Nash conditions on ω and μ are not always satisfied. First, when the brand awareness of the MNO is too strong regarding the MVNO's one (I_1/I_2 is large). Second, when the value added services the MVNO proposes are very attractive (V_2/V_1 is large) and the brand values of both operators are almost the same (I_1/I_2 is low). Actually, when the MVNO has a low brand appeal, it is more profitable to improve the value added services it offers.

Fig. 3 Nash condition on ω for $\alpha_1 = 0.8$ and $\alpha_2 = 0.2$

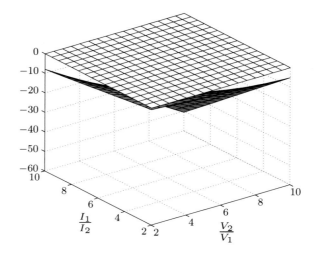

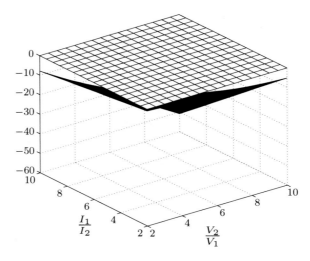

Fig. 4 Nash condition on μ for $\alpha_1 = 0.8$ and $\alpha_2 = 0.2$

Fig. 5 Nash condition on ω
for $\alpha_1 = 0.6$ and $\alpha_2 = 0.4$

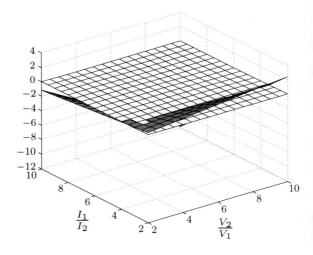

To summarize, the MNO may consider the MVNO as a new way to get more revenue from the market especially when this later proposes complementary products and target untapped segments. The MVNO may consider to invest in its services to improve its situation. The Fig. 7 is a summary for our results.

Fig. 6 Nash condition on μ
for $\alpha_1 = 0.6$ and $\alpha_2 = 0.4$

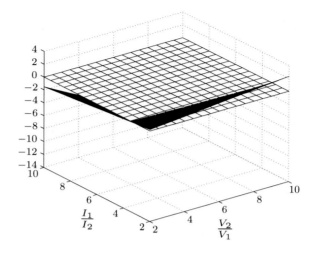

Fig. 7 Summarizing results

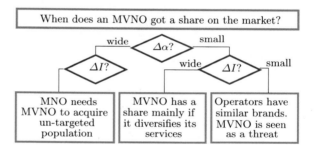

4 Conclusion

In this paper, we have considered a game between MNO and MVNO based on
market segmentation according to costumers sensitivity to brand appeal. We have
analyzed how MVNO's strategy depends on the MNO's one, but also on the
market's structure. We have concluded that the MNO accepts the MVNO only
when the MVNO does not present a real threat but provides customized services
that attract untapped segments. In an ongoing work, we investigate opportunities for
the MVNO to invest in its added services for a better positioning on the market.

References

1. Jeong-seok, P., Kyung-seok, R.: Developing MVNO market scenarios and strategies through a
   scenario planning approach. In: 7th International Conference on Advanced Communication
   Technology, pp. 137–140 (2005)

2. Kimiloglu, H., Kutlu, B.: Market analysis for mobile virtual network operators (MVNOs): The case of Turkey. In: The 6th International Journal of Business and Management, vol. 6, pp. 39–54 (2011)
3. Shin, D.H., Bartolacci, M.: A study of MVNO diffusion and market structure in the EU, US, Hong Kong, and Singapore. Telematics Inform. **24**, 86–100 (2007)
4. Kim, B.W., Seol, S.H.: Economic analysis of the introduction of the MVNO system and its major implications for optimal policy decisions in Korea. Telecommun. Policy **31**, 290–304 (2007)
5. Kalmus, P., Wiethaus, L.: On the competitive effects of mobile virtual network operators. ScienceDirect **5**, 262–269 (2010)
6. Thanh, D.V.: The mobile virtual network operator concept: Truth and myths. Telektronik, 3–6 (2001)
7. LeCadre, H., Bouhtou, M., Tuffin, B.: A pricing model for a MNO sharing limited resource with a MVNO. Netw. Econ. Next Gener. Netw. **5539**, 24–35 (2009). (LNCS)
8. LeCadre, H., Bouhtou, M., Tuffin, B.: Competition for subscribers between mobile operators sharing a limited resource. In: 9th International Conference on Game Theory for Networks, pp. 469–478 (2009)
9. Duan, L., Huang, J., Biying, S.: Investment and pricing with spectrum uncertainty: A cognitive operator's perspective. IEEE Trans. Mob. Comput. **10**, 1590–1604 (2011)
10. Duan, L., Huang, J., Biying S.: Cognitive mobile virtual network operator: Investment and pricing with supply uncertainty. In: IEEE International Conference on Computer Communications, pp. 1–9 (2010)
11. Debbah, M., Echabbi, L., Hamlaoui, C.: Market share analysis between MNO and MVNO under brand appeal based segmentation. In: 6th International Conference on Network Games, Control and Optimization, pp. 9–16 (2012)

A New Semi-analytical Performance Prediction of a Digital Communication System

Fatima Ezzahra Naamane, Mohamed Et-tolba and Mostafa Belkasmi

Abstract This paper proposes a new semi-analytical method for estimating the error probability for any digital communication system. We show that the problem of the error probability estimation is equivalent to estimating the conditional probability density function (pdfs) of soft receiver outputs. The proposed method uses Fast Fourier transform inversion for predicting the pdf. Furthermore, we applied Bootstrap criterion for selecting the optimum smoothing parameter which makes this method more accurate. Simulation results have shown that the obtained error probability with the new approach is close to that measured by Monte-Carlo simulation method.

Keywords Semi-analytical approach · BEP · Pdf · FFT · Bootstrap

1 Introduction

Monte Carlo (MC) simulation techniques are generally used to evaluate the error probability of a digital communication system once we cannot analytically compute its receiver performance. Unfortunately, it is well known that the MC method is very prohibitive in terms of computing time especially when a complex system receiver is considered. In order to overcome this problem, performance prediction methods have recently been proposed. The evaluation of the error probability has been made using semi-analytical approaches. In [1], the authors have shown that the problem of error probability is equivalent to estimate the conditional probability

F.E. Naamane (✉) · M. Belkasmi
Telecom and Embedded System Team, ENSIAS, Rabat, Morocco
e-mail: ezzahra.naamane@um5s.net.ma

M. Belkasmi
e-mail: m.belkasmi@um5s.net.ma

M. Et-tolba
Departement of Communications Systems, INPT, Rabat, Morocco
e-mail: ettolba@inpt.ac.ma

© Springer International Publishing Switzerland 2016
A. El Oualkadi et al. (eds.), *Proceedings of the Mediterranean Conference on Information & Communication Technologies 2015*, Lecture Notes in Electrical Engineering 380, DOI 10.1007/978-3-319-30301-7_23

density function (pdfs) of the observed soft samples at the receiver output. The method we have proposed in [2] considers the estimation of the pdf using Kernel estimator [3]. Furthermore, it has been shown that the accuracy of this method is very sensitive to the choice of the optimum smoothing parameter. Nevertheless, this technique can lead to inconsistent estimator and requires too much computing time. In this paper, we propose a new semi-analytical method based on Fast Fourier transform inversion for estimating the pdf. We applied Bootstrap criterion for selecting the optimum smoothing parameter, which makes the proposed method more accurate.

The remainder of the paper is organized as follows. In Sect. 2, we introduce the problem of error probability derivation. Section 3 details the probability density function estimation using Fast Fourier Transform Inversion. Simulations and numerical results are given in Sect. 4. Finally, the paper is concluded in Sect. 5.

2   Error Probability Derivation

Our main objective is to estimate the error probability of a digital communication system. The system is modeled as a classical communication chain, which consists of a source, a transmitter, a transmission channel and a receiver, as depicted in Fig. 1. The source is considered to be digital and delivers the information bits $b \in \{0, 1\}$. These bits are processed by a transmitter which can include channel coding, interleaving, and modulation. After that, the information symbols at the output of the transmitter are transmitted over a channel.

For simplicity, the channel is assumed to be Gaussian. The channel output is delivered to the receiver which tries to detect the information bits from a noisy signal by using a detector, a sampling process and a decision. Due to the channel effect, the receiver can make a wrong decision on information bits at its output $\tilde{b}$. So it is important to measure the communication system efficiency. The most popular mean to do this is the BEP (Bit error probability) evaluation. According to the system model, the bit error probability is written as:

$$
\begin{aligned}
P_e &= Pr[Error|b \, sent] \\
&= P_1 \cdot Pr(\tilde{b} = 0|b = 1) + P_0 \cdot Pr(\tilde{b} = 1|b = 0)
\end{aligned}
\tag{1}
$$

where $P_k, k = 0, 1$, is the probability that $b = k \cdot Pr(\cdot)$ is the conditional probability.

Fig. 1 General system model

If X is a random variable whose realization are the samples x observed by the receiver, then the P_e can be writhen as:

$$P_e = P_1 \cdot \int_{-\infty}^{0} f_X(x|b=1)\,dx + P_0 \cdot \int_{0}^{+\infty} f_X(x|b=0)\,dx \qquad (2)$$

where $f_X(x)$ denotes the pdf of x.

The conditional probability is evaluated by integrating the probability density functions of the random variable x. So, for predicting the error probability P_e, one has to estimate the probability density $f_X(x)$.

3 Probability Density Function and Error Probability Estimation

Due to the development in statistics theory in last years, several approaches are suggested to estimate the probability density function. In this paper, we use Fast Fourier transform inversion method to estimate the pdf.

The inverse fast Fourier transform of the function f_X is defined as:

$$\widetilde{f}(x;h) = \frac{1}{2\pi} \int_{-\infty}^{+\infty} e^{-jtx} \widetilde{\varphi}(t)\,dt \qquad (3)$$

where $\widetilde{\varphi}(t)$ is an estimator of the characteristic function of the pdf f which can be estimated as:

$$\varphi(t) = \mathbb{E}[e^{jtX}] \qquad (4)$$

For a given set of X of N received samples $X_1, X_2, \ldots, X_N$, it is natural to estimate the characteristic function as:

$$\varphi(t) = \frac{1}{N} \sum_{i=1}^{N} e^{jtX_i} \qquad (5)$$

Now, we can invert the characteristic function $\varphi(t)$ using the inverse Fourier transform to get the estimate of the probability density function f_X. One difficulty with applying this inversion formula is that it leads to a diverging integral since the estimate $\varphi(t)$ is unreliable for large t. To solve this problem, the estimator $\varphi(t)$ is multiplied by a damping function $\psi_h(t) = \psi(ht)$, which must be equal to 1 at the origin, and then falls to 0 at the infinity.

$$\tilde{\varphi}(t) = \frac{1}{N}\sum_{i=1}^{N} e^{jtX_i}\psi_h(t) \tag{6}$$

Now, we can invert $\varphi(t)$ using the inverse Fourier transform to get the estimate of the unknown density f:

$$\tilde{f}(x;h) = \frac{1}{Nh}\sum_{i=1}^{N} v(\frac{x - X_i}{h}) \tag{7}$$

where

$$v(x) = \frac{1}{2\pi} \int_{-\infty}^{+\infty} e^{-jtx}\psi(t)dt \tag{8}$$

where h is the smoothing parameter.

The most common choice for function ψ is a Gaussian function $\psi(ht) = e^{-\pi t^2}$. So, to get the expression of the estimated bit error probability P_e, we divide the set of observed samples into two subsets S_0 and S_1. The first subset contains N_0 observed samples which correspond to the transmission of $b = 0$. The second subset consists of N_1 observed samples when $b = 1$ is transmitted. In this manner, and by substituting the probability density $\tilde{f}(x;h)$, the estimated bit error probability is expressed as:

$$P_e = \frac{P_1}{N_1}\sum_{i=1}^{N_1} Q(\frac{(X_i)_1}{\sqrt{2\pi}h}) + \frac{P_0}{N_0}\sum_{i=1}^{N_0} Q(\frac{-(X_i)_0}{\sqrt{2\pi}h}) \tag{9}$$

where $(X_i)_0$ and $(X_i)_1$ are the observed samples correspnding to $b = 0$ and $b = 1$ respectively and $Q(:)$ denotes the complementary unit cumulative Gaussian distribution, that is:

$$Q(x) = \frac{1}{\sqrt{2\pi}} \int_{x}^{+\infty} e^{-t^2/2}dt \tag{10}$$

From (10), it is very clear that the accuracy of bit error probability estimation depends on the choice of the optimal smoothing parameter.

3.1 Bootstrap Criterion for Selecting the Optimum Smoothing Parameter

In [4], we have compared some criteria to make up for the optimum smoothing parameter choice. The first one is the minimum integrated squared error (MISE) [5], which exhibited a significant squared error between the true pdf and the estimated one. In the second method, the smoothing parameter is estimated using cross-validation (CV) method [6, 7]. Simulation study have concluded that the method called cross validation, outperforms the other criterion in terms of squared error. Nevertheless, this technique can lead to inconsistent estimator and requires too much computing time. In this work, we use Bootstrap criterion on the choice of the optimal smoothing parameter. Bootstrap procedures for selecting the smoothing parameter have been studied in the previous work [8–10]. The idea is to estimate the MISE using the bootstrap and then minimize over h. Let $\widetilde{f}_X(x; g)$ be the estimator of $f_X(x)$ obtained from $\{X_1, \ldots, X_N\}$, with a pilot smoothing parameter g. The straightforward approach to use the Bootstrap method would be to resample $\{X_1^*, \ldots, X_N^*\}$ from $\widetilde{f}_X(x; g)$ and then construct bootstrap estimates $\widetilde{f}_X^*(x; h)$ [11]. The bootstrap estimator of the MISE is given by:

$$E(R^*) = \varepsilon\left[\int_{-\infty}^{+\infty} [\widetilde{f}_X^*(x; h) - \widetilde{f}_X(x; g)]^2 dx\right] \tag{11}$$

Making a substitution followed by a Taylor series expansion and we assume that $h \to 0$ as $N \to \infty$ to get an asymptotic approximation [12]:

$$E(R^*) = \frac{1.074}{2Nh\sqrt{\pi}} + \frac{h^4}{4}\int \widetilde{f}_X^{iv}(x; g)\widetilde{f}_X(x; g)dx + O(h^6) \tag{12}$$

Using the condition that any probability density function satisfies:

$$[\widetilde{f}_X'''(x; g)\widetilde{f}_X(x; g)]_{-\infty}^{+\infty} = [\widetilde{f}_X''(x; g)\widetilde{f}_X'(x; g)]_{-\infty}^{+\infty} = 0 \tag{13}$$

The asymptotic expression for bootstrap estimator of MISE is:

$$E(R^*) = \frac{1.074}{2Nh\sqrt{\pi}} + \frac{h^4}{4}\int (\widetilde{f}_X''(x; g))^2 dx + O(h^6) \tag{14}$$

The easiest way to select the smoothing parameter is to find h value which minimizes $E(R^*)$:

$$h_{boot} = \arg_h \min(E(R^*)) \tag{15}$$

The optimal h value obtained is given as:

$$h_{boot} = \left(\frac{1.074}{2\sqrt{\pi} \int (\widetilde{f}_X''(x; g))^2 dx} \right)^{1/5} \cdot N^{-1/5} \tag{16}$$

As it can be seen from this equation, the optimal h_{boot} value depends on the second derivative of the estimate pdf $\int (\widetilde{f}_X''(x; g))^2 dx$, which depends on the choice of a pilot smoothing parameter g.

4 Simulation Results

To evaluate the performance of the proposed semi-analytical approach based on Fast Fourier Transform Inversion, we consider the framework of a SC-FDMA [13] system with binary phase-shift keying (BPSK) modulation and operate over an additive white Gaussian noise (AWGN) channel. One of the most efficient choices of the pdf is the inverse fourier transform method. In Fig. (2), we present both the theorical and the estimated conditional pdf of the observed soft sample using a smoothing parameter $h = 0.04$ and $N = 10^4$ observations. From the result, we prove that the IFFT method performs very well to recover the true density. The difference in the precision of the new approach and the direct computation method is very neglectable.

To see how accurate the Bootstrap smoothing parameter is, computer simulations are done in terms of squared error. The bootstrap performance is compared to Kernel based MISE criterion. Simulations obtained in Fig. 3 plot the evolution of the mean integrated squared error $MISE(h)$ and the Bootstrap Estimator $E(R^*)$. We prove that Bootstrap estimator tends rapidly to low values compared to the other estimators. From the results, it is clear that Bootstrap method outperforms ISE

Fig. 2 Pdf result from FFT inversion

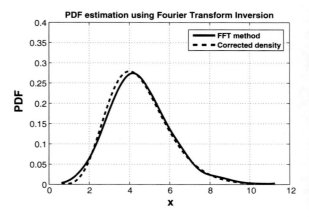

Fig. 3 Standard error
criterion comparison

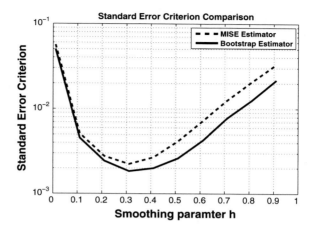

Fig. 4 BEP performance for
BPSK

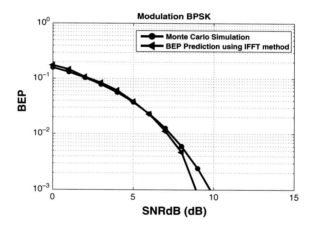

based method in terms of squared error and can be considered as a consistent
estimator.

After that, we validate the BEP prediction of a BPSK system over an AWGN
channel. We have shown that the new approach of BEP estimation (Fig. 4) provides
the same performance as the Monte Carlo method with a few numbers of data
observations. Also, in Fig. 5, computer simulations are run to assess the accuracy of
the new simulator considering a SC-FDMA system [13] over an AWGN channel.
The simulation are done with $N = 5000$ transmitted bits. From the result, it is
observed that the new simulator is accurate compared to Monte carlo simulation.
Moreover, a significant gain in terms of computing time is obtained.

Fig. 5 BEP performance for SC-FDMA system

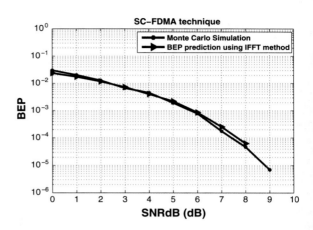

5 Conclusion

In this paper, we considered the problem of the error probability estimation for digital communication systems. We used a new semi-analytical approach of the BEP prediction, which is based on the pdf estimation using FFT. In the proposed method, we derived a new expression of the smoothing parameter, termed Bootstrap criterion. The results of this paper established that the obtained error probability with the new approach is close to that measured by Monte-Carlo simulation method.

References

1. Saoudi, S., Troudi, M., Ghorbel, F.: An iterative soft bit error rate estimation of any digital communication systems using a nonparametric probability density function. J. Wireless Commun. Netw. **2009** (2009)
2. Naamane, F.E., Et-tolba, M., Belkasmi, M.: Performance prediction of a coded digital communication system using cross-validation. Proc. IEEE PIMRC **2013**, 621–625 (2013)
3. Zucchini, W.: Applied smoothing technique, part 1: kernel density estimation (2003)
4. Naamane, F.E., Et-tolba, M., Belkasmi, M.: A semi-analytical performance prediction of turbo coded SC-FDMA. Proc. ICWMC **2013**, 146–151 (2013)
5. Sheather, S.J., Jones, M.C.: A reliable data-based bandwidth selection method for kernel density estimation. J. Roy. Stat. Soc. B **53**(3), 683–690 (1990)
6. Sain, S.R., Baggerly, K.A., Scott, D.W: Cross validation of multivariate densities. J. Am. Stat. Assoc. **89**, 807–817 (1992)
7. Savchuk, O.Y., Hart, J.D., Sheather, S.J.: An empirical study of indirect cross-validation. Submitted to Festschrift for Tom Hettmansperger, IMS Lecture Notes—Monograph Series
8. Faraway, J., Jhun, M.: Bootstrap choice of bandwidth for density estimation. J. Am. Stat. Assoc., 1119–1122 (1990)
9. Grund, B., Polzehl, J.: Bias corrected bootstrap bandwidth selection. J. Nonparametric Stat., 97–126 (1997)

10. Hall, P.: Using the bootstrap to estimate mean squared error and select smoothing parameter in nonparametric problems. J. Multivar. Anal., 177–203 (1990)
11. Marron, J.S.: Bootstrap bandwidth selection. In: LePage, R., Billard, L. (eds.) Exploring the Limits of Bootstrap, pp. 249–262 (1991)
12. Taylor, C.: Bootstrap choice of the tuning parameter in kernel density estimation. Biometrika, 705–712 (1989)
13. Myung, H.G., Goodman, D.J.: Single Carrier FDMA: a New Air Interface For Long Term Evolution. Wiley, New York (2008)

Network Higher Security of Intelligent Networks Platforms in Telecommunications Areas

Imane Maslouhi, Kamal Ghoumid, Jamal Zaidouni, Wiame Benzekri, El Miloud Ar-reyouchi and Ahmed Lichioui

Abstract The importance of intelligent network (IN) obliges each organism to secure continuously and efficient manner all equipment and data flowing in the network. This document proposes recommendations for telecommunications companies to improve the security level of its Intelligent Networks. These recommendations are justified by tests made by appropriate tools and implemented in an application in order to manage user's traceability. Therefore, the authors develop a traceability management solution to fight against inappropriate changes in network equipment and monitored.

Keywords Intelligent networks · Security network · Telecommunications · Protocol · Traceability management

1 Introduction

The security of the information system is considered one of the crucial challenges to establish. Each organization is led to define a security policy to succumb to its need and protect these resources and the secrets of its occupation. In view of the increased importance focused on intelligent networks (IN) which have a permanent financial source for the network operator [1], it seems obvious to implement on a regular security policy and audit in order to face up to all kinds of computer attacks. Network security is one of the most critical part of information system's security

I. Maslouhi · K. Ghoumid (✉) · J. Zaidouni
Electronics, Informatics and Telecommunications Department, ENSAO,
Mohammed I University, Oujda, Morocco
e-mail: ghoumid_kamal@yahoo.fr

W. Benzekri
ANOL Laboratory, FSO, Mohammed I University, Oujda, Morocco

E.M. Ar-reyouchi · A. Lichioui
Société Nationale de Radiodiffusion et de Télévision (SNRT), Rabat, Morocco

that deserves a specific monitoring and periodic audits. This audit concerns the IN Department of Nokia Siemens Networks company.

The document's aim is to carry out an audit of the IN security of platforms telecommunications companies, using the EBIOS (Expression of Needs and Identification of Security Objectives) method [2] audit, to develop recommendations for the company to improve its IN security level [3]. Thereafter, we will test these recommendations in order to justify our proposals and develop a traceability management solution to fight against inappropriate changes in network equipment and monitored.

2 Recommendations

The audit work that we have done in Nokia Siemens Networks Company was in the form of interview operations and platform tests. Thanks to these interviews we deduced several vulnerabilities in protocols level.

2.1 FTP Protocol

The File Transfer Protocol (FTP) is an insecure protocol, all the data exchanged with the server (upload or download) are transferred in plain text without any encryption, so it is inadvisable to use it. A common solution to this problem is to use the secure version of FTP (Secure File Transfer Protocol—SFTP). To do this we must put 'anonymous_enable' parameter at 'NO' in the configuration file vsftpd.conf.

2.2 TELNET Protocol

The Telnet protocol used between the administrator machine and administered equipment, transmits the login and the password in clear, they are easily intercepted. To remedy this flaw, we recommend to the company the use of an intermediary machine between the equipment and the administrator machine, It is required to use the Secure Shell protocol [4] (SSHv2) between the intermediate and the administrator machines, and Telnet protocol between equipment and this intermediate machine. Without forgetting, that we must apply the access-list at the equipment level to limit access to the intermediate machine.

2.3   RSH and Rlogin Protocol

The Remote Shell (RSH) protocol and The Remote Login (rlogin) protocol are both insecure protocol, so they should be replaced by SSHv2 protocol.

2.4   Vulnerability Detection

To be able to detect existing network vulnerabilities, the company should install Nessus software, which is a vulnerability scanner used to detect machine and network weaknesses.

2.5   Intrusion Detection

We consider that each telecommunication company needs an intrusion detection system for its IN. For this, we propose to implement SNORT 7 into the IN, to detect intrusions at machinery well as intrusions in throughout the network.

2.6   Traceability Management

We developed a traceability management application for existing network equipment in the IN telecommunications company. This application is intended to fight against frauds. It will be presented in the following.

2.7   HSRP Protocol

As indicated in [1] the Hot Standby Router Protocol (HSRP) has some vulnerability. To fix this problem, we suggest the use of the Virtual Router Redundancy Protocol (VRRP).

3   Experimental Section

In order to justify our proposals, we will illustrate our recommendations.

Fig. 1 Model of the test
solution proposed for the
Telnet protocol

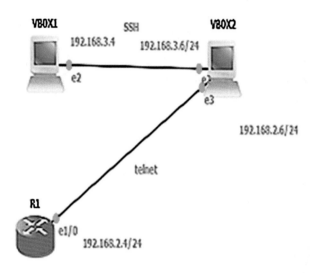

3.1   Testing the Telnet Protocol Recommendation

Figure 1 shows the test model (mock) that we used. As indicated, we use an
intermediary machine between the administrator machine and network equipment
while respecting the communication protocols between the equipment. Not forgetting that we will apply an Access Control List (ACL) at routers to limit access to
equipment, just to the staging machine.

As shown in Fig. 2, exchanged data between VBOX1 and VBOX2 is encrypted.

3.2   Test SFTP

In order to solve the problem of clear information exchange via FTP, we replaced it
by SFTP protocol which is based on Secure Sockets Layer (SSL)/Transport Layer
Security (TLS) (see Fig. 3). In this case the communication between machines will
be encrypted. The result after this modification is shown in Fig. 4.

Remark 1 To test the SFTP Protocol, we used gftp tool (Fig. 5).

3.3   Test SSH

The SSHv2 protocol enables to encrypt exchanged data between machines. As
shown in Figs. 6 and 7, the migration to this protocol will allow remedying the
vulnerabilities related to both rlogin protocols and to the SHR.

Fig. 2 Wireshark capture between VBOX and VBOX2

Fig. 3 Test SFTP network

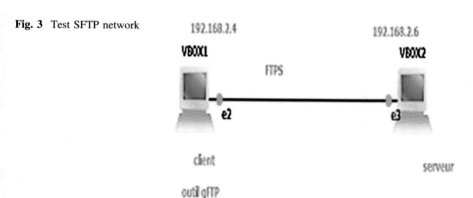

3.4 Protocol VRRP

To fix the problem related to the HSRP protocol, we suggested using VRRP protocol. One of the advantages of this protocol is that it is dedicated to all equipment. Indeed, according to Wireshark capture, the protocol allows to retrieve the maximum frame when the master router fails. But there is a possibility to retrieve the encrypted password due to the hash function based on the MD5 algorithm (Fig. 8).

According to this capture, we may find that the authentication is encrypted and the password does not appear clearly.

No.	Time	Source	Destination	Protocol	Info
59	36.391384	192.168.2.6	192.168.2.4	FTP	Response: \027\003\001\000I\201\324\024\217\217\272\23\
63	36.421066	192.168.2.4	192.168.2.6	FTP	Request: \027\003\001\000 \261\177r \221\3652g\l\025\
64	36.423882	192.168.2.6	192.168.2.4	FTP	Response: \027\003\001\000 \227\177
70	36.425409	192.168.2.6	192.168.2.4	FTP	Response: \027\003\001\000 #\300M\027\033\036\337Z\
71	36.425091	192.168.2.4	192.168.2.6	TCP	40903 > ftp [ACK] Seq=859 Ack=1935 Win=9104 Len=0 T
72	41.209074	192.168.2.4	192.168.2.6	FTP	Request: \027\003\001\000R\247\362C\311\022\273\22
73	41.218657	192.168.2.6	192.168.2.4	FTP	Response: \027\003\001\000\356\ae\262\v\355\320\36
74	41.221509	192.168.2.4	192.168.2.6	FTP	Request: \027\003\001\000 \005gM\257\3640\213\220\
75	41.222349	192.168.2.6	192.168.2.4	FTP	Response: \027\003\001\000\003\210\375\205\370\271\
79	41.224509	192.168.2.4	192.168.2.6	FTP	Request: \027\003\001\000B\205\262\255\017\353y\204
80	41.225220	192.168.2.6	192.168.2.4	FTP	Response: \027\003\001\000K\333s\222\353\367\246\06
85	41.238570	192.168.2.6	192.168.2.4	FTP	Response: \027\003\001\0000\003B\314\337M\333\241\22

Follow TCP Stream

Stream Content
.-*.M..
.....0..1.0...U....INI.0...U....rabat1.0...U....rabat1
3...U.
..INPT1
3...U....INPT1.0...U....souad1%b0#..*.M..
.....souad1ehla1i@gmail.com0..
1205211437142.
130521143714220..1.0...U....INI.0...U....rabat1.0...U....rabat1
3...U.
..INPT1
3...U....INPT1.0...U....souad1%b0#..*.H..
.....souad1ehla1i@gmail.com0...0
.-*.H.
..........0.......>@x.N......S......*q.............|0.dh.n.........F
5.....2[...2.2p..........@ZMO....aru....XF..~Q..w[.jS.;..|......C..}?[i.......PtN0...U......W.......|y*.0.f...

Fig. 4 Capture Wireshark of the manipulation of the SFTP

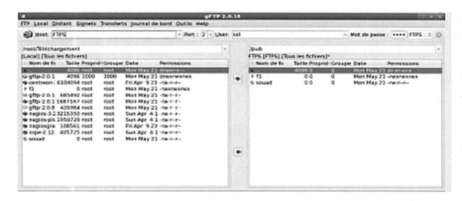

Fig. 5 Test with GFTP tool

Fig. 6 SSH network

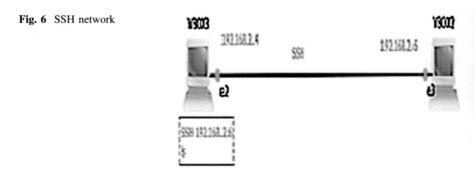

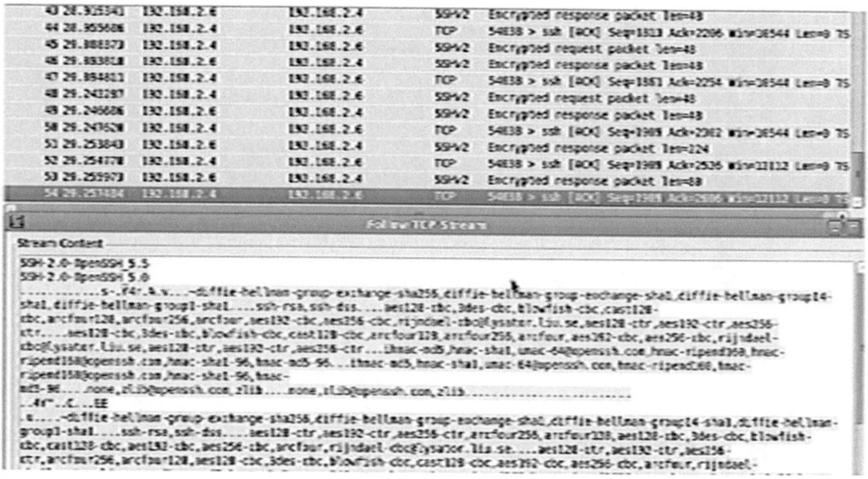

Fig. 7 Capture Wireshark of the SSH test

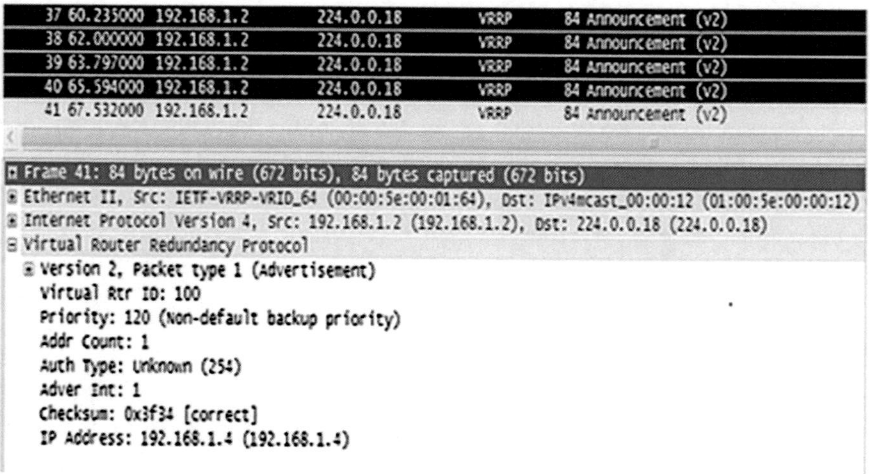

Fig. 8 Wireshark capture a VRRP packet

4 Realization and Model

The objective of the application is to have a traceability of actions made by users or administrators when they access the network equipment; router, switch or firewall, recording commands and modifications made at this level.

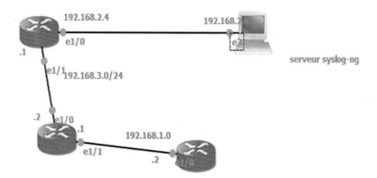

Fig. 9 Test of the network architecture

The application is based on Extract Transform and Load (ETL) data concept which aims to transferring data from production applications to decision systems.

4.1 Mockup Test

We test the application on the proposed architecture. As shown in Fig. 9 we used three routers belonging to different networks over a logging server (syslog-ng).

As logging server we proposed syslog-ng, it allows logs filtering from their source. So, we can specify log files for each device.

Remark 2 For each device, we redirect these log files to the logging server (through configurations made at server level).

4.2 Realization

The application is in the form of a web interface that enables access traceability and equipment configuration. The application menu provides three options:

The first one allows the addition of new equipment in the database in case of network extensions.

The second (traceability) gives an overview of all accesses made in equipment levels, by saving the IP address connected, the user login, executed commands and also a severity level. For example: the modification of the password or permissions. In this case a warning pop-up is displayed to inform the administrator that a password modification command is executed (Fig. 10).

Remark. From this list, the administrator can chooses the IP address that it wants to see its traceability.

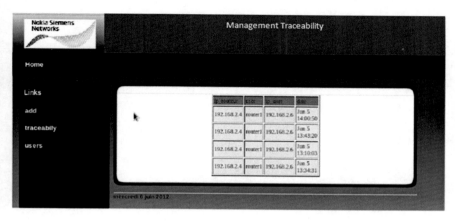

Fig. 10 Screenshot—Visualization of @ip of users

5 Conclusion

This study, has allowed us to remedy several flaws in the target system which provides a traceability of actions made by users or administrators when they access to the network equipment and modifications made at this level. This document lists recommendations made for the Nokia Siemens Network Company which are illustrated by tests and justifications; and present the traceability management application.

Acknowledgements This should always be a run-in heading and not a section or subsection heading. It should not be assigned a number. The acknowledgements may include reference to grants or supports received in relation to the work presented in the paper.

References

1. Anderson, J.: What's new in intelligent networking? In: IEE Third Tutorial Seminar on the Intelligent Network—The Next Generation, London, 17 May 1995, pp. 1–8. IEEE (1995)
2. Le bureau conseil de la DCSSI en collaboration avec club EBIOS, «ÉTUDE DE CAS @RCHIMED», 16/07/2004—27 p.
3. Porto de Albuquerque, J., Isenberg, H., Krumm, H., Lício de Geus, P.: Improving the Configuration Management of Large Network Security Systems, vol. 3775, pp. 36–47. Springer, Berlin (2005)
4. Bellare, Mihir, Kohno, Tadayoshi, Namprempre, Chanathip: Breaking and provably repairing the SSH authenticated encryption scheme: a case study of the encode-then-encrypt-and-mac paradigm. ACM Trans. Inf. Syst. Secur. **7**(2), 206–241 (2004)

Part V
Computer Vision

New Structural Method for Numeral Recognition

Soukaina Benchaou, M'Barek Nasri and Ouafae El Melhaoui

Abstract This paper proposes a new structural method of features extraction for handwritten, printed and isolated numeral recognition based on Freeman code method. It consists of extending the Freeman directions to 24-connectivity instead of 8-connectivity. This new technique shows its performances for the recognition in comparison with other techniques such as the classic Freeman code, the profile projection and the zoning. Numeral recognition is carried out in this work through k nearest neighbors. Results of simulations show that the proposed technique has obtained better results.

Keywords Structural approach · Handwritten and printed numeral recognition · Features extraction · Freeman code · Zoning · Profile projection · k nearest neighbors

1 Introduction

Postal sorting, bank check reading, order form processing, and others are among numerous technological applications where the recognition of characters, more particularly numerals, is hardly used and it knew an undeniable interest these last decades.

The recognition system goes through three main steps which are: preprocessing, features extraction and classification. The preprocessing phase is a current step in a recognition system. It consists first of discarding the imperfections by binarization using the global thresholding which consists on taking an adjustable and identical threshold for the entire image. Next, reducing the analyzed area by cropping the image

S. Benchaou (✉) · M. Nasri · O.E. Melhaoui
Laboratory MATSI, EST, University Mohammed First, 60000 Oujda, Morocco
e-mail: soukaina.benchaou@gmail.com

M. Nasri
e-mail: nasrihome@gmail.com

O.E. Melhaoui
e-mail: wafa19819@gmail.com

© Springer International Publishing Switzerland 2016
A. El Oualkadi et al. (eds.), *Proceedings of the Mediterranean Conference on Information & Communication Technologies 2015*, Lecture Notes in Electrical Engineering 380, DOI 10.1007/978-3-319-30301-7_25

to preserve only the numeral position and then normalizing the image in a predefined size to ensure that all processing on the images is performed on the same size.

The Numeral features extraction is a delicate process and is crucial [1] for a good numeral recognition. It consists of transforming the image into an attribute vector, which contains a set of discriminating characteristics for recognition, and also reducing the amount of information supplied to the system. These characteristics can be digital or symbolic. Mainly, we distinguish two approaches, statistical and structural.

In the literature, several works have been proposed for features extractions such as invariant moments [2], freeman coding [3], Zernike moments [4], Fourier descriptors [5], Loci characteristics [6], etc. Classification is the step of decision, which realizes the recognition. It consists of partitioning a set of data entity into separate classes according to a similarity criterion. Different methods are proposed in this context including neural networks [7], support vector machines [8], k nearest neighbors, k-means, etc.

In this study, a database of 600 numerals, printed and handwritten, provided by various categories of writers is used. The used database is split into two sets, 400 numeral images for learning and 200 numeral images for test. This work is organized as follows: Sect. 2 presents the different methods of features extraction. The k nearest neighbors technique of classification is discussed in Sect. 3. The proposed system for numeral recognition is presented in Sect. 4. The result of simulations and comparisons are introduced in Sect. 5. Finally, we give a conclusion.

2   Numeral Features Extraction

Features extraction is an important and essential task in any recognition system of forms. The main objective of features extraction is to represent objects by a vector of characteristics of fixed size and to keep only characteristics which contain the most relevant and the most discriminating information. The second objective of primitives selection is also to improve the performance of the used classifier in the recognition system.

In the literature, the feature extraction methods are classified according to two categories: statistical approach and structural approach. In our study, both approaches are used for the numeral recognition. For that purpose, we chose the zoning and the profile projection as statistical methods and Freeman code as structural method because of their efficiency and their robustness.

2.1   Zoning

Zoning is a statistical method of features extraction. Its widely used for characters recognition because of its simplicity and its good performance. This method consists of subdividing horizontally and vertically the pattern image into (m * n) zones of equal sizes (Fig. 1). For every zone i, the density d is calculated by dividing the number of pixels which represent the character on the total number of pixels in this zone. We obtain an attribute vector of m * n components for every pattern.

$$d(i) = \frac{Number\ of\ foreground\ pixels\ in\ zone\ i}{Total\ number\ of\ pixels\ in\ zone\ i} \tag{1}$$

Zoning method requires a phase of preprocessing. It consists of binarizing the numeral input image which is presented in grey level, and then we preserve the numeral position in image by cropping it. The next stage is to normalize the image in a predefined size and finally to skeletonize the resultant image.

2.2 Profile Projection

Profile projection is a statistical method. It consists of calculating the number of pixels between the left, bottom, right, top edge of the image and the first black pixel met on this row or column. The dimension of the obtained attribute vector is twice the sum of the number of rows and columns associated to the image of the numeral (Fig. 2).

This method requires the same preprocessing stage than the zoning method except the skeletonization.

2.3 Freeman Chain Codes

Freeman code issued to represent a shape boundary by a connected sequence of straight-line segments of specified length and direction. This representation is based

Fig. 1 Division of numeral 3 into 20 zones

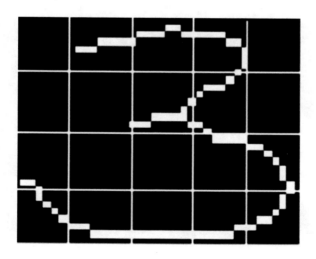

Fig. 2 The four profile
projections of numeral 4

on 4-connectivity or 8-connectivity of the segments. The first approach for representing the shape boundary was introduced by Freeman in 1961 using chain codes. This structural technique is based on contour detection, which is an important image processing technique. Freeman code can be generated by following a boundary of an object in a clockwise direction and coding the direction of each segment connecting every pair of pixels by using a numbering scheme.

The Freeman code of a boundary depends on the starting pixel $P(x_0, y_0)$. It was selected as being the first pixel encountered in the boundary of the object.

Our aim is to find the next pixel in the boundary or the first coding. There must be an adjoining boundary pixel at one of the eight locations surrounding the current boundary pixel. By looking at each of the eight adjoining pixels, we will find at least one that is also a boundary pixel.

Depending on which one it is, we assign a numeric code between 0 and 7 as already shown in Fig. 3. For example, if the pixel found is located at the right of the current location of pixel, a code 0 is assigned. If the pixel found is directly to the upper left, a code 3 is assigned. The process of locating the next boundary pixel and assigning a code is repeated until we return to the starting pixel.

Freeman code method requires a preprocessing stage. The applied operations are: transforming the input image in grey level, binarization, cropping,

Fig. 3 a 4-connectivity;
b 8-connectivity of Freeman
code

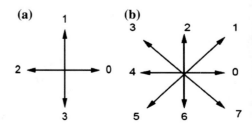

normalization, edge detection and edge closure by mathematical morphology operators, erosion and dilation.

Algorithm 1 presents the classic Freeman code.

Algorithm 1 Classic Freeman Code

---

1 : Research for the first pixel $P(x_0, y_0)$, called current pixel or starting pixel
2 : Research for the first coding pixel from the first pixel
3 : Affect the corresponding code to the first coding
4 : Save the first coding pixel
5 : New current pixel = coding pixel
6 : **while** *new current pixel* $\neq P(x_0, y_0)$ **do**
 | -Research for the following coding pixel surrounds 8 connectivity
 | -Affect the corresponding code (0 to 7) to the coding
 | -Save the coding
end

3 Classification Method: K Nearest Neighbors

K nearest neighbors (KNN) is a widely used method for data classification. Proposed in 1967 by Cover [9], it has been widely used in handwritten numerals recognition for its simplicity and its robustness [10]. KNN is a method, which was inspired from the closest neighbor rule. It is based on computing the distance between the test sample and the different learning data samples and then attributes the sample to the frequent class of their k nearest neighbors.

4 Proposed System

Figure 4 illustrates the three main steps of numeral recognition are: preprocessing, features extraction and classification.

In our case, the preprocessing stage is so fast in computation time. It consists of binarizing the numeral input image which is presented in grey level, setting out a threshold value against which the image pixels are compared, if the pixel value is greater than the threshold, it is set to one, it is set to zero otherwise. Then, we

Fig. 4 Scheme for numeral recognition

Fig. 5 Direction numbers for
24-directional chain codes

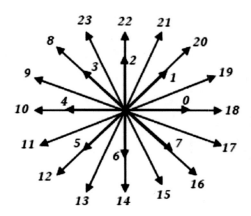

preserve only the numeral position in image by cropping it. The next stage is to normalize the image in a predefined size. After, we applied the edge detection, which can provide open boundary, so the closure must be done by morphological operators, which are erosion and dilation.

After the preprocessing, features extraction is carried out by a new technique based on Freeman codes method. Our contribution concerns the extension of the chain codes directions to 24-connectivity instead of 8-connectivity as shown in Fig. 5.

This contribution of ameliored Freeman code goes through steps presented in algorithm 2.

Algorithm 2 Ameliored Freeman Code

---

1 : Research for the first pixel $P(x_0, y_0)$, called current pixel
2 : Research for the first coding pixel from the first pixel
3 : Affect the corresponding code to the first coding
4 : Save the first coding
5 : New current pixel = coding pixel
6 : **while** *new current pixel* $\neq P(x_0, y_0)$ **do**
   -Research for the following coding pixel surround 8 connectivity
   -Affect the corresponding code (0 to 7) to the coding
   -Save the coding
   if *none following coding pixel met* **then**
     - Research for the following coding pixel surround 24 connectivity
     - Affect the corresponding code (8 to 23) to the coding
     - Save the coding
   end
end
7 : Return the chain code

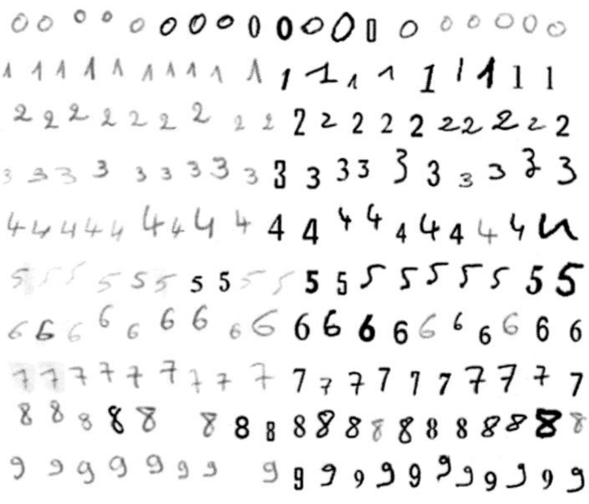

Fig. 6 A sample of handwritten and printed numerals

The process of locating the next boundary pixel and assigning a code is repeated until we return to the starting point. This type of coding helps to fill discontinuities that can persist at the preprocessing step.

This technique provides more information on the numeral and reduces the recognition error rate. The proposed technique is compared to the zoning, the profile projection and the classic Freeman code.

Numeral recognition is carried out through k nearest neighbors method.

In order to validate our contributions, we have used a database of 600 numerals, provided by different writers. A sample of the database is shown in Fig. 6. The database is divided into two sets, one set of 400 numerals is used for learning and the remaining 200 numerals are used for the test stage, the classes are equipropables.

5 Experimental Results and Comparative Study

To calculate the numeral recognition rate using the profile projection as feature extraction method and K nearest neighbors (KNN) as classification method, we have tested several sizes of normalization of the images and we found that the size 50 * 50 obtains the best recognition rate which is equal to 95 %.

In the same way, we used the zoning method. Our image is normalized with the size 50 * 50 with a division of 5 * 5 zones. The recognition rate obtained by this technique is equal to 91.5 %.

By using the classic Freeman code method of feature extraction and KNN classification method, the various simulations show that with the size 50 * 50, we obtained a recognition rate equal to 93.5 %.

As regards the new technique of feature extraction which consists of extending the Freeman directions to 24-connectivity, the obtained recognition rate is equal to 96.5 %.

Table 1 shows numeral recognition rates using four feature extraction methods presented in this work, and the K nearest neighbors as classification method.

Table 1 Recognition rate obtained by K nearest neighbors method

Features extraction methods	Recognition rate by KNN (%)
Proposed system	**96.5**
Classic Freeman code	93.5
Profile projection	95
Zoning	91.5

According to Table 1, we notice that the proposed system is more performant, it obtained a recognition rate equal to 96.5 % in fewer time, it equal 59 s. The classic Freeman code, the zoning and the profile projection have a lower recognition rate equal to 93.5, 91.5 and 95 % in 58, 11, and 37 s respectively. This results justify clearly the interest of extending the directions to 24-connectivity instead of 8-connectivity. It has provided more details for the attribute vector which has become more discriminative and has been able to differentiate between nearly similar numerals.

6 Conclusion

In this work, we have presented a new system for recognizing numerals, using the Freeman code. The image representing the numeral is initially preprocessed. It is binarized, cropped, brought to a fixed size and finally the edge detection and closure boundary were applied by morphological operators, which are erosion and dilation. The features extraction is carried out by a new technique based on Freeman code method. It consists of describing the boundary of an object by indicating the direction of a pixel given to its nearby pixel by one of the 24 possible directions instead of 8 directions. This process provides more information on the numeral and reduces the recognition error rate. The proposed technique was compared with classical Freeman code, zoning and profile projection, using the KNN classifiers. The simulations have obtained good results. A recognition rate of 96.5 % was achieved with the proposed method despite the different styles, fonts, scripts and writers.

References

1. Benne, R.G., Dhandra, B.V., Hangarge, M.: Tri-scripts handwritten numeral recognition: a novel approach. Adv. Comput. Res. 1, 47–51 (2009). ISSN 09753273
2. Kartar, S.S., Dhir, R.: Handwritten Gurumukhi character recognition using zoning density and background directional distribution feature. Int. J. Comput. Sci. Inform. Technol. 2, 1036–1041 (2011)
3. Nagare, A.P.: License plate character recognition system using neural network. Int. J. Comput. Appl. 25, 36–39 (2011)

4. Abu Bakar, N., Shamsuddin, S., Ali, A.: An integrated formulation of Zernik representation in character images. In: Trends in Applied Intelligent Systems, LNCS, vol. 6098, pp. 359–368. Springer, Heidelberg (2010)
5. Rajput, G.G., Mali, S.M.: Marathi Handwritten Numeral Recognition using Fourier Descriptors and Normalized Chain Code. IJCA Special Issue on Recent Trends in Image Processing and Pattern Recognition RTIPPR (2010)
6. El Melhaoui, O.: Arabic numerals recognition based on an improved version of the loci characteristic. Int. J. Comput. Appl. **24** (2011)
7. Benchaou, S., Nasri, M., El Melhaoui, O.: Neural network for numeral recognition. Int. J. Comput. Appl. **118**, 23–26 (2015)
8. Amrouch, M.: Reconnaissance de caractres imprims et manuscrits, textes et documents base sur les modles de Markov cach. Thse, Universit Ibn Zohr, Agadir, Maroc (2012)
9. Gil-Pita, R., Yao, X.: Evolving edited k nearest neighbor classifiers. Int. J. Neural Syst. **18**, 459–467 (2008)
10. Kuncheva, I.: Editing for the k-nearest neighbors rule by a genetic algorithm. Pattern Recogn. Lett. **16**, 809–814 (1995)

A Hybrid Feature Extraction for Satellite Image Segmentation Using Statistical Global and Local Feature

El Fellah Salma, El Haziti Mohammed, Rziza Mohamed and Mastere Mohamed

Abstract Satellite image segmentation is a principal task in many applications of remote sensing such as natural disaster monitoring and residential area detection... These applications collect a number of features of an image and according to different features of an object will detect the object from the image. This type of image (satellite image) is rich and various in content, the most of methods retrieve the textural features from various methods but they do not produce an exact descriptor features from the image. So there is a requirement of an effective and efficient method for features extraction from the image, some approaches are based on various features derived directly from the content of the image. This paper presents a new approach for satellite image segmentation which automatically segments image using a supervised learning algorithm into urban and non-urban area. The entire image is divided into blocks where fixed size sub-image blocks are adopted as sub-units. We have proposed a fusion of statistical features including global features based on the common moment of RGB image and local features computed by using the probability distribution of the phase congruency computed on each block. The results are provided and demonstrate the good detection of urban area with high accuracy and very fast speed.

Keywords Computer vision · Segmentation · Classification · Satellite image · Statistical feature · Phase gradient

E.F. Salma (✉) · E.H. Mohammed · R. Mohamed
LRIT, Associated Unit to CNRST (URAC 29), Faculty of Sciences,
Mohammed V-Agdal University, Rabat, Morocco
e-mail: elfellah.s@gmail.com

E.H. Mohammed
e-mail: melhaziti@yahoo.fr

R. Mohamed
e-mail: rziza@fsr.ac.ma

M. Mohamed
The National Institute for Urban and Territorial Planing, Rabat, Morocco
e-mail: mohamed.mastere@gmail.com

© Springer International Publishing Switzerland 2016
A. El Oualkadi et al. (eds.), *Proceedings of the Mediterranean Conference on Information & Communication Technologies 2015*, Lecture Notes in Electrical Engineering 380, DOI 10.1007/978-3-319-30301-7_26

247

1 Introduction

Remotely sensed images of the Earth that we can acquire through satellites are very large in number. The classification of data has long attracted the attention of the remote sensing community because classification results are the basis for many environmental and socioeconomic applications [1]. Scientists and practitioners have made much effort in developing advanced classification approaches and techniques for improving classification accuracy. Each image contains a lot of information inside it and can have a number of objects with characteristics related to the nature, shape, color, density, texture or structure. It is very difficult for any human to go through each image and extract and store useful patterns. An automatic mechanism is needed to extract objects from the image and then do classification. The problem of urban object recognition on satellite images is rather complicated because of the huge amount of variability in the shape and layout of an urban area, in addition to that, the occlusion effects, illumination, view angle, scaling, are uncontrolled [2]. Therefore, more robust methods are necessary for good detection of the objects in remotely sensed images.

Generally, most of existing segmentation approaches are based on frequency features and use images in gray levels. Texture based methods partition an image into several homogenous regions in terms of texture similarity. Most of the work has concentrated on pixel based techniques, [3]. The result of pixel-level segmentation is a thematic map in which each pixel is assigned a predefined label from a finite set. However, remote sensing images are often multispectral and of high resolution which makes its detailed semantic segmentation excessively computationally demanding task. This is the reason why some researchers decided to classify image blocks instead of individual pixels [4]. We also adopt this approach by automatically dividing the image into a single sub-image (block), and then evaluate classifiers based on a hybrid descriptors and support vector machines, which have shown good results in image classification.

The process of generating descriptions represents the visual content of images in the form of one or more features. They can be divided into; Local features computed over blocks and Global features computed over the entire image or the sub-image (block). In this paper, we focus on the emerging image segmentation method that is hybrid methods of feature extraction using statistical feature; in order to model global and local features.

The rest of the paper is organized as follows. Section 2 presents a review of some works related to our work, Sect. 3 describes the general framework of the proposed approach, while Sect. 4 shows the experimental results and finally Sect. 5 concludes the paper.

2   Related Work

There are many techniques for classification of satellite images. We briefly review here some of the methods that are related to our work:

Mehralian and Palhang [5] separate urban terrains from non-urban terrains using a supervised learning algorithm. Extracted feature for image description is based on principal components analysis of gradient distribution. Ilea et al. [6] considered the adaptive integration of the color and texture attributes and proposed a color-texture-based approach for SAR image segmentation. Fauquer et al. [7] classify aerial images based on color, texture and structure features; the authors tested their algorithm on a dataset of 1040 aerial images from 8 categories. Ma and Manjunath [8] use Gabor descriptors for representing aerial images. Their work is centered on efficient content-based retrieval from the database of aerial images and they did not try to automatically classify images to semantic categories. In this work we use either global or local features. Considering the incorporation of the two types of information, and believing that the combination of the two levels of features is beneficial on identifying image and more suitable to represent complex scenes and events categories, our feature vector extracted will be the result of the combination of both global and local feature.

3   Proposed Approach of Urban Terrain Recognition

In this paper, a new approach for satellite images segmentation is presented. The method comprises three major steps:

- Firstly, we start by splitting the image into blocks with size of (20×20)
- The second step consists in feature extraction, we calculate the computation of phase congruency map of each sub image (block) then, the statistical local features (mean, variance and skewness) are calculated from the probability distribution of the phase congruency; in addition the statistical global feature (mean, standard deviation and skewness) are computed from color moment of the sub image. These features are combined to construct a feature vector for each block. We use a hybrid of statistical global and local feature to characterize each sub image.
- Finally, these vectors of hybrid features are used as training and testing forward SVM to distinguish urban classes from non-urban classes. The tests show that the proposed method can segment images with high accuracy and low time complexity.

In what follows, it is assumed that satellite images are being analyzed for segmentation to urban and non-urban terrain. Below, we describe each of these steps.

3.1 Partitioning

For evaluation of the classifiers we used 800 × 800 pixel (RGB) image taken from Google Earth related to Larache city, Morocco, satellite images are sometimes very large and handling. We split this image into smaller blocks of 20 × 20 pixels, with an overlaps of 4 pixels at the borders. So we have in total 4493 blocks in our experiments. We classified all images into 2 categories, namely: urban and non-urban. Examples of sub-images from each class are shown in Fig. 1.

It should be noted that the distribution of images in these categories is highly different; this may be observed from the histogram graph in Fig. 2.

Fig. 1 Some sample blocks from satellite images; first row: non-urban areas, second row: Urban areas

Fig. 2 Histogram of urban and non-urban terrains

3.2 Feature Extraction

Before using a supervised learning process, the features must be defined from sub-images of urban and non-urban; positive and negative training samples.

Statistical local feature

To describe each sub-image, the statistical features of the 2D phase congruency histogram applied on each block and obtained values as principals used for features, so we have a 1D feature vector for each block, which will be used in training process.

The statistical features [9] provide information about the properties of the probability distribution. We use statistical features of the phase congruency histogram (PCH) as mean, variance and skewness which are computed by using the probability distribution of the different levels in the histograms of PCH. Let λi be a discrete random variable that represents different levels in a map and let $p(\lambda_i)$ be the respective probability density function. A histogram is an estimation of the probability of occurrence of values λj as measured by $p(\lambda_i)$. We content with three statistical feature of histogram:

- Mean (m): computes the average value. It is the standardized first central moment of the probability distribution in image.

$$m = \sum_{i=0}^{L-1} \lambda_i p(\lambda_i)$$

- Variance (σ): It's second central moment of the probability distribution, the expected value of the squared deviation from the mean.

$$\sigma = \sum_{i=0}^{L-1} p(\lambda_i)\lambda_i^2 - m^2$$

- Skewness (k): computes the symmetry of distribution. S gives zero value for a symmetric histogram about the mean and otherwise gives either positive or negative value depending on whether histogram has been skewed right or left to the mean.

$$k = \sum_{i=0}^{L-1} (\gamma_i - m)^3 p(\gamma_i)$$

After the calculation of these statistical features for each PCH, the feature vectors fPC of each block are constructed as:

$$fPC = \{mPCH, \sigma PCH, kPCH\} \tag{1}$$

Statistical Global feature

The basis of color moments is that can be interpreted as a probability distribution of color in an image. Probability distributions are characterized by a number of unique moments. The moments of distribution can be used as features to characterize the RGB image color. In our case we extract the color feature from RGB images using CM. Following equations define the mean, standard deviation and skewness of an image of size N. The corresponding calculation of the three color moments can then be defined as:

- *Mean*: can be understood as the average color value in the image.

$$mean(RGB) \approx \mu_i = \frac{1}{N}\sum_{j=1}^{N} P_{ij}$$

- *Standard Deviation*: The standard deviation is the square root of the variance of the distribution.

$$std(RGB) \approx \sigma_i = \left(\frac{1}{N}\sum_{j=1}^{N}(P_{ij} - \mu_i)^2\right)^{1/2}$$

- Skewness: computes the symmetry of distribution:

$$skw(RGB) \approx \gamma_i = \left(\frac{1}{N}\sum_{j=1}^{N}(P_{ij} - \mu_i)^3\right)^{1/3}$$

where P_{ij} the ith color channel at the jth image pixel, and N is the total number of pixels in the image. The feature vector of color $f(RGB)$ in each block is constructed as:

$$f(RGB) = \{mean(RGB), std(RGB), skw(RGB)\} \tag{2}$$

After construction of these feature vectors, a hybrid feature vector is constructed from each block by combining all of statistical local and global feature vectors (1) and (2) into vector of features:

$$f = \{mPCH, \sigma PCH, kPCH, f(RGB)\}. \tag{3}$$

The feature vectors of all the blocks images including urban and non-urban sub-image are constructed and stored to create a feature database. The feature vector of the testing image is also calculated in the same manner as described in Fig. 3.

It is time, after feature extraction, we use this vector feature to train and test SVM. Given a set of training examples, each marked as belonging to one of two

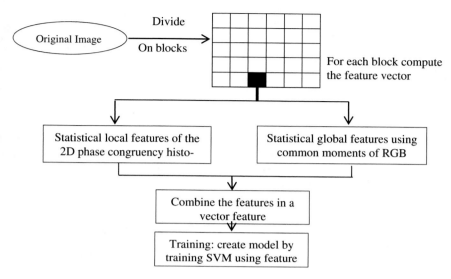

Fig. 3 Schematic diagram of the main step of the proposed method

categories. We used half of the images for training and the other half for testing, and after training process, the obtained model will be tested on test data. In the sections below the results of the test will be discussed.

4 Experimental Result of the Proposed Method

We experiment the approach on two sample of satellite image, taken both from Google Earth, related to Larache city, Morocco. They sizes are 800×800 pixel (RGB) image. We have split (with an overlaps of 4 pixels at the borders) these images into smaller sub-images of 20×20 pixels. In total, it occur 4493 sub images.

We characterize each sub-image by 1D statistical hybrid feature vector, which is based on the statistical local features (mean, variance and skewness) calculated from the probability distribution of the phase congruency; and the statistical global feature (mean standard deviation and skewness) computed from color moment of the block.

We used half of the features for training and the other half for testing, we label the training data manually with 1 and -1, which label 1 refers to urban category and the -1 refers to non-urban category, and the obtained model will be tested on test data. In the sections below the results of the test will be discussed.

Results of the proposed method compare three cases: the effectiveness between local, global and proposed hybrid features to improve the efficacy of hybrid feature in detection, and also we compare the accuracy between the sizes of blocks.

To examine the ability of proposed scheme, we have used accuracy and precision statistical measures.

Table 1 shows that Hybrid Feature gives 94 % accuracies for 20 of block size, which are the highest among all the three feature sets. And it is noticeable from the results that hybrid feature based block detection has superior results than both local and global feature.

The visual result of segmentation of the original image is shown in Fig. 4. The images segmented distinguish clearly the zone of urban terrain from non-urban terrain. The error in decision on category of blocks is due to the confusion with the streets. It may be an advantage for some application, if it selects block size smaller than 20, and then classify the block referring the streets as a non-urban terrain.

Table 1 Block detection results for local and hybrid feature with size of sub-image [20 × 20]

Metric	Local feature	Global feature	Hybrid feature
Accuracy	0.942	0.91	0.94
Precision	0.96	0.94	0.968
Time (ms)	19	17.16	62

Segmented image   Original image

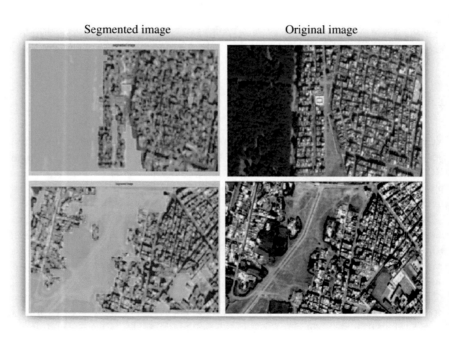

Fig. 4 Visual result of segmentation of the original images

5 Conclusion

In this work, we have focused on the different type of Feature Extraction Techniques, and we have proposed a hybrid features in the form of a novel fusion of statistical features including global feature based on common moments of RGB image, and local features in which we compute the probability of distribution of 2D phase congruency. We distinguish urban from non-urban terrain, the algorithm makes decision about image block (not a pixel) in both size (20×20), so each block are described by 1D vector features, then SVM are used for classification.

The results of hybrid approach given on test images yield good performance, better efficiency and high accuracy, comparing with global and local approach. As a future work, it can be considered for segmentation other types of terrains like forest, sea and desert; also it can select smaller block size than 20, and classify the streets as a non-urban terrain.

References

1. Blaschke, T.: Object based image analysis for remote sensing. ISPRS J. **65,** 2–16 (2010)
2. Sirmacek, B., Unsalan, C.: A probabilistic approach to detect urban regions from remotely sensed images based on combination of local features. In: 5th RAST'2011 Conference
3. Szummer, M., Picard, R.W.: Indoor-outdoor image classification. Proceedings of the IEEE ICCV Workshop, Bombay, India, pp. 42–51 (1998)
4. Pagare, R., Shinde, A.: A study on image annotation techniques. Int. J. Comput. Appl. **37**(6), 42–45 (2012)
5. Mehralian, S., Palhang, M.: Principal components of gradient distribution for aerial images segmentation. 11th Intelligent Systems Conference (2013)
6. Ilea, D.E., Whelan, P.F.: Image segmentation based on the integration of color texture descriptors—a review. Pattern Recogn. **44**, 2479–2501 (2011)
7. Fauqueur, J., Kingsbury, G., Anderson, R.: Semantic discriminant mapping for classification and browsing of remote sensing textures and objects. In: Proceedings of IEEE ICIP (2005)
8. Ma, W.Y., Manjunath, B.S.: A texture thesaurus for browsing large aerial photographs. J. Am. Soc. Inf. Sci. **49**(7), 633–648 (1998)
9. ShamikTiwari, V.P. Shukla, S.R. Biradar, Singh, A.K.: A blind blur detection scheme using statistical features of phase congruency and gradient magnitude. Adv. Electr. Eng. **2014,** 10 p Article ID521027 (2014)

A Multicriteria Approach with GIS for Assessing Vulnerability to Flood Risk in Urban Area (Case of Casablanca City, Morocco)

Abedlouahed Mallouk, Hatim Lechgar, Mohamed Elimame Malaainine and Hassan Rhinane

Abstract Flood risk management, since long based on the hazard control, is increasingly orientated towards an attempt to reduce the vulnerability of territories. In fact, the dangers of flood risk on the territory of Casablanca have been the subject of numerous analyzes, whereas studies on the vulnerability of Morocco's first economic city are very rare. This work aims to contribute to the development of methods for assessing the vulnerability of the urbanized watershed area of Oued Bouskoura facing the overflow of the latter in case of storm floods. The methodology of this study is the use of a geographic information system coupled with the hierarchical multi-criteria analysis developed by T. Saaty (Saaty 1980) for the identification and quantification of urban elements lying in the natural course of Oued Bouskoura and are at risk of flooding.

Keywords Flood risk · Geographic information system · Multi-criteria analysis · Vulnerability · Watershed · Bouskoura valley

1 Introduction

Since time immemorial, man is facing an enemy as powerful as unpredictable: the floods. In fact, floods are the most frequent natural disasters [1] and the most dangerous with their magnitude and suddenness [2]. Floods are a natural and

---

A. Mallouk (✉) · H. Lechgar · M.E. Malaainine · H. Rhinane
Faculté Des Sciences Ain Chock, Université Hassan II, Casablanca, Morocco
e-mail: abdelouahed.mallouk@gmail.com

H. Lechgar
e-mail: h.lechgar@gmail.com

M.E. Malaainine
e-mail: mohamed.malaainine@gmail.com

H. Rhinane
e-mail: h.rhinane@gmail.com

© Springer International Publishing Switzerland 2016
A. El Oualkadi et al. (eds.), *Proceedings of the Mediterranean Conference on Information & Communication Technologies 2015*, Lecture Notes in Electrical Engineering 380, DOI 10.1007/978-3-319-30301-7_27

seasonal phenomenon which corresponds to a flooding of an area usually out of the water.

The causes of these latter have evolved and their consequences are continuing to grow. Unfortunately, there are so far no means that can act upon the devastating effects of flooding and urban flooding, the only alternative is to take precautions at best by setting their knowledge at the forefront of policy priorities and intervention strategies [1].

During the last twenty-five years (1999–2014), Morocco has experienced a series of major floods, which affected more than 557,270 people by making 1192 deaths [3]. The damage caused by these floods hit three billions Dirhams [3]. The city of Casablanca, the economic capital of Morocco has not been spared. In fact, Heavy rainfall in 1996 as well as 2010 had caused extensive damage (collapse of dilapidated houses, electricity failure, paralysis of transport, school closures, and road accidents) and even losses of lives. The metropolis is characterized by the presence of the historical path of Oued Bouskoura that ran through, at the time, the city from east to west. The gradual occupation and urbanization of the major bed of Oued Bouskoura over the last decades and construction in the 80 s of the Eljadidaa road makes of this environment a vulnerable area for the population and infrastructure there.

The methodology of this study is the use of a geographic information system coupled with the hierarchical multi-criteria analysis developed by T. Saaty (Saaty 1980) for the identification and quantification of urban elements lying in the natural course of Oued Bouskoura and are at risk of flooding. The result is supplied as a synthetic cartography of the global indicator of vulnerability allowing a clear view of the risk associated with the flooding hazard. The result is an effective and useful tool for supporting decision-making for decision makers in the context of an integrated management for the areas of Casablanca subject to flooding.

1.1 Study Area

Casablanca is situated on the Moroccan Atlantic coast, west central Morocco. In geographic coordinates, the city is located about 33° 34' 42.44" North latitude and 7° 36' 23.89" West longitude. Casablanca is characterized by a semi-arid climate with normal average temperature (calculated from 1961 to 1990) ranging from 12.7 °C in winter to 21 °C in summer. The average annual rainfall is 427 mm. They may reach values lower than 200 mm or sometimes exceeding 800 mm [4].

Oued Bouskoura is undoubtedly one of the major threats to the city of Casablanca. The danger lies in the fact that the city is built on the natural bed of Oued Bouskoura. This gives particular importance to this area that covers three urban municipalities we have taken as a sample for the application of our model.

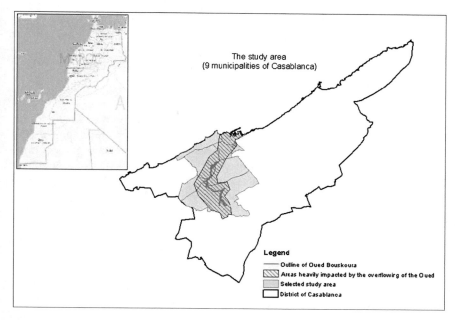

The study area
(9 municipalities of Casablanca)

Legend

—— Outline of Oued Bouskoura

▨ Areas heavily impacted by the overflowing of the Oued

▨ Selected study area

☐ District of Casablanca

Fig. 1 Study area

This being said, nothing prevents the application of the methodology of this study to be generalized across the area of Casablanca (Fig. 1).

1.2 Methodological Approach

The methodology of this study is the use of a geographic information system coupled with the hierarchical multi-criteria analysis developed by T. Saaty (Saaty 1980) for the identification and quantification of urban elements lying in the natural course of Oued Bouskoura and are at risk of flooding.

The elaborated assessment tool allows analyzing and evaluating local vulnerability to urban flooding in three types of issues: human issues, environmental issues and material issues. Each of these issues is the subject of an analysis and evaluation grid that enables the development of vulnerability indicators from the detailed lists of criteria. The result is supplied as a synthetic cartography of the global indicator of vulnerability allowing a clear view of the risk associated with the flooding hazard. The result is an effective and useful tool for supporting decision-making for decision makers in the context of an integrated management for the areas of Casablanca subject to flooding.

2 The Use of Multi-criteria Analysis Methods in the Vulnerability Assessment

2.1 Multicriteria Methods for Assessing the Vulnerability of a Territory

The methods of decision support or, more accurately, the multi-criteria methods for decision support are fairly recent techniques and under development [5]. Historically, these methods were used by managers of companies or project managers as decision support tools to deal with problems of choice of solution or alternative assessment. In fact, methods of decision support have been developed to facilitate the choice between a number of options available or evaluation in complex situations where several qualitative and quantitative criteria come into play [6]. More recently they have been applied repeatedly in environmental issues such as vulnerability assessment of a territory, in terms of flood risk, industrial risks or risks for transport of hazardous materials [7].

Today, there exist several decision support methods such as comparison criteria, weighted averages, the ordinal method, factor analysis or even multi-criteria approach to decision support AHP (Analytical Hierarchy Process) [6]. The latter, which is the methodological context of our study, was proposed by Thomas Saaty in the 1980s it consists in ordering solutions based on different criteria while examining the consistency and the logic of judgments about the weight of criterion considered and preferences of decision makers [8].

2.2 AHP Method: Principles and Computation

The process of the AHP method can be summarized in four steps: analyze the problem concerned, build a hierarchical decomposition by declining each objective comparison criteria, determine priorities for targets from expert judgment by checking the consistency of subjective assessments and aggregate at the end the experts' answers to get the vulnerability functions.

One of the most creative tasks in decision making is to choose the factors that are important for the decision [9]. Using the AHP method is initially based on the decomposition of the problem by performing a qualitative and quantitative accurate census of different criteria (vulnerable targets and vulnerability factors) [6]. This initial analysis is highly dependent on judgments, needs and knowledge of the participants in the process of decision making. Note also that the fineness of this decomposition depends on the "wealth" of the databases used (land use map, urban planning, population census, transportation networks, public facilities, etc.)

The second stage of the AHP method is to build a hierarchical tree composed of objectives, criteria, sub-criteria and targets in the form of father-son pair. This hierarchy provides a structured view of the problem studied in offering the user/decision maker better concentration on specific criteria [10].

Each criterion must be identified with its definition and intensity [8]. In fact, the weight of each item is determined using the pairwise comparison method. This involves comparing the different criteria son pairwise compared to the initial target (father criteria) to assess their importance. This comparison is done primarily through the collection of expert opinions through a questionnaire in the form of "evaluation matrix" and according to a so-called specific Saaty scale [11] (Table 1).

The evaluation matrices are square matrices, the order equal to the number of comparison criteria. The experts pronounce on the importance of criterion c_1 of the 1st line with criteria c_2, c_3, ..., c_n appearing successively at the top of each column. This operation is subsequently performed for each criterion of each line of the matrix. The sum of the weights of all criteria son of the same father criterion is 1. This weight rule is called "interdependent relationship."

In a decision problem, it is absolutely essential to measure the "degree" of consistency of judgments that are placed. In fact, the 3rd stage of the AHP method is to check the consistency of binary comparisons from expert judgment and included in the evaluation matrices. The consistency of the answers is estimated using a known ratio: Consistency Ratio (CR) [6]. This ratio is obtained from the eigenvalue of the comparison matrices and which should be less than 10 %. Calculating the Coherence Ratio is as follows.

Consider $C = (C_{ij})$ an evaluation matrix with C_{ij} the relative weight of the criterion C_i with respect to C_j criterion. C the matrix of order n with:

$$C_{ij} \neq 0, \; C_{ij} = 1/C_{ji} \text{ et } C_{ii} = 1. \tag{1}$$

The matrix is perfectly consistent if the weights C_{ij} are transitive, that is to say, for all i, j and k has the following equation:

$$C_{ij} = C_{ik} * C_{kj}. \tag{2}$$

In fact, for example, if an expert is pronounced for a greater relative weight for the target (I) relative to the target (K), and to a greater weight of the target (K) relative to (J) logically, the same expert will have to be pronounced for the greater weight of the target (I) relative to (J) [11].

Table 1 The AHP pairwise comparison scale [16]

Numeric value	Definition
1	Equal importance of both elements
3	Moderate importance of one element over other
5	Strong importance of one element over other
7	Very strong importance of one element over other
9	Extreme importance of one element over other
2, 4, 6, 8	Intermediate values between two neighboring appreciations

In this case of judgment perfectly transitive, the matrix C can be written as the following vector form: $C * V = \lambda * V$ where V is the eigenvector and λ is the eigenvalue of the matrix. In the case of a perfectly consistent matrix the λ value is equal to the order n and the rank of the matrix is equal to 1.

However, the weights established in the AHP process are based on human judgments that make it difficult to have a perfect Coherence Ratio (50). In this case, the previous equation is:

$$C * V = \lambda_{max} * V \quad \text{where} \quad \lambda_{max} \geq n. \tag{3}$$

The difference between λ_{max} and n indicates "the degree of inconsistency" of the judgment made [12].

3   Hierarchical Tree and Conception of a GIS Model

3.1   Development of Hierarchical Tree

The elaborated assessment tool developed within the framework of this study allows analyzing and evaluating local vulnerability to urban flooding in three types of issues: human issues, environmental issues and material issues. Each of these issues is the subject of an analysis and evaluation grid that enables the development of vulnerability indicators from the detailed lists of criteria. The suggested hierarchy is an overall view of the problem in terms of objective, criteria and alternatives.

Human vulnerable targets can be divided into three broad categories: resident population, non-resident population and population with assisted evacuation. Given the high density of population in the cities, using a small-scale measurement is paramount [13]. In our case, the following criteria were considered extremely important to be part of our evaluation criteria of human vulnerability: Total population (between 10 and 65, under 10 and over 65) population in the workplace and the reception centers (hotels, summer centers), and finally the population of hospital centers, schools and pretension centers.

The Material issues are six major types based on land use: residential buildings, industrial and commercial areas, administration and service, leisure and family gardens, transportation and public facilities. These elements are then divided into more detailed targets.

In addition to economic and social damage, floods in urban areas can also influence the environment and biodiversity of an ecosystem [14]. To assess the ecological risk of flooding, environmental vulnerability is represented by the following: potential contamination, the effect of overland flow on green landscaped and agricultural spaces, bare land, forests and zoos. Just like the material issues, these issues are subdivided and vulnerability factors can be assigned.

3.2 Data Structure According to Conceptual Model

Following the analysis and identification of needs, we identified the spatial data necessary for the completion of this study and completed the conceptual modeling (feature class, subclasses, entities, attributes, sub type of attributes). The data were later collected from different sources (cadastral data, road networks, population, water and sanitation networks, electricity network ...). This model was largely inspired from the Barczack and Grivault model in work "Geographical Information System for the assessment of vulnerability to urban surface runoff" [2].

4 Result of Analytic Hierarchical Process

4.1 Sectorization of the Study Area

According to the data dictionary of our conceptual model, the targets are in different geometric shapes and vectors (area, line, and point). Thus, they need to be standardized for their combination [6]. The commonly used generalization technique is the meshing of the study area [7, 15]. The choice of the size and number of meshes used are attributes of the degree of accuracy of the data and space studied itself [6]. We opted for a meshing of about 200 m from the side of each mesh. This makes a number of 4000 mesh to cover the entire study area. Note that we can use a narrower mesh (about 50 m) in a specific sector for further analysis needs.

4.2 Quantification of Targets

Quantification of human targets was based primarily on statistics of demography by the High Commission for planning municipality. The number of inhabitants per cell (H_m) was calculated from the population of the municipality (P_c) of the total number of housing in the municipality (N_{lc}) and the number of dwellings in the mesh (N_{lm}) using the following formula:

$$H_m = Nl_m * P_c / Nl_c. \qquad (4)$$

Unfortunately, there are no data on the employment potential of each municipality. The population at the workplace (P_{asm}) was estimated through the total active population in the city of Casablanca (P_a), the total number of economic entities by sector (N_{ec}), the percentage distribution of the workforce by sector (P_{as}) and the number of existing economic entities of the same sector in a mesh (N_{asm}):

$$P_{asm} = P_a * P_{as} * (N_{asm}/N_{ec}). \tag{5}$$

To quantify the economic and environmental targets, we determined for each areal target corresponding quantization factor (F_{qs}) using the surface of the target in the mesh (S_m) and the maximum surface of the target in a mesh (S_{max}) of the study area.

$$F_{qs} = S_m / S_{max}. \tag{6}$$

Similarly, for linear targets, quantization factors (F_{ql}) are derived from the length of the target in the mesh (L_m) and the maximum length of the target in a mesh.

$$F_{ql} = L_m/L_{max}. \tag{7}$$

The comparison matrices containing the expert judgments were the subject of a specific treatment to associate all assessments. Questionnaires that were retained are those with a coherence ratio less than 10 %. The calculation of the eigenvectors of each matrix was used to calculate the overall vulnerability function.

$$V_{global.} = 0.8064 * V_{humain} + 0.0322 * V_{materiel} + 0.1612 * V_{environnemental}. \tag{8}$$

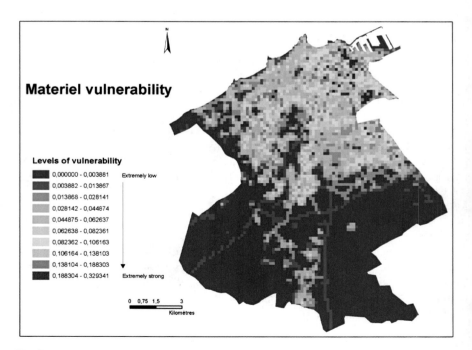

Fig. 2 Materiel vulnerability mapping

4.3 Discussion and Vulnerability Index Mapping

The analysis of the first results obtained from the evaluation matrices treated shows greater importance given to human vulnerability with 80 %, followed by environmental vulnerability of 17 % and at the end the materiel vulnerability with only 3 %. The presentation of the results with GIS data processing software leads to vulnerability maps as cited in the following example (Fig. 2).

5 Conclusion and Perspectives

The assessment of the vulnerability of urban land by using the AHP method is an exercise that requires a profound technical knowledge, specific resources and territorial and statistical data. Our research approach is to use a multi-criteria analysis method for the classification of the human, material and environmental stakes in the Casablanca area. The results obtained after using spatial analysis tool are represented as global vulnerability maps.

We must also consider the limitations of the proposed end approach to improve the relevance of results. Indeed, the AHP method was mainly criticized for too subjective aspect of judgments made by the experts. On the implementation side of the results obtained in a GIS, the mesh used as the unit of analysis is based on the assumption of a homogeneous distribution of the land use population within a islet. The lack of comprehensive, reliable and up to date on the Casablanca area is considered one of the major challenges we faced throughout the process.

This analysis will be enriched gradually by the integration of new matrices of judgment and the consideration of new issues that were not considered in this study. It would also be important to consider the temporal component, including the calculation of resident population and the workplace.

References

1. Abhas, K.J., Robin, B., Jessica, L.: Villes et inondations, Guide de gestion intégrée du risque d'inondation en zone urbaine pour le XXIe siècle. Résumé à l'intention des décideurs. Banque mondiale. 2012
2. Aleksandra, B., Camille, G.: Système d'information géographique pour l'évaluation de la vulnérabilité au risque de ruissellement urbain. NOVATECH 2007
3. The international disaster database EM-DAT. Centre for Research on the Epidemiology of Disasters (http://www.emdat.be)
4. Anthony, G.B., Stéphane, H., Salim, R., Asmita, T., Osama, H.: Adaptation au changement climatique et aux désastres naturels des villes côtières d'Afrique du Nord. Banque mondiale, Juin (2011)
5. Sami, B.M.: Introduction aux méthodes multicritères d'aide à la décision. Biotechnol. Agron. Soc. Environ. 4(2), p83–p93 (2000)

6. Renard, F., Chapon, P.-M.: Une méthode d'évaluation de la vulnérabilité urbaine appliquée à l'agglomération lyonnaise. L'Espace géographique 1/ 2010 (Vol. 39), pp. 35–50
7. Chloé, G.: Evaluation multicritère de la vulnérabilité des territoires aux risques de transport routier de matières dangereuses. XLIIIème Colloque de l'ASRDLF, 11–13 juillet 2007
8. Renaud, C.: Analyse multicritère : Étude et comparaison des méthodes existantes en vue d'une application en analyse de cycle de vie. CIRANO, Série scientifique, Montréal (2003)
9. Thomas, L.S.: How to make a decision: the analytic hierarchy process. Eur. J. Oper. Res. **48**, p9–p26 (1990)
10. Ouma, Y.O., Ryutaro, T.: Urban flood vulnerability and risk mapping using integrated multi-parametric AHP and GIS: methodological overview and case study assessment. Water (20734441) **6**(6), 1515 (2014)
11. Chloé, G.: Vulnérabilité et transport de matières dangereuses :une méthode d'aide à la décision issue de l'expertise de la Sécurité Civile. Cybergeo : European Journal of Geography [En ligne], Systèmes, Modélisation, Géostatistiques, document 361, mis en ligne le 12 janvier 2007
12. Geoff, C.: Practical Strategy. Open Access Material. AHP: THE ANALYTIC HIERARCHY PROCESS, Introduction. Pearson Education Limited 2004
13. Kubal, C., Haase, D., Meyer, V., Scheuer, S.: Integrated urban flood risk assessment adapting a multicriteria approach to a city. Nat. Hazards Earth Syst. Sci. **9**, 1881–1895 (2009)
14. Apel, H., Thieken, A.H., Merz, B., Bloschl, G.: Flood risk assessment and associated uncertainty. Nat. Hazards Earth Syst. Sci. **4**, 295–308 (2004)
15. Robert, D., Pascale, M.: La vulnérabilité territoriale: une nouvelle approche des risques en milieu urbain. *Cybergeo : European Journal of Geography* [En ligne], Dossiers, Vulnérabilités urbaines au sud, document 447
16. Berrittella, M., Certa, A., Enea, M., Zito, P.: An Analytic Hierarchy Process for The Evaluation of Transport Policies to Reduce Climate Change Impacts. Fondazione Eni Enrico Mattei, NOTA DI LAVORO (2007)

An Algorithm for Fast Computation of 3D Krawtchouk Moments for Volumetric Image Reconstruction

Abderrahim Mesbah, Mostafa El Mallahi, Hakim El Fadili, Khalid Zenkouar, Aissam Berrahou and Hassan Qjidaa

Abstract Discrete Krawtchouk moments are powerful tools in the field of image processing application and pattern recognition. In this paper we propose an efficient method based on matrix multiplication and symmetry property to compute 3D Krawtchouk moments. This new method is used to reduce the complexity and computational time for 3D object reconstruction. The validity of the proposed algorithm is proved by simulated experiments using volumetric image.

Keywords 3D Krawtchouk moments · Symmetry property · Matrix multiplication · Volumetric image reconstruction

1 Introduction

Image moments have been widely applied for solving different problems in pattern recognition and image analysis tasks [1–5]. Geometric moments and their translation, scaling and rotation invariants were introduced by Hu [6]. Teague in [7] proposed the concept of orthogonal continuous moments such as Legendre and

A. Mesbah (✉) · M. El Mallahi · A. Berrahou · H. Qjidaa
CED-ST, LESSI, Faculty of Sciences Dhar el Mehraz, Sidi Mohamed Ben
Abdellah University, Fez, Morocco
e-mail: abderrahim.mesbah@usmba.ac.ma

M. El Mallahi
e-mail: mostafa.elmallahi@usmba.ac.ma

H. El Fadili
Ecole Nationale des Sciences Appliquées, Sidi Mohamed Ben Abdellah University,
Fez, Morocco

K. Zenkouar
LRSI Laboratory, Faculty of Science Technique, Sidi Mohamed Ben
Abdellah University, Fez, Morocco

© Springer International Publishing Switzerland 2016
A. El Oualkadi et al. (eds.), *Proceedings of the Mediterranean Conference
on Information & Communication Technologies 2015*, Lecture Notes
in Electrical Engineering 380, DOI 10.1007/978-3-319-30301-7_28

267

Zernike moments to represent image with minimum amount of information redundancy. The major problem associated with these moments is the discretization error, which increases by increasing the moment order [1]. To solve the above problem of the continuous orthogonal moments, Mukundan et al. in [8] proposed a set of discrete orthogonal Tchebichef moments. Recently, Yap et al. [9] introduced a set of discrete orthogonal Krawtchouk moments. Their experimental results make Krawtchouk moments superior to Zernike, Legendre and Tchebichef moments in terme of image reconstruction.

The direct computation of Krawtchouk moments is time consuming process and the computational complexity increased by increasing the moment order [10]. Therefore, some algorithms have been developed to accelerate the computational time of Krawtchouk moments. A recursive algorithm based on Clenshaw's recurrence formula using a second order digital filter was proposed by P. Ananth et al. [11]. The authors developed both a direct recursive algorithm and a compact algorithm for the computation of Krawtchouk moment. The effective recursive algorithm for inverse Krawtchouk moment transform was also presented in [12]. However, this approach was developed for the case of 2D images and only few works were presented to compute 3D Krawtchouk moments.

Since moment computation from volumetric images is time consuming process, the reconstruction becomes even more longer, especially for large images and high orders. Some algorithms have been proposed to resolve this problem. Hosney in [13] introduced an algorithm for fast computation of 3D Legendre moments. Wu et al. [14] used an algorithm based on matrix multiplication to compute the scale invariants of 3D Tchebichef moments.

In this paper, we propose a new method for fast computation of 3D Krawtchouk moments. Symmetry property of the krawtchouk polynomials is employed to reduce the computational time. The set of 3D Krawtchouk moments are computed by using an algorithm based on matrix multiplication.

The rest of this paper is organized as follows: in Sect. 2, an overview of 3D Krawtchouk moments and object reconstruction are given. The proposed algorithm is presented in Sect. 3. Section 4 presents the simulation results on 3D image reconstruction. Finally, concluding remarks are presented in Sect. 5.

2   3D Krawtchouk Moments

In this section we will present the mathematical background behind the Krawtchouk moment theory including polynomials, moments computation and reconstruction.

2.1 Krawtchouk Polynomials

The discrete Krawtcouk polynomial of order n is defined as [9]:

$$K_n(x; p, N) = \sum_{k=0}^{N} a_{k,n,p} x^k = {}_2F_1\left(-n, -x; -N; \frac{1}{p}\right), \tag{1}$$

where $x, n = 0, 1, 2, \ldots N, N > 0, p \in (0, 1)$ and the function ${}_2F_1$ is the hypergeometric function which is defined as:

$$ {}_2F_1(a, b; c; z) = \sum_{k=0}^{\infty} \frac{(a)_k (b)_k}{(c)_k} \frac{z^k}{k!}. \tag{2}$$

The symbol $(a)_k$ in (2) is the Pochhammer symbol given by

$$(a)_k = (a)(a+1)(a+2)\ldots(a+k-1) = \frac{\Gamma(a+k)}{\Gamma(a)}. \tag{3}$$

In order to ensure the numerical stability of the polynomials, Yap et al. [9] introduced the set of weighted Krawtchouk polynomials, defined as:

$$\overline{K}(x; p, N) = K_n(x; p, N) \sqrt{\frac{w(x; p, N)}{\rho(n; p, N)}}, \tag{4}$$

where

$$w(x; p, N) = \binom{N}{x} p^x (1 - p)^{N-x}, \tag{5}$$

and

$$\rho(n; p, N) = (-1)^n \left(\frac{1-p}{p}\right)^n \frac{n!}{(-N)_n}. \tag{6}$$

The scaled weighted Krawtchouk polynomials satisfy the following recursive relation with respect to n:

$$p(n - N)\overline{K}_{n+1}(x; p, N) = A(Np + 2np + n - x)\overline{K}_n(x; p, N) \\ - Bn(1 - p)\overline{K}_{n-1}(x; p, N), \tag{7}$$

where $A = \sqrt{\frac{(1-p)(n+1)}{p(N-n)}}, B = \sqrt{\frac{(1-p)^2(n+1)n}{p^2(N-n)(N-n+1)}}$, with $\overline{K}_0(x; p, N) = \sqrt{w(x; p, N)}, \overline{K}_1(x; p, N) = \left(1 - \frac{x}{pN}\right)\sqrt{w(x; p, N)}$.

The recurrence relation can be utilized to calculate the weight function defined in (5) as follow:

$$w(x+1;p,N) = \left(\frac{N-x}{x+1}\right)\frac{p}{1-p}w(x;p,N),\qquad(8)$$

with $w(0;p,N) = (1-p)^N = e^{N\ln(1-p)}$.

2.2 3D Krawtchouk Moments

The 3D Krawtchouk moment of order $m+n+l$ of an image intensity function $f(x,y,z)$ are defined over the cube $[0,M-1]\times[0,N-1]\times[0,L-1]$ as:

$$Q_{mnl} = \sum_{x=0}^{M-1}\sum_{y=0}^{N-1}\sum_{z=0}^{L-1}\overline{K}_m(x;p_x,M-1)\overline{K}_n(y;p_y,N-1)\overline{K}_l(z;p_z,L-1)f(x,y,z).$$

$$(9)$$

2.3 Object Reconstruction Using 3D Krawtchouk Moment

Since, Krawtchouk polynomials $\overline{K}_n(x;p_x,N-1)$ forms a complete orthogonal basis set on the interval $[0,N-1]$ and satisfies the orthogonal property. The 3D image/object intensity function $f(x,y,z)$ can be expressed over cube $[0,M-1]\times[0,N-1]\times[0,L-1]$ as:

$$f_{rec}(x,y,z) = \sum_{m=0}^{M-1}\sum_{n=0}^{N-1}\sum_{l=0}^{L-1}\overline{K}_m(x;p_x,M-1)\overline{K}_n(y;p_y,N-1)\overline{K}_l(z;p_z,L-1)Q_{mnl}.$$

$$(10)$$

The direct method needs two separate steps. We first compute the Krawtchouk polynomials value and second, we evaluate the sum of Eq. (9). In the present case of 3D Krawtchouk moment the computation involved in Eq. (9) is highly expensive in terms of time computing. In fact, we have three sums, in spite of two in the 2D case, each of which necessitate the evaluation of three polynomials $\overline{K}_m(x;p_x,M-1)$, $\overline{K}_n(y;p_y,N-1)$ and $\overline{K}_l(z;p_z,L-1)$.

3 The Proposed Method

In order to speed up the total computation of Krawtchouk moment which is highly expensive in the 3D case, we use in this paper an algorithm based on matrix multiplication and symmetry property of Krawtchouk polynomials, this method can reduce the computation cost of moment and inverse transformations especially for high order moments and large size images.

3.1 Symmetry Property

The symmetry relation of the weighted Krawtchouk polynomials for p = 0.5 is given by:

$$\overline{K}_n(x; p, N - 1) = (-1)^n \overline{K}(N - 1 - x; p, N - 1). \tag{11}$$

Using this relation, the domain of $N \times N \times N$ image (when N is even) will be divided into eight parts. Only the first cube $(0 \le x, y, z \le N/2 - 1)$ will suffice to compute 3D Krawtchouk moments. The expression of Krawtchouk moments in Eq. (9) can be written as follow:

$$Q_{mnl} = \sum_{x=0}^{\frac{M}{2}-1} \sum_{y=0}^{\frac{N}{2}-1} \sum_{z=0}^{\frac{L}{2}-1} \overline{K}_m(x; p_x, M - 1)\overline{K}_n(y; p_y, N - 1)\overline{K}_l(z; p_l, L - 1)g(x, y, z),$$

$$\tag{12}$$

where

$$\begin{aligned} g(x, y, z) =& f(x, y, z) + (-1)^m f(M - 1 - x, y, z) + (-1)^n f(x, N - 1 - y, z) \\ &+ (-1)^l f(x, y, L - 1 - z) + (-1)^{m+n} f(M - 1 - x, N - 1 - y, z) \\ &+ (-1)^{m+l} \times f(M - 1 - x, y, L - 1 - z) + (-1)^{n+l} f(x, N - 1 - y, L - 1 - z) \\ &+ (-1)^{m+n+l} \times f(M - 1 - x, N - 1 - y, L - 1 - z). \end{aligned}$$

$$\tag{13}$$

3.2 Computation of 3D Krawtchouk Moments

The matrix form of Krawtchouk moments was proposed by Yap et al. [9] for 2D image

$$Q = K_1 A K_2^T, \tag{14}$$

where Q is an $m \times n$ matrix of Krawtchouk moments

$$Q = \{Q_{ij}\}_{i,j=0}^{i,j=N-1}, \qquad K_1 = \{\overline{K}_i(x; p_x, N-1)\}_{i=0,x=0}^{i=m-1,x=N-1},$$
$$K_2 = \{\overline{K}_j(y; p_y, N-1)\}_{j=0,y=0}^{j=n-1,y=N-1} \qquad \text{and} \quad A = \{f(x,y)\}_{x,y=0}^{x,y=N-1}.$$

For 3D case, inspired by the algorithm proposed by Wu et al. [14], we propose an improved approach to accelerate the computation by using the symmetry property. The order of the summation of Eq. (9) will firstly be exchanged as follow:

$$Q_{mnl} = \sum_{z=0}^{\frac{N}{2}-1} \overline{K}_l(z; p_z, N-1) \left\{ \sum_{x=0}^{\frac{N}{2}-1} \sum_{y=0}^{\frac{N}{2}-1} \overline{K}_m(x; p_x, N-1) \overline{K}_n(y; p_y, N-1) g(x,y,z) \right\}, \tag{15}$$

$$\begin{bmatrix} Q'_{00}(\frac{N}{2}-1) & Q'_{01}(\frac{N}{2}-1) & \dots & Q'_{0n}(\frac{N}{2}-1) \\ Q'_{10}(\frac{N}{2}-1) & Q'_{11}(\frac{N}{2}-1) & \dots & Q'_{1n}(\frac{N}{2}-1) \\ \dots & & & \\ Q'_{m0}(\frac{N}{2}-1) & Q'_{m1}(\frac{N}{2}-1) & \dots & Q'_{mn}(\frac{N}{2}-1) \end{bmatrix}$$

$$\begin{bmatrix} Q'_{00}(1) & Q'_{01}(1) & \dots & Q'_{0n}(1) \\ Q'_{10}(1) & Q'_{11}(1) & \dots & Q'_{1n}(1) \\ \dots & & & \\ Q'_{m0}(1) & Q'_{m1}(1) & \dots & Q'_{mn}(1) \end{bmatrix}$$

$$\begin{bmatrix} Q'_{00}(0) & Q'_{01}(0) & \dots & Q'_{0n}(0) \\ Q'_{10}(0) & Q'_{11}(0) & \dots & Q'_{1n}(0) \\ \dots & & & \\ Q'_{m0}(0) & Q'_{m1}(0) & \dots & Q'_{mn}(0) \end{bmatrix}.$$

These matrices will be rearranged, in the second step, and they are multiplied by the Krawtchouk polynomials matrices. Then we obtained the matrices of 3D Krawtchouk moments.

$$
\begin{bmatrix} Q_{000} & \cdots & Q_{0n0} \\ Q_{001} & \cdots & Q_{0n1} \\ \cdots & & \\ Q_{00l} & \cdots & Q_{0nl} \end{bmatrix} = \begin{bmatrix} \overline{K}_0(0;p_z,N-1) & \overline{K}_0(1;p_z,N-1) & \cdots & \overline{K}_0(\tfrac{N}{2}-1;p_z,N-1) \\ \overline{K}_1(0;p_z,N-1) & \overline{K}_1(1;p_z,N-1) & \cdots & \overline{K}_1(\tfrac{N}{2}-1;p_z,N-1) \\ \cdots & & & \\ \overline{K}_l(0;p_z,N-1) & \overline{K}_l(1;p_z,N-1) & \cdots & \overline{K}_l(\tfrac{N}{2}-1;p_z,N-1) \end{bmatrix}
$$

$$
\times \begin{bmatrix} Q'_{00}(0) & \cdots & Q'_{0n}(0) \\ Q'_{00}(1) & \cdots & Q'_{0n}(1) \\ \cdots & & \\ Q'_{00}(\tfrac{N}{2}-1) & \cdots & Q'_{0n}(\tfrac{N}{2}-1) \end{bmatrix},
$$

$$
\begin{bmatrix} Q_{100} & \cdots & Q_{1n0} \\ Q_{101} & \cdots & Q_{1n1} \\ \cdots & & \\ Q_{10l} & \cdots & Q_{1nl} \end{bmatrix} = \begin{bmatrix} \overline{K}_0(0;p_z,N-1) & \overline{K}_0(1;p_z,N-1) & \cdots & \overline{K}_0(\tfrac{N}{2}-1;p_z,N-1) \\ \overline{K}_1(0;p_z,N-1) & \overline{K}_1(1;p_z,N-1) & \cdots & \overline{K}_1(\tfrac{N}{2}-1;p_z,N-1) \\ \cdots & & & \\ \overline{K}_l(0;p_z,N-1) & \overline{K}_l(1;p_z,N-1) & \cdots & \overline{K}_l(\tfrac{N}{2}-1;p_z,N-1) \end{bmatrix}
$$

$$
\times \begin{bmatrix} Q'_{10}(0) & \cdots & Q'_{1n}(0) \\ Q'_{10}(1) & \cdots & Q'_{1n}(1) \\ \cdots & & \\ Q'_{10}(\tfrac{N}{2}-1) & \cdots & Q'_{1n}(\tfrac{N}{2}-1) \end{bmatrix},
$$

$$
\cdots
$$

$$
\begin{bmatrix} Q_{m00} & \cdots & Q_{mn0} \\ Q_{m01} & \cdots & Q_{mn1} \\ \cdots & & \\ Q_{m0l} & \cdots & Q_{mnl} \end{bmatrix} = \begin{bmatrix} \overline{K}_0(0;p_z,N-1) & \overline{K}_0(1;p_z,N-1) & \cdots & \overline{K}_0(\tfrac{N}{2}-1;p_z,N-1) \\ \overline{K}_1(0;p_z,N-1) & \overline{K}_1(1;p_z,N-1) & \cdots & \overline{K}_1(\tfrac{N}{2}-1;p_z,N-1) \\ \cdots & & & \\ \overline{K}_l(0;p_z,N-1) & \overline{K}_l(1;p_z,N-1) & \cdots & \overline{K}_l(\tfrac{N}{2}-1;p_z,N-1) \end{bmatrix}
$$

$$
\times \begin{bmatrix} Q'_{m0}(0) & \cdots & Q'_{mn}(0) \\ Q'_{m0}(1) & \cdots & Q'_{mn}(1) \\ \cdots & & \\ Q'_{m0}(\tfrac{N}{2}-1) & \cdots & Q'_{mn}(\tfrac{N}{2}-1) \end{bmatrix}.
$$

By using the same algorithm, we can calculate the intensity of the object by the matrix method. The image intensity function can be written in the matrix form for 2D case as follow [9]:

$$A = K_1^T Q K_2, \tag{16}$$

where A, K_1, Q and K_2 are defined above.

4 Numerical Simulation

In this section, the conducted numerical experiments are used to evaluate the efficiency of the proposed method which is compared with the corresponding methods. The reconstruction of volumetric image using Krawtchouk moments will be illustrated based on 3D MRI images. This experiment is performed by using the Mean-Square-Error (MSE) which is defined as:

$$MSE = \frac{1}{MNL} \sum_{x=0}^{M-1} \sum_{y=0}^{N-1} \sum_{z=0}^{L-1} (f_{rec}(x, y, z) - f(x, y, z))^2, \tag{17}$$

where $(M \times N \times L)$ is image size and f_{rec} is the reconstructed image function for different orders.

Figure 1 shows object reconstructed using matrix multiplication algorithm of Krawtchouk moments for different orders. A 3D MRI image was selected from [15]. The head image was resized at $128 \times 128 \times 128$ voxels. We illustrate different orders from 20 to 127. As it can be seen from the figures, the reconstructed images using Krawtchouk moments show progressively more visual resemblance with the original image in the early orders.

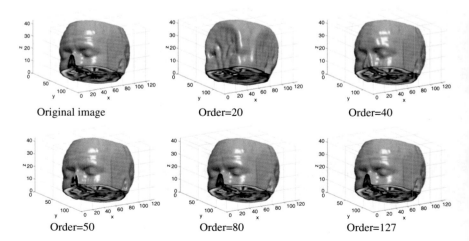

Fig. 1 Reconstruction of 3D MRI head image using Krawtchouk moments

Fig. 2 Comparative study of reconstruction error of Krawtchouk and Tchebichef moments of 3D MRI image

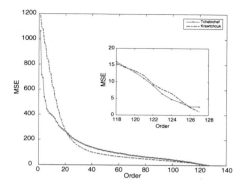

Fig. 3 Elapsed CPU time in second for 3D Krawtchouk moments computation

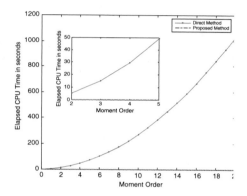

Figure 2 presents the plot of the reconstruction error of Krawtchouk and Tchebichef moments for different orders up to 127. We can see that the convergence is faster in terms of reconstruction error. However, Krawtchouk moments perform better as compared to Tchebichef moments.

The computational speed of the proposed algorithm is evaluated by using 3D MRI image. Figure 3 shows CPU elapsed times using our algorithm in comparison with the conventional algorithm. The CPU times is obtained using Matlab R2015a and implemented on MacBook Pro machine equipped with a processor of i5 (2.5 Ghz and 8 GB RAM). Figure 3 implies that our proposed method is very fast in terms of computation time compared with the classical method.

5 Conclusion

This paper proposed a new method for 3D Krawtchouk moments computation and its inverse transform by using an algorithm based on matrix multiplication combined with the symmetry property. This method avoids enormous computing caused by iteration. Our algorithm reduces significantly the computational complexity compared with the classical method. Simulated result clearly showed that the elapsed CPU time is significantly reduced.

References

1. Khotanzad, A., Hong, Y.: Invariant image recognition by Zernike moments. IEEE Trans. Pattern Anal. Mach. Intell. **12**(5), 489–497 (1990)
2. Belkasim, S., Shridhar, M., Ahmadi, M.: Pattern recognition with moment invariants: a comparative study and new results. Pattern Recogn. **24**(12), 1117–1138 (1991)
3. Flusser, J., Suk, T.: Pattern recognition by affine moment invariants. Pattern Recogn. **26**(1), 167–174 (1993)
4. Hsu, H.S., Tsai, W.H.: Moment-preserving edge detection and its application to image data compression. Opt. Eng. **32**(7), 1596–1608 (1993)
5. Zhu, H., Shu, H., Zhou, J., Luo, L., Coatrieux, J.L.: Image analysis by discrete orthogonal dual Hahn moments. Pattern Recogn. Lett. **28**(13), 1688–1704 (2007)
6. Hu, M.K.: Visual pattern recognition by moment invariants. IRE Trans. Inf. Theory **8**(2), 179–187 (1962)
7. Teague, M.R.: Image analysis via the general theory of moments. J. Opt. Soc. Am. **70**(8), 920–930 (1980)
8. Mukundan, R., Ong, S.H., Lee, P.A.: Image analysis by Tchebichef moments. IEEE Trans. Image Process. **10**(9), 1357–1364 (2001)
9. Yap, P.-T., Paramesran, R.: Image Analysis by Krawtcouk Moments. IEEE Trans. Image Process. **12**(11), 1367–1377 (2003)
10. Wang, G., Wang, S.: Recursive computation of Tchebichef moment and its inverse transform. Pattern Recogn. **39**(1), 47–56 (2006)
11. Venkataramana, A., Ananth Raj, P.: Recursive computation of forward Krawtchouk moment transform using Clenshaw's recurrence formula. Third National Conference on Computer Vision, Pattern Recognition, Image Processing and Graphics (2011)
12. Ananth Raj, P., Venkataramana, A.: Fast computation of inverse Krawtchouk moment transform using Clenshaw's recurrence formula. Third National Conference on Computer Vision, Pattern Recognition, Image Processing and Graphics(2011)
13. Hosny, K.M.: Fast and low-complexity method for exact computation of 3D Legendre moments. Pattern Recogn. Lett. **32**(9), 1305–1314 (2011)
14. Wu, H., Coatrieux, J.L., Shu, H.: New algorithm for constructing and computing scale invariants of 3D Tchebichef moments. Mathematical Problems in Engineering, pp. 1–8 (2013)
15. Fitzpatrick, J.: Retrospective image registration and evaluation project (RIRE). http://www.insight-journal.org/rire/index.php

Robust Face Recognition Using Local Gradient Probabilistic Pattern (LGPP)

Abdellatif Dahmouni, Karim El Moutaouakil and Khalid Satori

Abstract In this work, we propose a new local pattern for face recognition, named Local Gradient Probabilistic Pattern (LGPP). It is an extension of Local Gradient Pattern (LGP) that uses a very important result of probability theory; it is the law large numbers. In this direction, the distribution of the gray levels on a face image follows a law of probability, which is the sum of several normal laws. The current pixel will be evaluated by the confidence interval concept. The suggested model is merged with the most known algorithms of data analysis in the face recognition field, in particular the PCA, the LDA, the 2DPCA and the 2DLDA. The tests carried out on the ORL and YALE databases show the effectiveness of LGPP +2DPCA and LGPP+2DLDA systems. The experimental exactitude is of 96 %.

Keywords LBP · LGP · LGPP · 2DPCA · 2DLDA · Confidence interval

1 Introduction

The face recognition is a biometric modality that covers many applications such as biometric systems, computer security systems, access control systems, and surveillance systems. Many methods are developed in this field: Principal components analysis PCA [1] and 2DPCA [2, 3], Linear Discriminate Analysis LDA [4]

A. Dahmouni (✉) · K. Satori
LIIAN, Department of Mathematics and Computer Science,
Faculty of Sciences, Sidi Mohamed Ben Abdellah University,
Atlas-Fez, B.P 1796 Dhar-Mahraz, Morocco
e-mail: abdellatifdahmouni@gmail.com

K. Satori
e-mail: khalidsatorim3i@yahoo.fr

K.E. Moutaouakil
ENSAH, National School of Applied Sciences Al Hoceima,
Ajdir, BP 03 Al-Hoceima, Morocco
e-mail: karimmoutaouakil@yahoo.fr

© Springer International Publishing Switzerland 2016
A. El Oualkadi et al. (eds.), *Proceedings of the Mediterranean Conference on Information & Communication Technologies 2015*, Lecture Notes in Electrical Engineering 380, DOI 10.1007/978-3-319-30301-7_29

and 2DLDA [5, 6], Independent Component Analysis ICA [7], Artificial Neural Networks ANN [8], Hidden Markov Models HMM [9], Gabor Wavelets [10], and Local Binary Pattern LBP [11, 12, 13]. In this work, we propose a probabilistic approach for face recognition based on a new Local Gradient Probabilistic Pattern (LGPP) and on 2DPCA and 2DLDA methods. The LBP is a non-parametric operator for texture analysis, and extended for face recognition [13]. The original LBP uses a neighborhood of (3 by 3) pixels, centered on the current pixel to calculate its gray level value. Each of the eight neighboring pixels having a higher or equal value to the current pixel value is coded by 1, the others by 0 [11]. Because of its invariance to the monotonous lighting changes, and its low calculation complexity, the LBP became largely used. Several extensions of LBP are developed:

Ojala proposes a circular neighborhood; he noticed that more than 90 % of the texture information presents at most two transitions (01 or 10) it's the uniform variety $LBPU^2$ [12, 13]. Tan proposes the Local Ternary Pattern LTP, by using a three bits coding mode [14]. Ahonen introduces the Soft-LBP SLBP by using fuzzy logic [15], Liao proposes the multi-blocks' MB-LBP, he applies the LBP descriptor to the rectangular blocks intensities that consists to substitute the pixels of each locks by their average pixel [16]. Jabid proposes the Local Directional Pattern LDP, which uses the Kirsch masks as detectors of the maximum information directions [17].

The LGP operator introduced by Bongjin [18] uses the values of the gradients: $g_n = |i_n - i_c|$ of the eight neighboring pixels, and their average: $g_m = \frac{1}{P}\left(\sum_{n=1}^{P} g_n\right)$ to generate the mask: $M = [g_1, g_2, g_3, g_4, g_m, g_5, g_6, g_7, g_8]$. The LBP descriptor is applied to the mask M. The new gray level is generated by the following mathematical equation:

$$\text{LGP}_{P,R}(x_c, y_c) = \sum_{n=0}^{P-1} s(g_n - g_m) * 2^n \text{ Where : } s(x) = \begin{cases} 1, x \geq 0 \\ 0, x < 0 \end{cases}. \quad (1)$$

g_n and g_m, represent respectively the P edge values and the center value of the gradients mask of radius R.

The different varieties of LBP suffer from the loss of information related to the mode of hard thresholding. In this work, we proposed a new extension of the local gradient pattern LGP, called Local Gradient Probabilistic Pattern LGPP. This vision allows evaluating the current pixel while basing on the confidence interval concept. The proposed model is merged with the dimensionality reduction algorithms: PCA, LDA, 2DPCA and 2DLDA. Also, we compare the performances of the systems LBP+(ACP, LDA, 2DPCA and 2DLDA) and LGPP+(ACP, LDA, 2DPCA and 2DLDA) using the images extracted from the ORL and YALE databases. The obtained results show the effectiveness of the proposed approach compared to some existing face recognition techniques, especially for systems LGPP+2DPCA and LGPP+2DLDA.

The sequel of the paper consists of three sections followed by a conclusion. Section 2 presents in detail the LGPP descriptor variety based on the confidence interval. The hybrid systems based on the algorithms 2DPCA, 2DLDA and LGPP are examined in the Sect. 3. Section 4 presents the experimental results which evaluate the effectiveness of the suggested approach.

2 Local Gradient Probabilistic Pattern (LGPP)

As many distributions observed in reality, the distribution of face gray levels follows a sum of several normal laws. The normal law N (μ, σ) is defined by the probability density f and the distribution function F given by:

$$f(t) = \frac{1}{\sigma\sqrt{2\pi}}e^{-\frac{1}{2}\left(\frac{t-\mu}{\sigma}\right)^2} \forall t \in \mathbf{R} \text{ and } F(x) = \int_{-\infty}^{x} f(t)dt \ \forall x \in \mathbf{R}. \quad (2)$$

In the almost uniform areas, there exist small variations of intensity between neighboring pixels; we can remove these intensities variations by assimilating the distribution of the pixels in these areas to a normal law. Therefore, we pass from a thresholding mode: fixed and signed for LBP [11, 12], absolute and locally adapted for LGP [18], to a thresholding mode per a confidence interval based on the distribution of pixels around a current pixel.

We propose to generate a confidence interval [α_1, α_2], based on statistical moments in a neighborhood of (N = 5 by 5) pixels centered on current pixel. Each of the eight neighboring pixels having a value located inside the confidence interval is coded by 1, the others by 0. A byte of eight bits is generated, and then converted into gray level (Fig. 1). To evaluate the confidence interval, a random variable X is associated to the pixel gray level value i_n. The following statistical moments are calculated:

$$\text{Average}: \mu = \frac{1}{N}\sum_{n=1}^{N} i_n. \quad \text{Standard deviation}: \sigma = \sqrt{\frac{1}{N}\sum_{n=1}^{N}(i_n - \mu)^2} \quad (3)$$

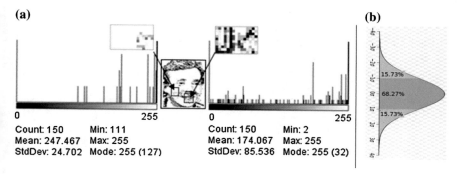

(a)

0   255

Count: 150   Min: 111
Mean: 247.467   Max: 255
StdDev: 24.702   Mode: 255 (127)

0   255

Count: 150   Min: 2
Mean: 174.067   Max: 255
StdDev: 85.536   Mode: 255 (32)

(b)

15.73%

68.27%

15.73%

Fig. 1 **a** Histograms of various areas obtained by LGPP. **b** Different confidence intervals

$$\text{Variation coefficient}: \delta = \frac{\sigma}{\mu} \quad \text{Symmetry coefficient}: S = \sum_{n=1}^{N} \frac{(i_{n}-\mu)^{3}}{\sigma^{3}}. \quad (4)$$

As it is known if the parameters δ and S fulfilling the condition:

$$S = 0 \text{ or } \delta < \beta \text{ Where, } \beta \text{ is a limit of validity of the dispersion } 0.1 < \beta < 0.2. \quad (5)$$

This random variable X follows the normal law N (μ, σ). In this case, the confidence interval [α_1, α_2] is:

$$P(\alpha_1 \leq X \leq \alpha_2) = \frac{1}{\sigma\sqrt{2\pi}} \int_{\mu-k\sigma}^{\mu+k\sigma} e^{-\frac{1}{2}\left(\frac{t-\mu}{\sigma}\right)^2} dt, \forall k \in \mathbf{R}. \quad (6)$$

$$[\alpha_1, \alpha_2] = [\mu - K, \mu + K], \text{ Where}: K = k\sigma.$$

The normal variable X is almost certainly located in the interval $[\mu - 3\sigma, \mu - 3\sigma]$. One defines a convergence limit of the normality in the confidence degree to 99.99 % for the interval $[\mu - 4\sigma, \mu - 4\sigma]$ therefore P ($\alpha_1 \leq X \leq \alpha_2$) ≈ 1. For any X not fulfilling (5) one takes:

$$K = \begin{cases} \sigma & \text{if } \beta < \delta < \beta + 0.1 \\ (\beta + 0.1)\mu & \text{other} \end{cases} \quad (7)$$

Such choices permit to correct the asymmetry and kurtosis problems for X.

To evaluate the confidence interval associated to LGPP descriptor, a random variable G is associated to gradient value g_n. According to Eqs. (6) and (7), we have:

$$|i_n - \mu| \leq K \Rightarrow |(i_n - i_c) - (\mu - i_c)| \leq K$$
$$\Rightarrow ||i_n - i_c| - |\mu - i_c|| \leq K$$
$$\Rightarrow -K + |\mu - i_c| \leq |x_n - x_c| \leq K + |\mu - i_c|$$

From where : $\max(0, -K + g_m) \leq g_n \leq K + g_m$ Where, $g_m = |\mu - i_c|$. (8)

The LGPP confidence interval is : $[\gamma_1, \gamma_2] = [\max(0, -K + g_m), K + g_m]$. (9)

The mathematical formulation of (LGPP) is given by:

$$\text{LGPP}_{P,R,N}(x_c, y_c) = \sum_{n=0}^{P-1} s(g_n) * 2^n \text{ Where}: s(x) = \begin{cases} 1, & \gamma_1(N) \leq x \leq \gamma_2(N) \\ 0, & \text{others} \end{cases}.$$

$$(10)$$

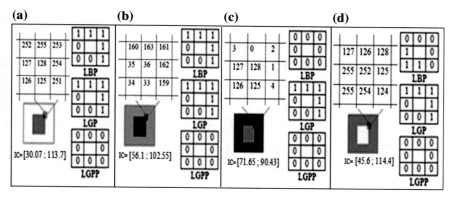

Fig. 2 LBP, LGP and LGPP for different intensity changes: (**a**) original, (**b**) Monotonic illumination change, (**c**) Background change, and (**d**) Foreground change of *gray* levels

Unlike the LBP and LGP, which are deterministic models, our approach considers the face as distributions of pixels which follow a law of probability. We thus pass to a probabilistic and statistical description where the notion of the confidence interval allows assigning values near 255 at all pixels of almost homogeneous areas (forehead, cheeks....). And the values near 0 at all corners. The nose, eyes and mouth which having strong curves have different values that can be easily located (Fig. 1). Like LGP the LGPP is robust to local and global intensity variations. However, in the Fig. 2b where the background and the foreground together change (global), LBP, LGP and LGPP codes are the same as the (Fig. 2a). On the other side, when the background and the foreground separately change (local), the LGPP code remains invariant, but LBP code changes (Fig. 2c, d).

3 Recognition Systems Based on LGPP

According to our knowledge the LGP is generally used in face detection. In this work, we use the variety LGPP for face recognition. Generally, the face recognition based on LBP consists in concatenation of histograms Iso-blocks. To profit from the advantages of the local and global approaches, LGPP is combined with the PCA, LDA, 2DPCA and 2DLDA global algorithms. In contrast, the conventional algorithms PCA and LDA; 2DPCA and 2DLDA represent original images by 2D matrix. The covariance matrix will be of small size (n, n) instead of (n^2, n^2). Consequently, 2DPCA and 2DLDA have important advantages compared to PCA and LDA. First of all, it is easier to evaluate with precision the covariance matrix, less time is necessary to determine the eigenvectors. In addition, 2DLDA dominates the singularity problem (Size of database can generate a covariance matrix not diagonalizable). The steps of 2DPCA and 2DLDA are presented below:

Let R be a vector of dimension n and Γ a matrix of size (m, n). The linear transformation $Y = \Gamma R$ is a projection of Γ on R. The projected vector Y of dimension m is called characteristic vector of Γ. For M training images $[\Gamma_1,\ldots, \Gamma_M]$, one defines: For 2DPCA the covariance matrix C:

$$C = \frac{1}{M}\sum_{j=1}^{M}(\Gamma_j - \psi_{mean})^t(\Gamma_j - \psi_{mean}) \text{ Where}: \frac{1}{M}\sum_{j=1}^{M}\Gamma_j. \tag{11}$$

For 2DLDA the intra-class S_w and inter-Class S_b dispersion matrices:

$$S_w = \sum_{i=1}^{P}\sum_{\Gamma_k \in C_i}(\Gamma_k - \psi_{C_i})(\Gamma_k - \psi_{C_i})^t. \tag{12}$$

$$S_b = \sum_{i=1}^{P} q_i(\psi_{moy} - \psi_{C_i})(\psi_{moy} - \psi_{C_i})^t. \tag{13}$$

Where the average image of class ``i'' is : $\psi_{C_i} = \frac{1}{q_i}\sum_{k=1}^{q_i}\Gamma_k. \tag{14}$

The matrix of optimal projection $R_{opt} = \text{argmax}(J(R))$ is obtained by maximization of generalized variance criterion: $J(R) = R^tCR$ for 2DPCA, and Fisher criterion: $J(R) = R^tS_wR/R^tS_bR$ for 2DLDA. R_{opt} is composed by the Eigenvectors $[R_1\ldots R_d]$ corresponding to «d» greater Eigen values of the C or $S_w^{-1}S_b$ specials matrix. Then we obtain for each image Γ a characteristic matrix of dimension m * d: $Y = [Y_1 \ldots Y_d]$, whose components $Y_k = \Gamma R_k (k = 1\ldots d)$. Finally, the similarity between the test and training images is measured by the distance: dist = argmin $|Y_{test} - Y_{train}|$.

4 Experimentations

We propose a methodology which uses the obtained image from (LGPP) to direct input of algorithms PCA, LDA, 2DPCA and 2DLDA (see Fig. 3).

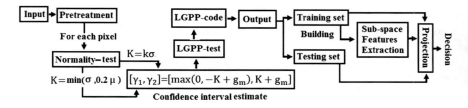

Fig. 3 Design block of our face recognition system

To evaluate the performances of the proposed approach, we have used ORL and YALE databases. The ORL database is made of 40 subjects having each one 10 different views [19]; the images in gray levels have the same size (92 * 112) pixels. Some images were taken under various conditions. The YALE database is made up of 165 images of 15 subjects representing 11 conditions of size (320 * 243) [20], and scaled to size (112 * 92) pixels. One pre-treatment with histogram equalization is necessary to harmonize the gray levels distribution.

As it is shown, in the Fig. 3, formed by images respectively obtained by application of LBP, LGPP (k = 1), LGPP (k = 2), LGPP (k = 3) and LGPP (k = 4). The pixels of the almost homogeneous areas, and the pixels of the peaks areas are perfectly separate in the case of LGPP (k = 3) and LGPP (k = 4).

Initially, we compare the performances (rate and time recognition) of 2DPCA and PCA for different dimensions of Eigen-space projection, for a number of 5 training images per class of ORL database.

Table 1 shows that LGPP+2DPCA and LGPP+2DLDA algorithm reaches its maximum rate for a small Eigen-base size: LGPP+2DPCA: (dim = 8; rate = 96; Time = 7.1) and LGPP+2DLDA: (dim = 8; rate = 96; Time = 9.22), differently to LGPP+PCA: (dim = 18; rate = 90; Time = 11.65).

ORL database: for a set of 5 training images and 5 testing images by class, and for optimal base of projection; the results, illustrated in Table 2, show that the association of algorithms 2DPCA, 2DLDA and LGPP (k > 2) gives the best recognition rate 2DPCA (96), 2DLDA (96).

Table 1 Rate and time recognition for different Eigen space dimensions on database ORL

Dimensions		6	8	12	16	18	20
Rate(%)	LGPP+PCA	71	81.5	88	89	90	89.5
	LGPP+2DPCA	94	96	95	94.5	94	94
	LGPP+2DLDA	94.5	96	95	95	94.5	94
Time(s)	LGPP+PCA	10.17	10.45	10.39	10.61	11.65	11.73
	LGPP+2DPCA	6.31	7.09	8.44	9.86	11.33	12.73
	LGPP+2DLDA	8.11	9.22	10.59	11.45	12.27	13.34

Table 2 Rate recognition for different methods applied on database ORL

Methods	Only	LBP	LGPP1	LGPP2	LGPP3	LGPP4
PCA	89	75	87	90	90	90
LDA	90	82.5	88	90	90	90
2DPCA	94	88	95	96	96	96
2DLDA	94.5	90.5	95	95.5	96	96

Table 3 Rate recognition for different methods applied on database YALE

Methods	Only	LBP	LGPP1	LGPP2	LGPP3	LGPP4
PCA	79.67	77.33	78.67	86.67	86.67	85.33
LDA	81.33	84	88	92	92	92
2DPCA	82.67	82.67	85.33	96	96	96
2DLDA	89.33	82.67	86.67	94.67	94.67	94.67

YALE database: for a set of 6 training images and 5 testing images by class, and for optimal base of projection, the results illustrated in Table 3 show that the association of algorithms 2DPCA, 2DLDA and LGPP (k > 2) gives the best recognition rate 2DPCA (96), 2DLDA (94.67).

The Fig. 4, expose, the recognition rates for different number of training images from the ORL and Yale databases. The number of erroneous testing images depends on the training images number, on used database, on global methods selected, and on the LGPP confidence interval size (Fig. 5).

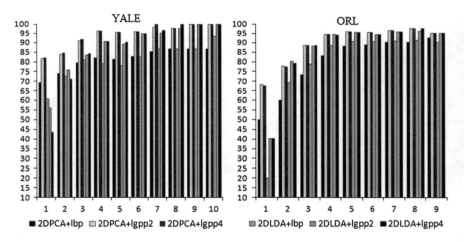

Fig. 4 Recognition rates for different number of training images on ORL and Yale

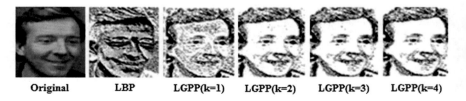

Fig. 5 Obtained ORL images by: LBP and LGPP (k = 1, 2, 3, 4) descriptors

5   Conclusion

In this paper, we have proposed a new model for local characteristics extraction, named LGPP. This probabilistic LGP variety is based on the idea that the face pixels distribution follows a probability law. The obtained results by the face recognition system based on fusion the LGPP with 2DPCA and 2DLDA methods prove that the approach based on the LGPP is robust than that based on LBP and LGP. The suggested system reach their best performances for the confidence interval of k > 2. In future work, we will propose a new face recognition system based on the artificial intelligence concepts.

References

1. Turk, M., Pentland, A.: Eigenfaces for recognition. J. Cogn. Neurosci. **3**, 71–86 (1991)
2. Yang, J., Zhang, D., Frangi, A.F., Yang, J.-Y.: Two dimensional PCA: A new approach to appearance-based face representation and recognition. IEEE Trans. Patter Anal. Mach. Intell. **26**, 131 (2004)
3. Yang, W., Sun, C., Ricanek, K.: Sequential row–column 2DPCA for face recognition. Neural Comput. Appl. **21**(7), 1729–1735 (2012)
4. Etemad, K., Chellappa, R.: Discriminant analysis for recognition of human face images. J. Optical Soc. Am. **14**, 1724–1733 (1997)
5. Zhou, D., Yang, X., Peng, N., Wang, Y.: Improved-LDA based face recognition using both facial global and local information. Pattern Recogn. Lett. **27**, 536–543 (2006)
6. Yang, W., Yan, X., Zhang, L., Sun, C.: Feature extraction based on fuzzy 2DLDA. Neurocomputing **73**(10), 1556–1561 (2010)
7. Bartlett, M.S., Movellan, J.R., Sejnowski, T.J.: Face recognition by independent component analysis. IEEE Trans. Neural Netw. **13**, 1450–1464 (2002)
8. El-Bakry, H.M., Mastorakis, N.: A novel model of neural networks for fast data detection. WSEAS Trans. Comput. **5**(8), 1773–1780 (2006)
9. Kim, D.J., Chung, K.W., Hong, K.S.: Person authentication using face, teeth, and voice modalities for mobile device security. IEEE Trans. Consum. Electr. **56**, 2678 (2010)
10. Tan, X., Triggs, B.: Fusing Gabor and LBP feature sets for kernel-based face recognition. In: Proceedings of the 3rd International Conference on Analysis and Modeling of Faces and Gestures (2007)
11. Ojala, T., Pietikainen, M., Harwood, D.: A comparative study of texture measures with classification based on feature distributions. Pattern Recogn. **29**(1), 51–59 (1996)
12. Ojala, T., Pietikäinen, M., Mäenpää, T.: Multiresolution gray-scale and rotation invariant texture classification with local binary patterns. IEEE Trans. Pattern Anal. Mach. Intell. **24**, 971–987 (2002)
13. Ahonen, T., Hadid, A., Pietikainen, M.: Face description with local binary patterns: Application to face recognition. IEEE Trans. Pattern Anal. **28**, 2037 (2006)
14. Tan, X., Triggs, B.: Enhanced local texture feature sets for face recognition under difficult lighting conditions. IEEE Trans. Image Process. **19**(6), 1635–1650 (2007)
15. Ahonen, T., Pietikäinen, M.: Soft histograms for local binary patterns. In: Proceedings of the Finnish Signal Processing Symposium, FINSIG 2007, vol. 1, p. 14 (2007)
16. Liao, S., Zhu, X., Lei, Z., Zhang, L., Li, S.Z.: Learning multi-scale block local binary patterns for face recognition. In: ICB, pp. 828–837 (2007)

17. Jabid T, Kabir, M.H., Chae, O.S.: Local directional pattern for face recognition. In: Proceeding of the IEEE International Conference of Consumer Electronics (2010)
18. Jun, B., Kim, D.: Robust face detection using local gradient patterns and evidence accumulation. Pattern Recogn. **45**, 3304 (2012)
19. The ORL face database at the AT&T http://www.uk.research.att.com/facedatabase.html
20. The Yale Face Database, http://cvc.yale.edu/proiects/yalefaces/yalefaces.html

Comparative Study of Mesh Simplification Algorithms

Abderazzak Taime, Abderrahim Saaidi and Khalid Satori

Abstract Many applications in the field of computer graphics are becoming more complex and require more accurate simplification of the surface meshes. This need is due to reasons of rendering speed, the capacity the backup and the transmission speed 3D models over networks. We presented four basic methods for simplifying meshes that are proposed in recent years. The result obtained by the implementation of these methods will be the subject of a comparative study. This study aims to evaluate these methods in terms of preserving the topology and speed.

Keywords Computer graphics · Simplification · Surface meshes

1 Introduction

The use of 3D models, represented as mesh, is ever increasing in many applications and are also used in very different fields, such as video games, animation, archeology, medical imaging, scientific visualization, augmented reality, and computer aided design. These models are finely detailed to satisfy an expectation of increasingly

A. Taime (✉) · A. Saaidi · K. Satori
LIIAN, Department of Mathematics and Computer Science Faculty of Sciences,
Dhar-Mahraz Sidi Mohamed Ben Abdellah University, B.P 1796 Atlas- Fez, Morocco
e-mail: taime35550@gmail.com

K. Satori
e-mail: khalidsatorim3i@yahoo.fr

A. Saaidi
LSI, Department of Mathematics, Physics and Computer Science Polydisciplinary
Faculty of Taza, Dhar-Mahraz Sidi Mohamed Ben Abdellah University,
B.P. 1223 Taza, Morocco
e-mail: abderrahim.saaidi@usmba.ac.ma

© Springer International Publishing Switzerland 2016
A. El Oualkadi et al. (eds.), *Proceedings of the Mediterranean Conference on Information & Communication Technologies 2015*, Lecture Notes in Electrical Engineering 380, DOI 10.1007/978-3-319-30301-7_30

strong realism in computer graphics. However, the increased complexity of the data still surpasses the improvements in hardware performance, which requires the simplification of meshes while preserving the topology. In this context, several approaches and strategies are adopted for controlled simplification of meshes:

- The first Involves the removal of geometric entities, typically the vertices [1] or triangles [2] and the remesh the holes thus formed.
- Some strategies have tried to agglomerate geometrically the nearby vertices to replace an representative [3–8].
- Another type of method is to merge coplanar facets [9–11].
- Other approaches have tried optimizing an energy function [12, 13].
- Wavelets [14, 15], in a single structure, provide a mathematical framework to be several representations more or less detailed of the same object.

Another type of method called remeshing consists of randomly positioning the points on the surface and move them with attraction/repulsion forces by forcing them to stay on the surface [16].

We can still classify these methods on the basis of local and global approaches: in the first case, mesh modifications are operated upon a local optimization criterion (e.g. simplification envelopes [17] and other approaches decimation). In the second, a global optimization process is applied to the whole mesh. (e.g. energy optimization approaches [12, 13], Multiresolution decimation [18, 19], and multiresolution analysis [20]).

This work presents a brief introduction to mesh simplification methods. Its main objective is to analyze and compare some methods and main approaches on the basis of two criteria, a criterion for minimizing the error or difference between the simplified mesh and the original mesh and a criterion of run times.

2   Error Evaluation

The approximation error is managed in many various manners by the different simplification approaches, but there are two principal manners between them:

- Approaches that are based on the locally bounded errors, i.e. the approximation accuracy is known in each surface unit (e.g. methods of mesh decimation [1, 21]).
- Approaches which only support globally bounded approximation errors, i.e. the approximation accuracy is known around each surface entity (e.g. the simplification envelopes method [17], Agglomeration of the vertices [3, 4, 8] and energy optimization approaches [12, 13]).

3 Simplification Approaches Reviewed

To quickly view a scene and allow it to move, the mesh is simplified by reducing the number of triangles. The strategy for simplifying a mesh differ from one approach to another, On the other hand, the majority of them degrade the mesh approximation precision to reduce the number of vertices/triangles. Here is an overview of approaches and the most commonly used methods for simplifying meshes we presented in this work:

- Vertex decimation: This algorithm iteratively removes a vertex of the mesh and all the adjacent faces. In each step of decimation process, all vertices are evaluated according to their importance. The least important vertex is selected for removal and all the facets adjacent to that vertex are removed from the model and the resulting hole is triangulated (see [1] for more details). Since the triangulation requires a projection of the local surface onto a plane, these algorithms are generally limited to manifold surfaces. Vertex decimation methods preserve the mesh topology as well as a subset of original vertices.
- Energy function optimization: the two methods most usually used within this framework are those proposed by Hoppe [12, 13]. The principle of these methods consists in assigning an energy function to three characteristics of the mesh: the number of nodes, the approximation error and the regularity of the mesh (determined from the length of the edges). These methods seek to minimize the energy in order to solve the mesh optimization problem. This approach can produce much better simplification results, but it increases a huge amount of computational cost.
- Agglomeration of the vertices: the methods based on this principle are distinguished by how to choose the vertices to be agglomerated. There are essentially two ways to do it. The first uses a three-dimensional grid, possibly adaptive, to replace all of the vertices belonging to a cell of the grid with a single vertex. [3] The second approach uses a metric for determining the nodes to be merged [4] Fig. 1.

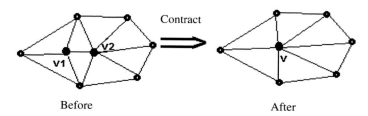

Before   After

Fig. 1 Edge contraction. The highlighted edge is contracted into a single point. The *shaded triangles* become degenerate and are removed during the contraction

In this work we present a comparative study of four methods [1, 3, 4, 12] representative of the approaches mentioned above. These methods are most commonly used in the literature and a source of inspiration for many works published in recent years.

4 Evaluation of the Simplified Meshes

In order to give a theoretical assessment but also objective of the methods examined, we adopted an empirical approach based on four implementations in which the approximation error introduced in the simplification process is calculated. For this purpose, we have based on a metric of the approximation error used by Garland [4], capable of measuring the average squared distance between the approximation and the original model. The error approximation Ei = E (M, E) is defined as:

$$E_i = \frac{1}{|X_n| + |X_i|} \left(\sum_{\vartheta \in X_n} d^2(\vartheta, M_i) + d^2(\vartheta, M_n) \right) \tag{1}$$

where X_n and X_i are sets of points sampled on the models M_n and M_i respectively. The distance $d(\vartheta, M) = \min_{p \in M} ||\vartheta - p||$ is the minimum distance from v to the closest face of M.

Moreover, On the basis of the maximum error like metric associated to the faces/vertices [21], in order to have a more precise idea of these methods.

5 Empirical Evaluation

To validate the effectiveness of these methods, we choose a set of data representing a broad category (Fig. 2): the meshes obtained with standard CAD system—SOCKET, available at http://www1.cs.columbia.edu/~cs4162/models/.

The four methods are executed on a computer with the following features: Core 2 Duo processor, 1.60 GHz, 2 GB RAM.

Table 1 shows the numerical results obtained by the implementation of the four methods. The results relative to the evaluation of the approximation error are also summarized in the Figs. 3 and 4 (see also Fig. 5): the Fig. 3 shows the maximal error of the various methods and sizes. Similarly, Fig. 4: on the top graph, we plot the error curve (EAVG) for different approaches and sizes. For good visualization, we plot, in the lower graph, the error curve (EAVG) for three approaches and

Fig. 2 A CAD triangulated mesh: SOCKET original—Meshing: 836 vertices

Table 1 Comparison of various simplification algorithms: SOCKET mesh, (errors are measured as percentages of the datasets bounding box diagonal; times are in milliseconds)

SOCKET (826 vertices, 1696 triangles, bounding box: 4.8 × 5.7 × 2.8)

N_{Vert}	N_{Triang}	E_{Max}	E_{Avg}	Time
Mesh decimation [1]				
634 (75 %)	1302	0.00092	1.081e-5	218.05
412 (50 %)	852	0.00532	5.091e-5	239.89
200 (25 %)	424	0.11553	0.00367	250.23
106 (12 %)	218	0.59810	0.01339	254.66
37 (4 %)	71	1.67380	0.04800	262.08
Vertex clustering [3]				
634 (75 %)	1302	2.18201	0.03201	43.05
413 (50 %)	852	3.19371	0.16009	41.89
202 (25 %)	424	5.00101	0.54008	45.23
107 (12 %)	218	7.98261	1.80041	44.66
36 (4 %)	N/A	N/A	N/A	N/A
Mesh optimization [12]				
635 (75 %)	1302	0.22921	0.00237	69,760
415 (50 %)	852	0.30975	0.00270	76,827
203 (25 %)	424	0.41219	0.00295	79,876
109 (12 %)	218	0.40520	0.00348	79,754
37 (4 %)	71	0.45591	0.00597	83,645
Quadric error metrics [4]				
635 (75 %)	1302	0.00009	3.001e-5	298.05
413 (50 %)	852	0.00067	6.073e-5	327.79
200 (25 %)	424	0.00442	0.00022	341.13
106 (12 %)	218	0.03746	0.00173	344.50
37 (4 %)	71	0.09253	0.00571	351.18

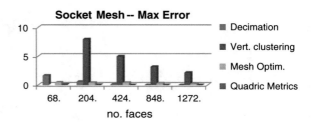

Fig. 3 The graphs show the performance of four approaches (E.MAX)

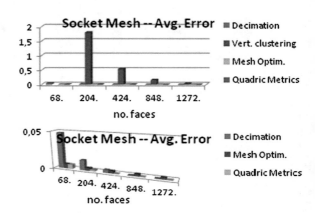

Fig. 4 The graphs show the performance of four approaches (EAVG)

different sizes. The size of the simplified mesh is represented on the X axis and the error on the Y axis.

In particular, the method of Vertex clustering [3] does not reach a high level of simplification. Moreover, she recorded in general the worst result in terms of error, but the maximum error curve showed a relative improvement; this depends on the strategy adopted in the removal of the vertices. By cons, it is interesting to note that this method produces the best results when speed is needed.

As expected, the good results in terms of average error are given by the Mesh Optimization method and Quadric Error Metrics. However, the method of Mesh decimation is less reliable in terms of accuracy. On the other hand, methods [3, 4, 12] which only support globally bounded approximation errors and which use a global optimization process at mesh modifications produce better results when we consider the maximal error.

It is interesting to note that the method of QEM and Mesh decimation are produced the better results in terms of speed. In contrast, the method of Mesh Optimization is the slowest.

Mesh decimation   ROSSIGNAC 93

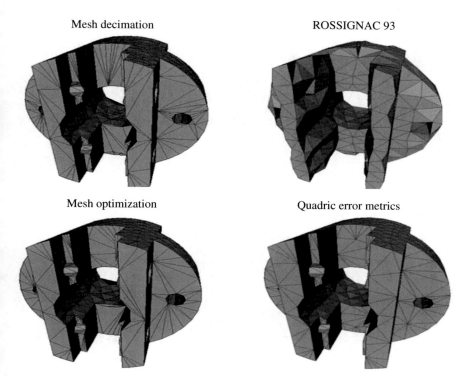

Mesh optimization   Quadric error metrics

Fig. 5 Simplified Socket meshes ($\sim$852 faces)

6 Conclusion

This work presented a brief overview of mesh simplification methods proposed in recent years. We presented the different characteristics of the four basic methods that are based on the simplification strategy. We also discussed and classified these methods according to their ability to preserve topology and manage the errors. In addition, the results of an empirical comparison have been presented. Four implementations have been performed on a set of data. We compared the computational time and accuracy of simplified meshes.

From the point of view of accuracy, the results showed that approaches QEM and Mesh Optimization gives the best results in terms of error.

Finally, all tested solutions have a common weakness: they are defined to work on a single mesh. This is not the case of CAD models or virtual reality sessions where we may need to simplify the complex objects or scenes with a non-topological clean composition between components. New solutions are required for these applications to provide increased generality and robustness.

References

1. Schroeder, W.J., Zarge, J.A., Lorensen, W.E.: Decimation of triangle meshes. ACM siggraph computer graphics, pp. 65–70. ACM, New York (1992)
2. Taubin, G., Guéziec, A., Horn, W., Lazarus, F.: Progressive forest split compression. In: Proceedings of the 25th annual conference on Computer Graphics and Interactive Techniques. ACM, New York, pp. 123–132 (1998)
3. Rossignac, J., Borrel, P.: Multi-resolution 3D approximations for rendering complex scenes, pp. 455–465. Springer, Heidelberg (1993)
4. Garland, M., Heckbert, P.S.: Surface simplification using quadric error metrics. In: Proceedings of the 24th annual Conference on Computer Graphics and Interactive Techniques. ACM Press/Addison-Wesley Publishing Co. pp. 209–216 (1997)
5. Kanaya, T., Teshima, Y., Kobori, K.I., Nishio, K.: A topology-preserving polygonal simplification using vertex clustering. In: Proceedings of the 3rd International Conference on Computer Graphics and Interactive Techniques. ACM Australasia and South East Asia, pp. 117–120 (2005)
6. Boubekeur, T., Alexa, M.: Mesh simplification by stochastic sampling and topological clustering. Comput. Graph. 33(3), 241–249 (2009)
7. Li, Y., Zhu, Q.: A new mesh simplification algorithm based on quadric error metrics. In: International Conference on Advanced Computer Theory and Engineering, ICACTE'08, IEEE, pp. 528–532 (2008)
8. Li, G., Wang, W., Ding, G., Zou, Y., Wang, K.: The edge collapse algorithm based on the batched iteration in mesh simplification. In: IEEE/ACIS 11th International Conference on Computer and Information Science (ICIS), 20, IEEE, pp. 356–360 (2000)
9. Dehaemer, M.J., Zyda, M.J.: Simplification of objects rendered by polygonal approximations. Comput. Graph. 15(2), 175–184 (1991)
10. Hinker, P., Hansen, C.: Geometric optimization. In: Proceedings of the 4th Conference on Visualization'93. IEEE Computer Society, pp. 189–195 (1993)
11. Kalvin, A.D, Haddad, B., Noz, M.E.: Constructing topologically connected surfaces for the comprehensive analysis of 3-D medical structures. In: Med. Imaging V. Image Process. Int. Soc. Opt. Photonics, 247–258 (1991)
12. Hoppe, H., Derose, T., Duchamp, T., McDonald, J., Stuetzle, W.: Mesh optimization. In: Proceedings of the 20th Annual Conference on Computer Graphics and Interactive Techniques. ACM, New York, pp. 19–26 (1993)
13. Hoppe, H.: Progressive meshes. In: Proceedings of the 23rd Annual Conference on Computer Graphics and Interactive Techniques. ACM, New York, pp. 99–108 (1996)
14. Eck, M., Derose, T., Duchamp, T., et al.: Multiresolution analysis of arbitrary meshes. In: Proceedings of the 22nd Annual Conference on Computer Graphics and Interactive Techniques. ACM, New York, pp. 173–182 (1995)
15. Wang, W., Zhang, Y.: Wavelets-based NURBS simplification and fairing. Comput. Methods Appl. Mech. Eng. 199(5), 290–300 (2010)
16. Turk, G.: Re-tiling polygonal surfaces. ACM SIGGRAPH Comput. Graph. 26(2), 55–64 (1992)
17. Cohen, J., Varshney, A., Manocha, D., Turk, G., Weber, H., Agarwal, P., Wright, W.: Simplification envelopes. In: Proceedings of the 23rd Annual Conference on Computer Graphics and Interactive Techniques. ACM, New York, pp. 119–128 (1996)
18. Ciampalini, A., Cignoni, P., Montani, C., Scopigno, R.: Multiresolution decimation based on global error. Visual Comput. 13(5), 228–246 (1997)
19. Mocanu, B., Tapu, R., Petrescu, T., Tapu, E.: An experimental evaluation of 3D mesh decimation techniques. In: 10th International Symposium on Signals, Circuits and Systems (ISSCS), IEEE, pp. 1–4 (2011)

20. Cohen, A., Dyn, N., Hecht, F., Mirebeau, J.M.: Adaptive multiresolution analysis based on anisotropic triangulations. Math. Comput. **81**(278), 789–810 (2000)
21. Ciampalini, A., Cignoni, P., Montani, C., Scopigno, R.: Multiresolution decimation based on global error. Visual Comput. **13**(5), 228–246 (1997)

Human Tracking Based on Appearance Model

Khadija Laaroussi, Abderrahim Saaidi, Mohamed Masrar and Khalid Satori

Abstract The Mean Shift algorithm for tracking the location of an object has recently gained considerable interest because of its speediness and efficiency. However, the appearance description using only color features cannot provide enough information when the target and its background have similar colors. In response to this problem, an improved human tracking system based on Mean Shift algorithm is incorporated in this paper. The proposed method combines color-texture features and background information to find the most distinguished features between the target and background for target representation. The experimental results show that the proposed method presents a good compromise between computational cost and accuracy; its performance is compared with recent state-of-the-art algorithm on Benchmark dataset and it achieved excellent results.

Keywords Human tracking · Mean shift algorithm · Local binary pattern · Color histogram

K. Laaroussi (✉) · A. Saaidi · M. Masrar · K. Satori
LIIAN, Department of Mathematics and Computer Science, Faculty of Science Dhar El Mahraz, Sidi Mohamed Ben Abdellah University, P.O 1796, Atlas, Fez, Morocco
e-mail: khadija.laaroussi3@usmba.ac.ma

M. Masrar
e-mail: masrar.m@hotmail.com

K. Satori
e-mail: khalidsatorim3i@yahoo.fr

A. Saaidi
LSI, Laboratory of Engineering Sciences, Polydisciplinary Faculty of Taza, Sidi Mohamed Ben Abdellah University, P.O 1223, Taza, Morocco
e-mail: abderrahim.saaidi@usmba.ac.ma

© Springer International Publishing Switzerland 2016
A. El Oualkadi et al. (eds.), *Proceedings of the Mediterranean Conference on Information & Communication Technologies 2015*, Lecture Notes in Electrical Engineering 380, DOI 10.1007/978-3-319-30301-7_31

297

1   Introduction

Object tracking in image sequences is a critical task in computer vision applications such as human-computer interaction, robotics, traffic flow monitoring and video surveillance. Many tracking algorithms [1–5] have been proposed to overcome the difficulties arising from noise, illumination changes, occlusions, camera motion, changes of viewpoint, background clutters and so on. Tracking is to estimate the locations of the target in each image of a video sequence. The localization process is based on the object recognition from a set of visual features such as color, shape, texture, speed, etc. Among various object tracking algorithms, the Mean Shift approach [5] which has recently gained much attention due to its characteristics of simple, robust, and real-time performance. Mean Shift method, iteratively seeks to maximize a local measure of similarity between the target model and target candidate [5]. Currently, a widely used form of target representation is the color histogram [5] which could be viewed as the discrete probability density function (PDF) of the target. However, using only color features in Mean Shift algorithm has some limitations. First, the spatial information of the target is lost. Second, when the target and background have similar colors, the appearance description using only color features cannot provide enough information to distinguish them. To improve the traditional Mean Shift algorithm, several attempts have been made by many researchers [4]. The main improvements include scale adaptation, background information, kernel selection, on-line model updating, feature selection and mode optimization [4].

In this paper, a new approach will be discussed; it is an approach of human tracking based on improved Mean Shift algorithm. The proposed method integrates color-texture features and background information to find more rich features for target representation. The mains contributions of this paper are to use the Local Binary Pattern (LBP) operator [6] in order to represent texture features which reflect the spatial information of the target. Also, to reduce the interference of background in target localization a background weighted histogram (BWH) is adopted to decrease the weights of both prominent background color and texture features similar to the target. This new approach has several advantages: the joint color-texture features eliminates smooth background and reduces noise in the tracking process. The experimental results show that the proposed approach presents a good compromise between computational cost and accuracy; it's also successfully coped with background clutter and interference, camera motion, partial occlusion, deformation, fast motion and illumination changes.

The paper is organized as follows. Section 2 briefly introduces the traditional Mean Shift algorithm. Section 3 analyzes the LBP operator and background information, then we present the proposed method. Experimental results are presented and discussed in Sect. 4. Section 5 concludes the paper.

2 Mean Shift Algorithm

2.1 Target Representation

The appearance of the target is represented by an m-bin RGB color histogram computed from a rectangle region. Let $\left\{z_i^*\right\}_{i=1\cdots n}$ be the normalized pixel locations in the target model centred at location 0. The probability of the feature $v = 1\cdots m$ in the target model is then computed as:

$$\hat{e}_v = S\sum_{i=1}^{n} k\left(\left\|z_i^*\right\|^2\right)\theta\left[c\left(z_i^*\right) - v\right] \tag{1}$$

where θ is the Kronecker delta function [5] and S is a normalized constant. The function $c : R^2 \rightarrow \{1\cdots m\}$ associates to the pixel at location z_i^* the index $c\left(z_i^*\right)$ of its bin in the quantized feature space, $k(z)$ is an isotropic kernel profile [5].

Similarly, let $\{z_i\}_{i=1\cdots n_h}$ be the normalized pixels locations of the target candidate, centered at x in the current frame and S_h is a normalized constant. Using the same kernel profile $k(z)$, but with bandwidth h, the probability of the feature $v = 1\cdots m$ in the target candidate model is given by:

$$\hat{f}_v(x) = S_h\sum_{i=1}^{n_h} k\left(\left\|\frac{x - z_i}{h}\right\|^2\right)\theta\left[c(z_i) - v\right] \tag{2}$$

2.2 Target Localization

To find the location corresponding to the target in the current frame, the distance between $\hat{f}(x)$ and $\hat{e}$ should be minimized as a function of x:

$$dist\left[\hat{f}(x), \hat{e}\right] = \sqrt{1 - \rho\left[\hat{f}(x), \hat{e}\right]} \tag{3}$$

where $\rho\left[\hat{f}(x), \hat{e}\right] = \sum_{v=1}^{m}\sqrt{\hat{f}_v(x)\hat{e}_v}$ is the Bhattacharyya coefficient. In the iterative process, the estimated target moves from the current location $\hat{x}_0$ to the new location $\hat{x}_1$ according to the relation:

$$\hat{x}_1 = \frac{\sum_{i=1}^{n_h} z_i\varphi_i g\left(\left\|\frac{\hat{x}_0 - z_i}{h}\right\|^2\right)}{\sum_{i=1}^{n_h} \varphi_i g\left(\left\|\frac{\hat{x}_0 - z_i}{h}\right\|^2\right)} \tag{4}$$

where

$$\varphi_i = \sum_{v=1}^{m} \sqrt{\frac{\hat{e}_v}{\hat{f}_v(\hat{x}_0)}} \theta[c(z_i) - v] \tag{5}$$

In the final step, the tracking window center will converge in the real target location with limited iterations in a current frame using (4). In this work, $k(z)$ is the Epanechnikov profile [7] and $g(z) = -k'(z)$.

3   Proposed Tracking Algorithm

3.1   Target Representation Based on Color-Texture Features and BWH

The Local Binary Pattern (LBP) [6] operator is a gray-scale invariant texture primitive statistic; it is a powerful means of texture description. The original version of the LBP operator works in a 3×3 pixel block of an image. The pixels in this block are thresholded by its center pixel value, multiplied by powers of two and then summed to obtain a label for the center pixel. The LBP operator is defined as follows:

$$LBP_{P,R}(x_i, y_i) = \sum_{n=0}^{P-1} b(g_n - g_i)2^n, \quad b(x) = \begin{cases} 1 & x \geq 0 \\ 0 & x < 0 \end{cases} \tag{6}$$

where P indicates target area of pixels, R represents the target pixel radius, g_i is the gray value of the center pixel (x_i, y_i) and g_n is the gray values of the adjacent pixels around center point g_i in the region. The texture model derived by (6) has only gray-scale invariance. The grayscale and rotation invariant LBP texture model [6] is obtained by:

$$LBP_{P,R}^{riu2} = \begin{cases} \sum_{n=0}^{P-1} b(g_n - g_i) & \text{if } U(LBP_{P,R}) \leq 2 \\ P + 1 & \text{otherwise} \end{cases} \tag{7}$$

where

$$U(LBP_{P,R}) = |b(g_{P-1} - g_i) - b(g_0 - g_i)| + \sum_{n=1}^{P-1} |b(g_n - g_i) - b(g_{n-1} - g_i)|$$

Superscript ^{riu2} reflects the use of rotation invariant "*uniform*" patterns that have U value of at most 2. By definition, exactly $P + 1$ "*uniform*" binary patterns can occur in a circularly symmetric neighbor set of P pixels. In this paper, we adopt the modified thresholding strategy in the LBP operator presented in [8] and employ $LBP_{8,1}^{riu2}$ to describe the target texture features. In target representation, the micro-textons such as edges, line ends and corners, by name of "*major uniform patterns*", represent the main features of target, while spots and flat areas, called "*minor uniform patterns*", are minor textures. Thus, we extract the main uniform patterns of the target by the following equation:

$$
LBP_{8,1}^{riu2} =
\begin{cases}
\sum_{n=0}^{7} b(g_n - g_i + a) \; U\left(LBP_{8,1}\right) \leq 2 & \text{and} \\[2mm]
\sum_{n=0}^{7} b(g_n - g_i + a) \; \in \{2, 3, 4, 5, 6\} \\[2mm]
0 & \text{otherwise}
\end{cases}
\tag{8}
$$

In [5], the background weighted histogram (BWH) is represented as $\{\hat{b}_v\}_{v=1 \cdots m}$ (with $\sum_{v=1}^{m} \hat{b}_v = 1$) and $\hat{b}^*$ be its smallest nonzero entry. The weights

$$
\left\{ \alpha_v = \min\left(\frac{\hat{b}^*}{\hat{b}_v}, 1\right) \right\}_{v=1 \cdots m}
\tag{9}
$$

These weights are only used to define a transformation between the representations of target model and target candidate. The transformation diminishes the importance of those features which have low α_v, i.e., are prominent in the background. The new target model and target candidate are then defined by:

$$
\begin{cases}
\hat{e}'_v = M' \alpha_v \sum_{i=1}^{n} k\left(\left\|z_i^*\right\|^2\right) \theta\left[c\left(z_i^*\right) - v\right] \\[2mm]
\hat{f}'_v(x) = M'_h \alpha_v \sum_{i=1}^{n_h} k\left(\left\|\frac{x-z_i}{h}\right\|^2\right) \theta[c(z_i) - v]
\end{cases}
\tag{10}
$$

3.2   Proposed Method

We use the RGB color space and the LBP patterns extracted by (8) to jointly represent the target and embed it into the Mean Shift tracking framework. The entire procedure of the proposed modified Mean Shift tracking algorithm is summarized as follows:

Algorithm: Proposed method with joint color-texture histogram and BWH

INPUT: Image sequence $I_0, I_1, \cdots, I_n$ and the initial object location $\hat{x}_0$.
OUTPUT: Target locations $c_0, c_1, \cdots, c_n$ at tracking over times, in frames $I_0, I_1, \cdots, I_n$

Begin
  1. Compute the color-texture distribution of the target model $\hat{e}$ in the initial position $\hat{x}_0$ in I_0 according to (1) and (8), in which $v = 16 \times 16 \times 16 \times 5$, then, compute the background-weighted histogram $\{\hat{b}_v\}_{v=1\cdots m}$ and $\{\alpha_v\}_{v=1\cdots m}$ by (9) and the transformed target model $\hat{e}'$ by (10).
  2. Initialize the location of the target in the current frame with $\hat{x}_0$. Compute the color-texture distribution $\hat{f}_v'(\hat{x}_0)$ according to (10) and evaluate $\rho[\hat{f}'(\hat{x}_0), \hat{e}'] = \sum_{v=1}^m \sqrt{\hat{f}_v'(\hat{x}_0)\,\hat{e}'}$.
  3. Calculate the new weights formula $\varphi_i' = \sum_{v=1}^m \sqrt{\frac{\hat{e}_v'}{\hat{f}_v'(x)}}\theta[c(z_i) - v]$.
  4. Find the next location $\hat{x}_1$ of the target candidate according to (4).
  5. Compute $\hat{f}_v'(\hat{x}_1)_{v=1\cdots m}$ and evaluate $\rho[\hat{f}'(\hat{x}_1), \hat{e}'] = \sum_{v=1}^m \sqrt{\hat{f}_v'(\hat{x}_1)\,\hat{e}'}$.
  6. While $\rho[\hat{f}'(\hat{x}_1), \hat{e}'] < \rho[\hat{f}'(\hat{x}_0), \hat{e}']$
       Do $\hat{x}_1 \leftarrow \frac{1}{2}(\hat{x}_0 + \hat{x}_1)$
       Evaluate $\rho[\hat{f}'(\hat{x}_1), \hat{e}']$
  7. If $\|\hat{x}_1 - \hat{x}_0\| < \varepsilon$ (default value $\varepsilon = 0.1$) Stop.
     Otherwise Set $\hat{x}_0 \leftarrow \hat{x}_1$ and go to (3).
end

---

4 Experimental Results

This section describes a series of experiments that have been carried out using the proposed Mean Shift tracking algorithm. Two Benchmark sequences were used to access the tracking performances captured by moving cameras [9]. All the algorithms are run on a PC (Intel® Core ™ 2 Duo CPU T5800 with 2.00 GHz CPU and 3 GB RAM). We selected RGB color space as the feature space and it was quantized into $16 \times 16 \times 16$ bins. The tracking results are compared with the Mean Shift algorithm with background weighted histogram (MS-BWH) [5]. To further evaluate the performance of the proposed method and to compare with the existing methods, the ground truth of the target location for all these sequences is available [9].

The first experiment is the "David" sequence with 252 frames of spatial resolution 640×480 where the tracked target is the walking man. There are various factors that make the tracking challenging: occlusion, deformation, out-of-plane rotation and background clutters. Refer to (Fig. 1) which shows the qualitative tracking results of our proposed method and the MS-BWH approach, because of the

(a)

(b)

Fig. 1 Results for "David" sequence for frames 21, 88, 142 and 169 are displayed. **a** Tracking results using the MS-BWH method [5]. **b** Tracking results using the proposed method

changing background, partial occlusion during tracking process, the MS-BWH method loses the target very quickly (Fig. 1a), while the proposed approach track exactly the moving man in the whole sequence (Fig. 1b). The quantitative results in Table 1 show that the proposed method has more accurate localization accuracy than the MS-BWH approach, because the former truly exploits the features in the target representation.

The second experiment is the "Human" sequence with 305 frames of spatial resolution 320×240. This sequence consists of environments of variant light, scale change, fast motion, non-rigid object deformation and motion blur. From the tracking results shown in (Fig. 2a, b) we can see that the proposed algorithm can track the target perfectly over the whole sequence when the target has a fast and complex motion, however, the MS-BWH approach loses the target due to the fast camera motion. Refer to the statistics results presented in Table 1 the proposed approach achieves higher target localization accuracy than the MS-BWH model.

Table 1 Quantitative performance evaluation: average error, average number of iterations and average processing time for each method

Video sequence	Target representation	Average error	Mean shift iteration		Average processing time (s/frame)
			Total iterations	Average iterations	
David	MS-BWH	56.05	1501	5.95	2.06
	Proposed method	**21.26**	1636	6.49	2.30
Human	MS-BWH	70.6	2175	7.13	1.73
	Proposed method	**11.85**	2230	7.31	1.43

(a)

(b)

Fig. 2 Results for "Human" sequence for frames 26, 149, 174 and 225 are displayed. **a** Tracking results using the MS-BWH method [5]. **b** Tracking results using the proposed method

Fig. 3 The average error for the two methods on "David" and "Human" sequences

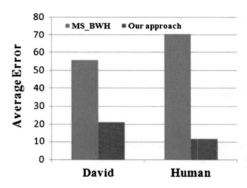

From the obtained results of this sequence, we find that the proposed improvement into Mean Shift is invariant in several real world complex situations such as target rotating, partial occlusion, fast motion and background clutter, and this due to the use of color description enhanced by discriminate features. The graphical tracking result of the average error for the two methods on the "David" and "Human" sequences is plotted in Fig. 3.

5 Conclusion

In this paper, we have proposed an enhanced human tracking algorithm based on Mean Shift procedure. The novelty of this work is to combine color-texture features and background information to reduce the computational cost and improve the robustness of target representation. The proposed method makes the classical Mean Shift algorithm more consistent and robust against very complex conditions, such as similar target, background appearance, partial occlusion and camera motion, where the obtained results are very spectacular. Experimental results indicate that the proposed method performs much better than other state-of-the-art algorithm with smallest average error localization and fewer iteration numbers, so it is sufficient to track human in real time, which is a crucial requirement in visual surveillance.

References

1. Li, X., Hu, W., Shen, C., Zhang, Z., Dick, A.R., Van den Hengel, A.: A survey of appearance models in visual object tracking. ACM Trans. Intel. Syst. Technol. **4**(4), Article 58 (2013)
2. Laaroussi, K., Saaidi, A., Masrar, M., Satori, K.: Video-surveillance system for tracking moving people using color interest points. World Appl Sci J **32**(2), 289–301 (2014)
3. Laaroussi, K., Saaidi, A., Satori, K.: People tracking using color control points and skin color. J. Emerg. Technol Web Intell **6**(1), 94–100 (2014)
4. He, L., Xu, Y., Chen, Y., Wen, J.: Recent advance on mean shift tracking: A survey. Int. J. Image Graph. **13**(3) (2013)
5. Comaniciu, D., Ramesh, V., Meer, P.: Kernel-based object tracking. IEEE Trans. Pattern Anal. Mach. Intell. **25**(5), 564–577 (2003)
6. Ojala, T., Pietikäinen, M., Maenpaa, T.: Multiresolution gray-scale and rotation invariant texture classification with local binary patterns. IEEE Trans. Patt. Anal. Mach. Intell. **24**(7), 971–987 (2002)
7. Qingchang, G., Xiaojuan, C., Hongxia, C.: Mean-shift of variable window based on the Epanechnikov Kernel. In: Proceedings of the International Conference on Mechatronics and Automation, pp. 2314–2319 (2007)
8. Heikkilä, M., Pietikäinen, M.: A texture-based method for modeling the background and detecting moving objects. IEEE Trans. Patt. Anal. Mach. Intell. **28**(4), 657–662 (2006)
9. http://www.visual-tracking.net/

Towards a Conversational Question Answering System

Fatine Jebbor and Laila Benhlima

Abstract More and more people resort to the web for searching and sharing information about their health using search engines and forums. These tools have many drawbacks such as: the lack of concise answers, difficulties of presentation of the complex needs of information...In this paper, we look at a new way of delivering information to the users via coherent conversations. After completing a comparative study on recent question answering (QA) systems, the major motivation is to find a solution to address their weaknesses. The solution that we propose is a QA system which takes into account the context of research and allows interaction with the user to understand its purpose, by using different types of data sources.

Keywords Information extraction · Question answering system · Conversational system

1 Introduction

Information extraction is a complex process allowing the identification of previously unknown and potentially useful original information from big masses of data. Among its domains of application, we find question answering (QA) systems. A QA system takes as input a natural language question and a collection of sources, and produces the appropriate natural language answer. To build the answer, it gathers relevant data and relations from the sources and then fuses or summarizes them into a single answer as appropriate [1].

Several QA systems exist such as: Quantum [2], Piquant [3] and Queri [4]. The most popular is IBM Watson [5], but it is a commercial system. Each of these

F. Jebbor (✉) · L. Benhlima
Analysis, Modeling and Integration of Processes and Systems Team,
Computer Science Department, Mohammadia Engineering School,
Mohammed V University, Rabat, Morocco
e-mail: fatine.jebbor@gmail.com

L. Benhlima
e-mail: benhlima@emi.ac.ma

© Springer International Publishing Switzerland 2016
A. El Oualkadi et al. (eds.), *Proceedings of the Mediterranean Conference on Information & Communication Technologies 2015*, Lecture Notes in Electrical Engineering 380, DOI 10.1007/978-3-319-30301-7_32

systems has limitations. In fact, we first find the restriction of data sources: use of unstructured data sources only (free text), whereas structured sources are their natural complement insofar as they cover a range more restricted of questions but are more accurate on this range. This restriction is due to the problem of translating the question in natural language to a structured query. The second limitation is the absence of interaction with the user: building answers might need a conversation with the user in order to obtain clarification or refine the question and the answer.

To illustrate this, consider the following question: *"How to treat skin rash?"*. After having analyzed this question and detected keywords, the system realizes that there are several cures for skin rash depending on their location on the body. Knowing that the user did not mention this information in his initial question, so how the system will be able to choose the efficient cure? To do this, the system needs to know the concerned part of the body. Hence, it is necessary to interact with the user to ask him for clarification in order to provide a relevant answer. To meet this need, we propose as solution a conversational QA system using different types of data sources (structured, semi-structured, and unstructured). Our system is generic and is dedicated to the open domain.

This article is organized as follows, we start by giving a brief overview of some existing systems. Then we present our solution by describing the general architecture of our system and each of its components. Next, we detail the functioning of our system via a medical scenario. Finally, we end with a conclusion and perspectives.

2   Related Work

We'll present in this article three QA systems that we have studied during the review of the literature, namely: Quantum, System Based on Linked Data and IBM Watson.

2.1   Quantum

Quantum is a QA system developed in collaboration between the laboratory RALI (Recherche Appliquée en Linguistique Informatique) of the Montreal university, the laboratory ClaC (Computational Linguistics at Concordia) of Concordia university and the BCE company (Bell Canada Enterprises). The purpose is to find accurate answers to the questions by searching in a large collection of unstructured documents.

This system is composed of three modules: Question analysis, Research of relevant passages and Answer extraction [2]. It is characterized by a cross-linguistic architecture that consists in translating only the question and the answer to the desired language, and keeping initial system and documentary base [6].

2.2 System Based on Linked Data (SBLD)

The authors in [7] presented a method for mapping the questions expressed in natural language into structured queries based on ontologies. This to extract direct answers from open knowledge bases (linked data). SBLD translates the questions into RDF triple patterns by using the dependency tree of text. In addition this method uses the relational patterns extracted from the web [6].

The functioning of this system goes through three stages: the first one is extraction of triple patterns by using Stanford CoreNLP library [8] which provides natural language analysis tools. The second step is extraction of all DBpedia elements corresponding to the triples of the previous stage. The final step is answer extraction that consists of building all the possible queries to extract the answer from objects and the data properties obtained previously.

2.3 IBM Watson

Watson is a QA system designed by IBM and dedicated to the open domain. In 2011, it participated in the Jeopardy competition and beats the best two champions in real time. It is based on DeepQA (Deep Question Answering) which is an extensible architecture that combines and deploys a wide range of algorithmic techniques in the open domain of question answering [5]. The process goes through five steps: Content acquisition, Question analysis, Hypothesis generation, Scoring, and Merging and final ranking [6]. For the extraction, Watson uses PRISMATIC [5] which is the implementation and the result of the knowledge extraction approach adopted.

Several components of DeepQA use structured data sources (databases, knowledge bases...) to generate the candidate answers or to find evidence. Indeed, the extraction of evidence from structured sources has a great importance when it comes to geospatial and temporal constraints.

2.4 Criticism and Motivation

Concerning the structure dimension, the first two systems we have introduced are basically based on information retrieval (IR) and use just unstructured data sources, while Watson takes advantage of structured data to provide evidence to support and estimate the confidence of the candidate answers [9]. This dimension is ignored by the majority of the QA systems although the use of structured sources can help to reduce the resources to the most appropriate to the information need.

And concerning the interaction with the user, none of these systems support it. They follow the paradigm where the interaction is limited to a search in a single direction, while research in the real world is naturally interactive. Indeed, the user may play a role of paramount importance in improving the research results. He is able to reformulate the query based on his knowledge domain and interests. Moreover, he may interrogate the system in an iterative manner until he is sure that the system has understood his question.

Hence the need to develop systems that combine the different types of sources to provide effective retrieval of appropriate answers to a wide range of questions, and support the interactive QA (dialog) by allowing rich interactions with the human user in order to clarify and explore the subject of the question.

3   Our Solution

We present in this section our solution, a conversational QA system that meets the following requirements.

3.1   Overview

Question Side. The system focuses on the natural language questions under the open domain. It allows the identification of more precise and relevant answers.

Dialog Side. The conversation with the user is done in natural language. The goal is to refine the understanding of the issue and improve the quality of the answer. In other words, dialog comes to fill inevitable gaps in the knowledge of the system.

Answer Side. The answer is extracted from the corpus. It has several levels of granularity depending on the question: for some questions, a named entity for example is sufficient. Other questions require passages or documents as a response. Answers must meet the requirements of the question and be specific as possible.

3.2   Architecture

Our system's architecture consists of three components as shown in Fig. 1: Question processing, Extraction and Conversational components.

We will detail in what follows each of the three components.

Question Processing Component. The objective is the analysis of the question and then its transformation into a query. This goes through the following stages.

Question Classification. This consists of identifying the question type (definition, description…) using classification techniques as the Bayesian classification.

Focus and LAT (Lexical Answer Type) Detection. The focus is the part of the question which, if we replace it with the response, we will have a stand-alone sentence. It contains important information about the answer.

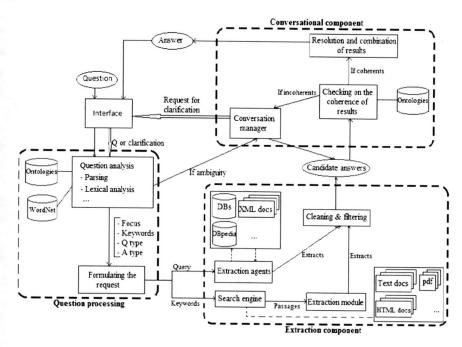

Fig. 1 System's architecture

The LAT is a word or noun phrase, contained in the question that specifies the type of the answer without understanding its semantics.

Detection of Relations among the Question Entities. This stage is about looking for the relations that link the different words or entities constituting the question. Consider the following example: *"What type of vitamin is used to treat Influenza?"* The relation contained in this question is represented as: (Influenza, ?x, vitamin)

During the question processing, the system uses several techniques such as:

Tokenizing. This is about breaking down the text into words, keywords, phrases, symbols and other elements called tokens.

Morphological Analysis. This technique uses a list of regular expressions as well as a set of rules concerning these expressions.

Named Entities Tagging. The system uses pre-defined lists of common names (indicators of classes to which belong the words), organizations, titles, dates… in order to identify and tag the words in the text.

Extraction Component. Extraction from structured sources consists on executing the query generated during the question analysis phase on available sources (databases, knowledge bases…) by using a set of agents.

Indeed, this component contains a set of intelligent agents which run in parallel to extract answers from data sources. Each agent gives out several possible answers that are processed and filtered to produce a list of candidate answers.

For the extraction from unstructured data sources, it goes through two stages:

Passage Retrieval. Our system uses a search engine that receives the question keywords and returns the paragraphs or passages that contain all these keywords. The system evaluates the quality of returned paragraphs and assigns them scores. *Answer Extraction.* The system extracts candidate answers from the paragraphs of the previous step. This involves the use of a parser to identify the named entities. The parser assigns semantic tags and returns candidate answers contained in paragraphs. Finally, our system evaluates the candidate answers by calculating their scores.

Conversational Component. At this stage, the sub-component 'Checking on the coherence of results' evaluates the coherence of the results (answers candidates) given by the Extraction component by using ontologies.

If these results are coherent, they pass to the component 'Resolution and combination of results' to combine and reconcile the different proposed candidate answers and collect the elements that will constitute the final answer.

If the system detects an inconsistency in the results, it raises a conversation with the user in order to solve the incoherence problem by asking him for clarification.

The proposed conversational QA system allows the end-users to vaguely express and gradually refine their information needs using only natural language questions or statements as input.

Once the user responds to the system, its answer that can be in the form of a clarification or reformulation of the initial question is treated from the start as if it was a new question. If the extraction results for this second question are coherent, they will be resolved and combined with the selected results of the original question. Otherwise, the system raises again a conversation, and so on until the user is satisfied.

4   Medical Scenario

Consider the following medical scenario that describes the straight-through processing of the question: *"What is the gene involved in Aniridia?"*

This involves the following steps.

4.1   Question Analysis

First of all, the system analyzes the question and extracts the following elements:

- Question Type: According to the Bloom taxonomy [10], the use of the word "What" means that the user seeks knowledge. So the type of this question is Knowledge.

- Answer Type: The user looks for the name of a gene, so according to WordNet, the type of the expected answer is Noun.
- Focus: In this question, the focus is the word "Gene".
- Keywords: The keywords of this question are: gene, involved, Aniridia.

4.2 Query Formulation

After extracting the previous elements, our system built the corresponding query. There are many languages that query RDF data, we opted for SPARQL because it is a W3C standard that is implemented by a lot of programs, so this interoperability will be useful.

The SPARQL query corresponding to our question is illustrated by Fig. 2.

Our system passes the keywords extracted previously to the search engine, belonging to the Extraction Component, which will return all documents, web pages... that may contain the answer. Then these resources go to the extraction module.

4.3 Extraction Module

In this step, our system extracts all relevant passages contained in the resources given by the search engine. This is achieved by applying some passage extraction techniques based on a set of criteria such as the average distance between the keywords and the focus that figure in the passage. Once a set of passages is selected, our system assigns them scores in order to classify them. Later it extracts the possible answers that will go to the sub-component 'Cleaning & filtering'.

4.4 Extraction Agents

The Extraction component contains a set of extraction agents. Every single agent executes the SPARQL query (Fig. 2) on the corresponding data source. The results

Fig. 2 The SPARQL query corresponding to the considered question

```
SELECT  ?gene   ?name
WHERE {
        ?gene   rdf :type   onto :Gene
        ?gene   onto: Noun   ?name
        ?gene   onto: isResponsableOf   onto: Aniridia
        }
```

of this stage (which are gene names for our question) go to the sub-component 'Cleaning & filtering' to provide the candidate answers.

4.5 Conversational Component

The sub-component 'Checking on Coherence' receives the candidate answers, checks their consistency and triggers a conversation with the user if necessary. For our question, the system did not find any ambiguity, so the set of candidate answers goes to the sub-component 'Resolution and combination' to construct the final answer that is: PAX6 gene.

5 Conclusion and Perspectives

We presented in this paper a solution that combines the advantages of both types of systems: IR and QA. It ensure an effective extraction of suitable answers for a wide range of questions expressed in natural language, and it incorporates a rich user-system dialog component to better understand the question and refine answers.

The innovative aspects of our system that distinguish it from other QA systems are:

First it goes beyond the document retrieval approach: it returns specific answers and not documents. Then it takes advantage of the complementarity of different sources: parallel use of structured and unstructured data sources. Besides it allows to the user to interact in real time to provide more details or reformulate his question.

Our future work will focus on the Extraction and the Conversational components. In other words, our short-term objective is developing the extraction agent, as well as defining the characteristics and functionalities of the extraction module. Our mid-term objective is developing the conversational strategies that our system will use.

References

1. Allan, J.: Challenges in information retrieval and language modeling. Workshop held at the Center for Intelligent Information Retrieval, University of Massachusetts, Amherst (2002)
2. Plamondon, L.: Le système de question-réponse Quantum. Mémoire de maîtrise, université de Montréal, Canada (2002)
3. Chu-Carroll, J., Czuba, K., Prager, J., Blair-Goldensohn, S., Ittycheriah, A.: IBM's PIQUANT II in TREC 2004. IBM T.J. Watson Research Center and Columbia University (2004)

4. Merdaoui, B.; Frasson, C.: QUERI: Un système de question-réponse collaboratif et interactif. Département d'informatique et de recherche opérationnelle, université de Montréal, Canada (2004)
5. Ferrucci, D.A.: Introduction to this is Watson. IBM Journal of Research and Development. IBM Research Division, Thomas J. Watson Research Center, Yorktown Heights, NY, USA (2012)
6. Jebbor, F.; Benhlima L.: Overview of knowledge extraction techniques in five question-answering systems. In: 9th International Conference on Intelligent Systems: Theories and Applications (SITA-14), IEEE, pp. 1–8 (2014)
7. Sherzod, H., Hakan, T., Marlen, A., Erdogan, D.: Semantic question answering system over linked data using relational patterns. TOBB University of Economics and Technology, Ankara (2013)
8. Toutanova, K., Klein, D., Manning, C.D., Singer, Y.: Feature-rich part-of-speech tagging with a cyclic dependency network. In: Proceedings of the 2003 Conference of the North American Chapter of the Association for Computational Linguistics on Human Language Technology, vol. 1, pp. 173–180 (2003)
9. Kalyanpur, A., Boguraev, B.K., Patwardhan, S., Murdock, J.W., Lally, A., Welty, C., Prager, J.M., Coppola, B., Fokoue-Nkoutche, A., Zhang, L., Pan, Y., Qiu, Z.M.: Structured data and inference in DeepQA. IBM J. Res. Dev. **56**(3/4), (Paper 10) (2012)
10. Bloom's Taxonomy. Available online: http://www.cbv.ns.ca/sstudies/links/learn/1414.html. Accessed 17 Apr 2015

Part VI
Optimization and Modeling in Wireless Communication Systems

Clonal Selection Algorithm
for the Cell Formation Problem

Bouchra Karoum, Bouazza Elbenani
and Abdelhakim Ameur El Imrani

Abstract The cell formation problem attempts to group machines and part families in dedicated manufacturing cells such that the inter-cell movement of the products are minimized while the machine utilization are maximized. In this paper, a clonal selection algorithm is proposed for solving this problem. This algorithm introduces theories of clonal selection, hypermutation and receptor edit to construct an evolutionary searching mechanism which is used for exploration. A local search mechanism is integrated to exploit local optima. In order to demonstrate the effectiveness of the proposed algorithm, most widely used benchmark problems are solved and the obtained results are compared with different methods collected from the literature. The results demonstrate that the proposed algorithm is a very effective and performs well on all test problems.

Keywords Cell formation · Grouping efficacy · Local search · Clonal selection · Receptor editing

1 Introduction

Group technology is a disciplined approach based on grouping parts with similar design and manufacturing characteristics into part families in order to improve the productivity of the manufacturing system. Cellular manufacturing is one of the most

B. Karoum (✉) · B. Elbenani
Research Computer Science Laboratory (LRI), Faculty of Science,
Mohammed V University of Rabat, B.P. 1014 Rabat, Morocco
e-mail: bouchra.karoum@gmail.com

B. Elbenani
e-mail: elbenani@fsr.ac.ma

A.A. El Imrani
Conception and Systems Laboratory (LCS), Faculty of Science,
Mohammed V University of Rabat, B.P. 1014 Rabat, Morocco
e-mail: elimrani@fsr.ac.ma

© Springer International Publishing Switzerland 2016
A. El Oualkadi et al. (eds.), *Proceedings of the Mediterranean Conference on Information & Communication Technologies 2015*, Lecture Notes in Electrical Engineering 380, DOI 10.1007/978-3-319-30301-7_33

319

important applications of the group technology that attempts to convert a production system into several mutually separable production cells then assigns parts and machines to these cells. Its aim is to reduce setup and throughput times, minimize the material handling costs, increment the flexibility, etc. The cell formation problem (CFP), which is a non-polynomial hard optimization problem [1], is considered the first issue faced in the designing of cellular manufacturing systems. Extensive research has been devoted to solve this problem. Therefore, many solution methods have been proposed. These include cluster analysis approaches (e.g. [2]), exact algorithms (e.g. [3]) and metaheuristic algorithms (e.g. [4, 5]), among others.

The objective of this paper is to introduce a procedure to construct part-machine groupings when the manufacturing system is represented by a binary machine-part incidence matrix. The approach combines a clonal selection (CS) algorithm with a local search (LS) heuristic to intensify the search towards promising regions. The LS method is based on the one introduced by Goncalves and Resende [4]. The numerical experimentations indicate that the proposed algorithm performs well on all test problems in a reasonable computational time.

The rest of this article is structured as follow: Sect. 2 describes the CFP; Sect. 3 introduces the improvement carried out on the CS algorithm to solve the CFP. The results of numerical experiments on a set of benchmark problems are reported in Sect. 4. Finally, the conclusion of this paper is provided in Sect. 5.

2 Problem Formulation

The input parameter of the CFP is the machine-part incidence matrix A. Where, each row represents a machine and each column represents a part. To formulate this problem, a mathematical model similar to the one used in [3] is adopted.

Parameters:

i, j, k: Index of machines, parts and cells, respectively.
M, P, C: Number of machines, parts and cells, respectively.
A: Machine-part incidence matrix $A = \lfloor a_{ij} \rfloor$.
a: Total number of entries equal to 1 in the matrix A.
$a_{ij} = 1$ if machine i process part j; 0 otherwise.
$x_{ik} = 1$ if machine i belongs to cell k; 0 otherwise.
$y_{jk} = 1$ if part j belongs to cell k; 0 otherwise.

The objective function of the CFP can be presented as follow:

$$\min ComEff(x, y) = \frac{a + \sum_{k=1}^{C} \sum_{i=1}^{M} \sum_{j=1}^{P} (1 - 2a_{ij}) x_{ik} y_{jk}}{a + \sum_{k=1}^{C} \sum_{i=1}^{M} \sum_{j=1}^{P} (1 - a_{ij}) x_{ik} y_{jk}} \tag{1}$$

Subject to:

$$\sum_{k=1}^{C} x_{ik} = 1 \quad i = 1, \ldots, M \tag{2}$$

$$\sum_{k=1}^{C} y_{jk} = 1 \quad j = 1, \ldots, P \tag{3}$$

$$\sum_{i=1}^{M} x_{ik} \geq 1 \quad k = 1, \ldots, C \tag{4}$$

$$\sum_{j=1}^{P} y_{jk} \geq 1 \quad k = 1, \ldots, C \tag{5}$$

$$x_{ik} = 0 \text{ or } 1 \quad i = 1, \ldots, M; k = 1, \ldots, C \tag{6}$$

$$y_{jk} = 0 \text{ or } 1 \quad j = 1, \ldots, P; k = 1, \ldots, C \tag{7}$$

Constraint (2) ensures that each machine is assigned to exactly one cell. Constraint (3) guarantees that each part must be assigned to only one cell. Inequalities (4) and (5) ensure that each cell includes at least one machine and one part. Finally, constraints (6) and (7) denote that the decision variables are binary. In the numerical experimentation, the number of cells C is fixed for each problem to its value in the best-known solution reported in the literature.

3   The Proposed Algorithm

The CS algorithm is a multi-model function optimization algorithm inspired by the biological immune system. It was formally introduced by De Castro et al. [6]. In this paper, the proposed method encodes solutions as antibodies, which are evaluated by affinity function. Each iteration begins by selecting a percentage b of the best antibodies with the highest affinities. Then, the selected antibodies proliferate into several clones according to their affinities, i.e. the higher affinity, the higher number of clones. The generated antibodies undergo a one-point crossover then a local search mechanism that ensures each individual antibody is the best one among its neighbors. Finally, a receptor editing process, in which the worst antibodies in the actual population are replaced by the best generated antibodies, is performed. An overview of the method is depicted in Fig. 1.

Fig. 1 Flowchart of the
proposed CS algorithm

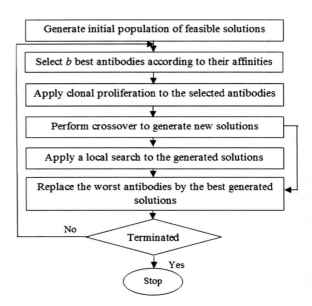

Scheme for Coding The problem variable is represented as a vector with length of $P + M$: $(P_1, \ldots, P_P | M_1, \ldots, M_M)$. Where P_j is the index of the cell including part j and M_i is the index of the cell including machine i. The chosen encoding is similar to the one used by Elbenani et al. [5].

Initial Population The initial solutions are generated randomly. Each machine i and part j are assigned randomly to a cell k. To fix infeasibilities that may result from an empty cell (cell without parts or machines), a repair process is applied to reassign one part or machine to this cell, inducing the smallest decrease of the grouping efficiency.

Clonal Proliferation Clonal proliferation is the process of producing exact copies of the parent. The number of clones of an antibody r is calculated based on its affinity value $(Affinity(r))$, the population size $(popsize)$ and the sum of affinities of the whole population $\left(\sum_{t=1}^{popsize} Affinity(t) \right)$:

$$Number\ of\ clones = \frac{Affinity(r)}{\sum_{t=1}^{popsize} Affinity(t)} \times popsize \qquad (8)$$

Affinity The affinity value of an antibody r is calculated as:

$$Affinity(r) = \frac{1}{Objective\ function(r)} \qquad (9)$$

4 Computational Result

The proposed algorithm is tested on several benchmark problems [5] with different sizes in order to evaluate the scalability of this method. For each problem, its size, the minimum, the average and the maximum solutions obtained, the computational time in second required to find the best solution are reported in Table 1 respectively.

Table 1 The numerical results of the proposed algorithm

No.	M	P	C	Min. Sol.	Ave. Sol.	Max. Sol.	CPU. Time (s)
P1	5	7	2	82.35	82.35	82.35	0.06
P2	5	7	2	69.57	69.57	69.57	0.06
P3	5	18	2	79.59	79.59	79.59	0.10
P4	6	8	2	76.92	76.92	76.92	0.07
P5	7	11	5	60.87	60.87	60.87	0.18
P6	7	11	4	70.83	70.83	70.83	0.14
P7	8	12	4	69.44	69.44	69.44	0.18
P8	8	20	3	85.25	85.25	85.25	0.29
P9	8	20	2	58.72	58.72	58.72	0.15
P10	10	10	5	75.00	75.00	75.00	0.24
P11	10	15	3	92.00	92.00	92.00	0.21
P12	14	24	7	72.06	72.06	72.06	5.13
P13	14	24	7	70.59	71.17	71.83	4.62
P14	16	24	8	51.09	52.17	53.26	8.62
P15	16	30	6	63.43	67.09	69.53	8.15
P16	16	43	8	57.32	57.43	57.53	32.82
P17	18	24	9	56.84	57.46	57.73	11.10
P18	20	20	5	42.45	42.94	43.45	2.77
P19	20	23	7	50.81	50.81	50.81	7.01
P20	20	35	5	71.93	77.31	77.91	10.49
P21	20	35	5	57.22	57.90	57.98	10.96
P22	24	40	7	100.0	100.0	100.0	33.04
P23	24	40	7	85.11	85.11	85.11	41.07
P24	24	40	7	73.51	73.51	73.51	34.60
P25	24	40	11	52.94	53.20	53.29	94.99
P26	24	40	12	47.89	48.29	48.61	97.39
P27	24	40	12	44.90	45.79	46.21	87.63
P28	27	27	5	52.85	54.42	54.8	7.50
P29	28	46	10	44.94	46.39	47.08	163.68
P30	30	41	14	62.24	62.80	63.31	276.54
P31	30	50	13	58.86	59.57	59.77	295.12
P32	30	50	14	50.00	50.44	50.83	355.11

(continued)

Table 1 (continued)

No.	M	P	C	Min. Sol.	Ave. Sol.	Max. Sol.	CPU. Time (s)
P33	36	90	17	45.66	46.493	47.71	2489.81
P34	37	53	3	59.41	59.41	59.41	18.28
P35	40	100	10	84.03	84.03	84.03	1527.35

Preliminary testing indicates that the following values seem to be appropriate: a number of iterations equal to 200 iterations, a population size of 20 solutions and a percentage b of the selected antibodies for cloning process equal to 25 %.

For each problem, 10 independent runs of the algorithm with these parameters are carried out. The algorithm is coded using Java, and the numerical tests are completed on a Personal Computer equipped with Pentium® Dual Core CPU T4500 clock at 2.30 GHz and 2 GB of RAM.

As shown in Table 1, the proposed method generates very good solutions in a short computational time. Besides, the obtained results indicate that the performance of this method is rather constant on all types of problems ranging from small to large.

To evaluate the performance of the proposed algorithm, the obtained results are compared with the results of several algorithms developed in the literature for the CFP. Table 2 reports the results of these algorithms and the best known solution found for each problem (Bk Sol.).

Table 2 The numerical results of the proposed algorithm

No.	GRAFICS	EA	SA	GRASP	GA-LNS	HGBPSO	Proposed method	Bk Sol.
P1	73.68	73.68	82.35	73.68	82.35	82.35	82.35	82.35
P2	60.87	62.50	69.57	62.50	69.57	69.57	69.57	69.57
P3	–	79.59	79.59	79.59	79.59	79.59	79.59	79.59
P4	–	76.92	76.92	76.92	76.92	76.92	76.92	76.92
P5	53.12	53.13	60.87	53.13	60.87	60.87	60.87	60.87
P6	–	70.37	70.83	70.37	70.83	70.83	70.83	70.83
P7	68.30	68.29	69.44	69.44	69.44	69.44	69.44	69.44
P8	85.24	85.25	85.25	85.25	85.25	85.25	85.25	85.25
P9	58.33	58.72	58.72	58.72	58.72	58.72	58.72	58.72
P10	70.59	70.59	75.00	70.59	75.00	75.00	75.00	75.00
P11	92.00	92.00	92.00	92.00	92.00	92.00	92.00	92.00
P12	64.36	69.86	72.06	69.86	72.06	72.06	72.06	72.06
P13	65.55	69.33	71.83	69.33	71.83	71.83	71.83	71.83
P14	45.52	52.58	53.26	51.96	53.26	53.41	53.26	53.26
P15	67.83	67.83	68.99	67.83	69.53	68.99	69.53	69.53
P16	54.39	54.86	57.53	56.52	57.53	57.53	57.53	57.53
P17	48.91	54.46	57.73	54.46	57.73	57.73	57.73	57.73

(continued)

Table 2 (continued)

No.	GRAFICS	EA	SA	GRASP	GA-LNS	HGBPSO	Proposed method	Bk Sol.
P18	38.26	42.94	**43.45**	42.96	**43.45**	43.26	**43.45**	**43.45**
P19	49.36	49.65	**50.81**	49.65	**50.81**	**50.81**	**50.81**	**50.81**
P20	75.14	76.22	**77.91**	76.54	**77.91**	**77.91**	**77.91**	**77.91**
P21	–	58.07	57.98	58.15	**57.98**	**57.98**	**57.98**	**57.98**
P22	**100.0**	**100.0**	**100.0**	**100.0**	**100.0**	**100.0**	**100.0**	**100.0**
P23	**85.11**	**85.11**	**85.11**	**85.11**	**85.11**	**85.11**	**85.11**	**85.11**
P24	**73.51**	**73.51**	**73.51**	**73.51**	**73.51**	**73.51**	**73.51**	**73.51**
P25	43.27	51.88	**53.29**	51.97	**53.29**	**53.29**	**53.29**	**53.29**
P26	44.51	46.69	**48.95**	47.37	**48.95**	**48.95**	48.61	**48.95**
P27	41.67	44.75	**47.26**	44.87	46.58	**47.26**	46.21	**47.26**
P28	47.37	54.27	**54.82**	54.27	**54.82**	**54.82**	**54.82**	**54.82**
P29	32.86	44.37	**47.23**	46.06	47.06	–	47.08	**47.23**
P30	55.43	58.11	**63.31**	59.52	63.12	**63.31**	**63.31**	**63.31**
P31	56.32	59.21	59.77	60.00	**60.12**	59.77	59.77	**60.12**
P32	47.96	50.48	**50.83**	50.51	**50.83**	–	**50.83**	**50.83**
P33	39.41	42.12	47.14	45.93	**47.75**	–	47.71	**47.75**
P34	52.21	56.42	**60.64**	59.85	60.63	**60.64**	59.41	**60.64**
P35	83.92	**84.03**	**84.03**	**84.03**	**84.03**	–	**84.03**	**84.03**

The comparator algorithms are GRAFICS method [2], evolutionary algorithm (EA) [4], simulated annealing (SA) [7], GRASP heuristic [8], genetic algorithm and large neighborhood search (GA_LNS) [5] and HGBPSO method [9]. The results show that the proposed algorithm performs very well since its results are equal or better than those of all comparator algorithms except for six problems where it performs less than others algorithms.

Regarding problem P14, the optimal solution is already reached using an exact method [3] and it is equal to 53.26 (the same value attained by our method), which is conflicting with the one obtained by HGBPSO method. Also, the data reported for problem P21 in EA and GRASP methods are inconsistent with the original data in the literature.

5 Conclusion

The cell formation problem is one of the first issues faced in the designing of cellular manufacturing systems. In this paper, a new solution approach based on the clonal selection algorithm is presented for solving the problem with the aim of maximizing the grouping efficacy. To intensify the search, the proposed method is refined by a local search mechanism.

Numerical results confirm the effectiveness of the proposed method. Indeed, the comparison with six algorithms developed in the literature indicates that on 29 out of 35 problems (82.86 %) the results of the proposed approach are among the best results. These results justify the superior performance of the method compared to other recently proposed algorithms.

References

1. Dimopoulos, C., Zalzala, A.M.: Recent developments in evolutionary computation for manufacturing optimization: Problems, solutions, and comparisons. IEEE Trans. Evol. Comput. **4**, 93–113 (2000)
2. Srinivasan, G., Narendran, T.T.: GRAFICS—A nonhierarchical clustering algorithm for group technology. Int. J. Prod. Res. **29**, 463–478 (1991)
3. Elbenani, B., Ferland, J.A.: Cell formation problem solved exactly with the Dinkelbach algorithm. CIRRELT, pp. 1–14 (2012)
4. Goncalves, J., Resende, M.: An evolutionary algorithm for manufacturing cell formation. Comput. Ind. Eng. **47**, 247–273 (2004)
5. Elbenani, B., Ferland, J.A., Bellemare, J.: Genetic algorithm and large neighbourhood search to solve the cell formation problem. Expert Syst. Appl. **39**, 2408–2414 (2012)
6. De Castro, L.N., Von Zuben, F.J.: Learning and optimization using the clonal selection principle. IEEE Trans. Evol. Comput. **6**, 239–251 (2002)
7. Ying, K.C., Lin, S.W., Lu, C.: Cell formation using a simulated annealing algorithm with variable neighbourhood. Eur. J. Ind. Eng. **5**, 22–42 (2011)
8. Diaz, J.A., Luna, D., Luna, R.: A GRASP heuristic for the manufacturing cell. Top **20**, 679–706 (2012)
9. Kashan, A.H., Karimi, B., Noktehdan, A.: A novel discrete particle swarm optimization algorithm for the manufacturing cell formation problem. Int. J. Adv. Manuf. Technol. **73**, 1543–1556 (2014)

A Comparative Study of Three Nature-Inspired Algorithms Using the Euclidean Travelling Salesman Problem

Yassine Saji and Mohammed Essaid Riffi

Abstract Recently, the nature has become a source of inspiration for the creation of many algorithms. A great research effort has been devoted to the development of new metaheuristics, especially nature-inspired one to solve numerous difficult combinatorial problems appearing in various industrial, economic, and scientific domains. The nature-inspired algorithms offer additional advantages over classical algorithms; they seek to find acceptable results within a reasonable time, rather than an ability to guarantee the optimal or sub-optimal solution. The travelling salesman problem (TSP) is an important issue in the class of combinatorial optimization problem and also classified as NP-hard problem and no polynomial time algorithm is known to solve it. Based on three nature-inspired algorithms, this paper proposes a comparative study to solve TSP. The proposed algorithms are evaluated on a set of symmetric benchmark instances from the TSPLIB library.

Keywords Travelling salesman problem · NP-hard problem · Nature-inspired algorithms · Combinatorial optimization problem

1 Introduction

The travelling salesman problem (TSP) belongs to class of NP-hard combinatorial optimization problems [1]. The aim of this problem is to find the shortest possible route that visits a list of cities exactly once, and returns to the origin city, while minimizing the total cost of round-trip. The TSP problem is simple to describe but very hard to solve; for example, we assume that we have a list of n cities, so the number of possible tours for TSPs is equal to $(n - 1)!/2$, in this case if one tour is

Y. Saji (✉) · M.E. Riffi
LAROSERI Laboratory, Department of Computer Science, Faculty of Science, Chouaïb Doukkali University, Route Ben Maachou, 24000 El Jadida, Morocco
e-mail: yassine.saji@gmail.com

M.E. Riffi
e-mail: said@riffi.fr

© Springer International Publishing Switzerland 2016
A. El Oualkadi et al. (eds.), *Proceedings of the Mediterranean Conference on Information & Communication Technologies 2015*, Lecture Notes in Electrical Engineering 380, DOI 10.1007/978-3-319-30301-7_34

calculated in 1 ns, the selection of the optimal (shortest) route among 6.08E+16 routes will take about 2 years for 20 given cities.

In general, there are two different kinds of TSP problem, the Symmetric TSP and the Asymmetric TSP. In the symmetric variation known as STSP, the distance or the cost between cities A and B is equal to the distance or the cost between cities B and A. However, in asymmetric TSP form known as ATSP, often there is more than one way between two adjacent cities; in this case, it is possible to have two different costs or distances between two cities. In this study, we only consider the STSP in our simulations. Mathematically, the TSP is formulated as to find a permutation $\pi = \{1, 2, ..., n\}$ that minimizes:

$$C(\pi) = \sum_{i=1}^{n-1} d_{\pi(i)\pi(i+1)} + d_{\pi(n)\pi(1)} \tag{1}$$

Assuming that $d_{i,j}$ is the Euclidean distance or the travelling cost from city i to city j.

Generally, the optimization problems can be divided into two main classes continuous and discrete. An optimization problem with real decision variables is known as a continuous optimization problem. Conversely, it is called discrete optimization problem if decision variables take discrete (usually integer) values. Moreover, in discrete problems, the set of feasible solutions is discrete or can be reduced to a discrete one by the discretization of the continuous space.

The remainder of this paper is organized as follows: Sect. 2 presents a brief review on three discretization methods. Section 3, describes three nature-inspired algorithms to solve travelling salesman problem. The experimental results obtained from applying the proposed algorithms on some benchmark instances of the TSP are given in Sect. 4 and the conclusion is given in Sect. 5.

2   Discretization Methods

In literature, there exist many discretization techniques to move from continuous to discrete space; some of them are presented below.

2.1   *Sigmoid Function*

The sigmoid function (SF) is one of the most popular discretization methods [2] used to transform a continuous space value into a binary one through the following equations:

Initial population :

-3.028	0.8045	0.0363	3.5790	-0.53	-1.24

Probability transformation :

0.0462	0.690	0.5090	0.9728	0.370	0.2245

Random number σ:

0.20	0.45	0.62	0.85	0.11	0.30

Binary population:

0	1	0	1	1	0

Fig. 1 Binary encoding scheme

$$S(x_i) = \frac{1}{1 + e^{-x_i}} \tag{2}$$

$$x_i' = \begin{cases} 1 & \text{if } S(x_i) > \sigma \\ 0 & \text{otherwise} \end{cases} \tag{3}$$

where $S(x_i)$ denotes the probability that the bit x_i, takes the value 1 and σ is a random number taken from the interval $[0, 1[$.

Figure 1 summarizes the basic steps of binary encoding with SF.

2.2 Random-Key

The Random-Key (RK) encoding scheme [3] is used to transform the position in RK continuous space to an integer/combinatorial space by encoding the solution vector. A vector in RK space consists of real numbers drawn uniformly from $[0, 1)$. To decode the position as well as a solution, this real vector is sorted in an ascending order with an integer series according to the weights of real numbers. The Random-Key encoding mechanism can be illustrated as shown in Fig. 2.

2.3 Modified Position Equation

The first use of the Modified Position Equation (MPE) was in Particle Swarm Optimization algorithm (PSO) to solve the travelling salesman problem [4]. The particle swarm optimization is a population based stochastic optimization approach developed by Eberhart and Kennedy in 1995 [5]; this algorithm was inspired by the

Fig. 2 Random-key encoding scheme

Random key:

0.2	0.4	0.6	0.8	0.7	0.3

Decodes as:

1	3	4	6	5	2

coordinate movement of fish schools and bird flocks. The PSO approach engages a set of agents to solve a given problem. These agents called swarms are consisted of a set of particles. The movement of each particle in n-dimensional search space is characterized by two vectors, its position $x_i = (x_{i1}, x_{i2}, \ldots, x_{in})$ and its velocity $v_i = (v_{i1}, v_{i2}, \ldots, v_{in})$. During the search process, the particle tends to fly towards the best positions (solutions) found. At each iteration, and after finding two best values, each particle updates its velocity and positions based on experience accumulated during previous iterations of the search process according to the two following equations:

$$v_i^t = w * v_i^{t-1} + c_1 * rand_1 \times \left(P_{i_{best}} - x_i^{t-1}\right) + c_2 * rand_2 \times \left(g_{best} - x_i^{t-1}\right) \quad (4)$$

$$x_i^t = x_i^{t-1} + v_i^t, \quad (5)$$

where w is the inertia factor; P_{ibest} is the best position reached by the i-th particle; g_{best} represents the global best position; c_1 and c_2 are the weights determining the influence of P_{ibest} and g_{best}; $rand_1$ and $rand_2$ are two random numbers in $[0, 1]$.

To shift from the original PSO algorithm to the discrete search domain, it was necessary to modify the original position update and velocity Eqs. (4) and (5). The traditional addition and subtraction operators in Eq. (4) cannot have their usual meaning and should be redefined as follows:

- The operator + in (4) is used to concatenate two permutations; it means velocity + velocity return velocity vector in which elements consist of the two velocities.
- The—operator in (4) returns a set of permutations that allow passing from a position to another.
- The operator × in (4) is a multiplication of permutation by a positive real (k) and we can distinguish the following cases: if $0 < k < 1$ we truncate from V a set of N elements with $N = E(k * |V|)$. E is the integer part and $|V|$ is the number of elements. If k is an integer then we repeat the permutation k times. Finally, if k is a real upper to one, we separate the integer part and the decimal part and we refer to the previous two steps.
- The + operator in (5) can apply to sort a set of permutations, it means position + velocity return a new position sorted by elements of the velocity.

3 Nature-Inspired Algorithms for TSP

3.1 Binary Particle Swarm Optimization

Basically, in continuous optimization problems, each particle moves in search space towards continuous valued position, but many problems are, however, defined for discrete valued spaces where the domain of the variables is finite. Li et al. [6]

introduced in 2006, a binary version of particle swarm optimization to solve the travelling salesman problem by using sigmoid function. The binary PSO saves the original concept of the PSO algorithm except that each particle in a swarm consists of binary matrix representing a particle's position in the search space. It uses the velocity as a probability of making the transition from a city to a non-visited city. The main changes will be in Eq. (5) by using (6), however (4) remains unchanged.

$$x_{jk}^i = \begin{cases} 1 & \text{if the path goes from city } j \text{ to city } k \\ 0 & \text{otherwise} \end{cases} \tag{6}$$

$$S\left(v_{jk}^i\right) = \frac{1}{1 + e^{-v_{jk}^i}} \tag{7}$$

where $S\left(v_{jk}^i\right)$ represents the probability of bit x_{jk}^i takes the value 1.

3.2 Random-Key Cuckoo Search

The cuckoo search algorithm is a nature-inspired population based metaheuristic originally developed by Yang and Deb in 2009 [7] for solving optimization problems. This algorithm mimics the brood parasitism of some cuckoo species that lay their eggs in the nests of host birds of other species.

The RK encoding scheme was utilized to transform the continuous search space in the original Cuckoo Search (CS) into a discrete solution space, especially for travelling salesman problem and named Random-Key Cuckoo Search (RKCS) [8]. As the CS is typically a population-based stochastic global search algorithm, in RKCS algorithm [8] a simple local search procedure was employed to enhance the exploitation capability for CS. Furthermore, the RKCS algorithm was based on Levy's flight to manipulate the search strategy of exploitation and exploration. The research process of generating a TSP solution using random keys as described in [8] can be presented in Fig. 3.

Cuckoos chose random location, according to their real values in [0, 1). Each cuckoo is associated with an integer number (index city) denotes its ascending order

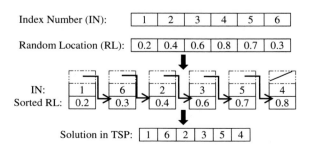

Fig. 3 Random-keys steps to TSP solution

among other cuckoos in the linked list. Then the real numbers are sorted in an increasing order, their corresponding indices (treated as cities) in the sorted order yield an initial solution of the TSP instance.

3.3 Modified Bat Algorithm

Bat-inspired Algorithm (BA) is a population-based stochastic optimization technique developed by Yang in 2010 [9]. This algorithm is inspired by the echolocation behaviour of microbats to find their prey in complete darkness. Initially, the fundamental BA algorithm has been developed to optimize continuous non-linear functions [9]. In the basic bat algorithm developed by Xin-She Yang [9], the movement of the virtual bats is given by updating their frequencies f_i, velocities v_i^{t+1} and positions x_i^{t+1} at time step $t + 1$ using Eqs. (8), (9) and (10), as follows:

$$f_i = f_{\min} + (f_{\max} - f_{\min})\beta \tag{8}$$

$$v_i^{t+1} = v_i^t + (x_i^t - x_*)f_i, \tag{9}$$

$$x_i^{t+1} = x_i^t + v_i^{t+1}, \tag{10}$$

where $\beta \in [0, 1]$ is a random vector generated from uniform distribution and x_* denotes the current global best solution.

However, it had been impossible to exploit this basic version of BA to solve discrete problems. In 2012, Nakamura et al. [10] proposed a first binary version of BA for feature selection. In this paper, we present a discrete BA to solve travelling salesman problem as described in [11]. The standard position update and velocity equations can be redefined by using two following equations:

$$v_i^{t+1} = \chi(x_i^t, x_*, f_i), \tag{11}$$

$$x_i^{t+1} = \phi(x_i^t, v_i^{t+1}), \tag{12}$$

The function χ is a crossover function that has three input arguments (two solutions and one integer) and returns a set of permutations reached by applying 2-exchange crossover mechanism described in [11]. The integer denotes the number of elements saved from x_i^t when the latter crosses x_*.

The function ϕ is a function to sort x_i^t while taking into account the set of permutations of v_i^{t+1}.

4   Numerical Results and Discussion

Three typical nature-inspired population-based algorithms are used to make a comparison involving the Modified Bat Algorithm (MBA) [11], Hybrid Discrete Particle Swarm Optimization (HDPSO) [6] and Random Key Cuckoo Search (RKCS) [8] to solve the travelling salesman problem. The described algorithms are applied to Euclidean symmetric instances benchmark taken from TSPLIB [12]. Each instance provides cities with their coordinates, in addition to name, type and dimension of the instance. Table 1 summarizes the simulation results of MBA and HDPSO algorithms over 10 independents runs and the optimal solutions found by these algorithms are shown in bold. In Table 1, the column "best" denotes the best-so-far optimal solution quality, the column "worst" denotes the worst solution found over 10 runs and the column "PDav(%)" means percentage deviations of solution found from optimal solution known. To ensure a fair comparison among MBA and HDPSO, we used six available and typical data sets, Att48, Eil51, St70, Eil76, Gr96 and Kro100 as a test material (the instance named Att48 as the other instances, the number 48 in the name represents the number of provided cities). Besides, Table 2 displays the experimental results of RKCS algorithm as described in [8] compared to MBA over 30 runs.

Table 1 Comparison of experimental results of MBA algorithm and HDPSO [6] over 10 runs

Instance	Opt	HDPSO			MBA		
		Best	Worst	PDav(%)	Best	Worst	PDav(%)
att48	10,628	10,842	11,070	2.80	**10,628**	**10,628**	**0.00**
eil51	426	442	442	3.75	**426**	**426**	**0.00**
st70	675	694	704	3.11	**675**	**675**	**0.00**
eil76	538	550	568	3.27	**538**	541	0.09
gr96	55,209	60,263	63,087	11.07	**55,209**	**55,209**	**0.00**
kroA100	21,282	22,829	23,867	8.24	**21,282**	**21,282**	**0.00**

Table 2 Comparison of experimental results of MBA algorithm and RKCS [8] over 30 runs

Instance	Opt	RKCS			MBA		
		Best	Worst	PDav(%)	Best	Worst	PDav(%)
eil51	426	**426**	430	0.21	**426**	**426**	**0.00**
berlin52	7542	**7542**	**7542**	**0.00**	7542	7542	**0.00**
st70	675	**675**	684	0.34	**675**	**675**	**0.00**
Pr76	108,159	**108,159**	109,085	0.04	**108,159**	**108,159**	**0.00**
eil76	538	**538**	541	0.20	**538**	540	0.02
kroA100	21,282	**21,282**	21,343	0.03	**21,282**	**21,282**	**0.00**
eil101	629	**629**	636	0.33	**629**	643	0.55
bier127	118,282	**118,282**	120,773	0.43	**118,282**	118,693	0.11
pr136	96,772	97,046	98,936	0.96	**96,772**	97,892	0.27
pr144	58,537	**58,537**	58,607	0.03	**58,537**	**58,537**	**0.00**
ch130	6110	6126	6210	0.87	**6110**	**6168**	0.28

According to the results displayed in Table 1, we can see that the MBA out-performs HDPSO with a significant difference among the results reported. For all compared instances in this table, the MBA can reach the optimal solutions with a success rate of 100 %. Furthermore, the average differences from the optimal solutions obtained by MBA and HDPSO regarding to all tested instances are, respectively, 0.015 and 5.37 %.

Besides, as seen from the results presented in Table 2, we can easily find that MBA algorithm is better than RKCS on the most tested benchmark instances with a significant average difference from the optimal solutions equal to 0.11 % in front of 0.30 % for RKCS, which means that the solutions found by MBA are nearer to optimal solutions than those found by RKCS.

5 Conclusion

The paper has presented a comparative study of three nature-inspired algorithms adapted to travelling salesman problem by using the most common discretization methods and encoding strategies proposed in the literature. These algorithms demonstrate how continuous algorithms can be useful in some discrete problems. Furthermore, the empirical results presented in this study show that the BA is a promising approach to solve TSP problem as a typical case, but we cannot give a single conclusion about which one has the best performance in solving other combinatorial problems. This depends upon the problem tackled, the programming languages, the machine configurations and the complexities of algorithms. Nevertheless, the approximate comparison offered here could help future authors to choose the suitable method and discretization strategy for their problems.

References

1. Arora, S.: Polynomial time approximation schemes for Euclidean traveling salesman and other geometric problems. J. ACM (JACM) **45**, 753–782 (1998)
2. Palit, S., Sinha, S.N., Molla, M.A., Khanra, A., Kule, M.: A cryptanalytic attack on the knapsack cryptosystem using binary Firefly algorithm. In: 2nd International Conference on Computer and Communication Technology (ICCCT), IEEE, pp. 428–432 (2011)
3. Bean, J.C.: Genetic algorithms and random keys for sequencing and optimization. ORSA J. Comput. **6**, 154–160 (1994)
4. Wang, K.-P., Huang, L., Zhou, C.-G., Pang, W.: Particle swarm optimization for traveling salesman problem. In: International Conference on Machine Learning and Cybernetics, IEEE, pp. 1583–1585 (2003)
5. Kennedy, J.: Particle swarm optimization. Encyclopedia of machine learning, pp. 760–766. Springer, Berlin (2010)
6. Li, X., Tian, P., Hua, J., Zhong, N.: A hybrid discrete particle swarm optimization for the traveling salesman problem. Simulated evolution and learning, pp. 181–188. Springer, Berlin (2006)

7. Yang, X.-S., Deb, S.: Cuckoo search via Lévy flights. In: World Congress on Nature & Biologically Inspired Computing, NaBIC 2009, pp. 210–214. IEEE (2009)
8. Ouaarab, A., Ahiod, B., Yang, X.-S.: Random-key cuckoo search for the travelling salesman problem. Soft Comput., 1–8 (2014)
9. Yang, X.-S.: A new metaheuristic bat-inspired algorithm. Nature inspired cooperative strategies for optimization (NICSO 2010), pp. 65–74. Springer, Berlin (2010)
10. Nakamura, R.Y., Pereira, L.A., Costa, K., Rodrigues, D., Papa, J.P., Yang, X.-S.: BBA: A binary bat algorithm for feature selection. In: 25th SIBGRAPI Conference on Graphics, Patterns and Images (SIBGRAPI), pp. 291–297. IEEE (2012)
11. Saji, Y., Riffi, M.: A novel discrete bat algorithm for solving the travelling salesman problem. Neural Comput. Appl., 1–14 (2015)
12. Reinelt, G.: TSPLIB—A traveling salesman problem library. ORSA J Comput. **3**, 376–384 (1991)

A Compact Planar Monopole Antenna with a T-Shaped Coupling Feed for LTE/GSM/UMTS Operation in the Mobile Phone

Lakbir Belrhiti, Fatima Riouch, Jaouad Terhzaz, Abdelwahed Tribak and Angel Mediavilla Sanchez

Abstract In this paper a novel compact planar monopole antenna with a T-shaped coupling feed for LTE/GSM/UMTS operation in the mobile phone is presented and studied. The proposed antenna has a simple structure, a planar configuration and can be directly printed on the system circuit board of the mobile phone. The antenna size is 40×20 mm^2 it consisting of T-shaped driven strip and a coupled radiating structure. In addition two wide bands can be generated by the designed antenna for the LTE/GSM/UMTS operation in the mobile phone. The proposed antenna is validated by using two electromagnetic softwares CST-MWS and Ansoft HFSS. Operating principle of this antenna and details of the various antenna parameters are also studied and discussed in this paper. The simulated results including return loss, radiation patterns and current distributions are presented for the proposed antenna.

Keywords Planar monopole antenna · LTE/GSM/UMTS · Coupling feed · Wide band · Mobile phone · CST-MWS · HFSS

L. Belrhiti (✉) · F. Riouch · A. Tribak
National Institute of Posts and Telecommunications, Rabat, Morocco
e-mail: belrhiti@inpt.ac.ma

F. Riouch
e-mail: riouch@inpt.ac.ma

A. Tribak
e-mail: tribak@inpt.ac.ma

J. Terhzaz
Regional Center for Trades Education and Training (CRMEF),
Casablanca, Morocco
e-mail: terhzazj@yahoo.fr

A.M. Sanchez
DICOM, University of Cantabria, Santander, Spain
e-mail: media@dicom.unican.es

© Springer International Publishing Switzerland 2016
A. El Oualkadi et al. (eds.), *Proceedings of the Mediterranean Conference
on Information & Communication Technologies 2015*, Lecture Notes
in Electrical Engineering 380, DOI 10.1007/978-3-319-30301-7_35

1   Introduction

In recent years wireless communications and internet services have been penetrating into our society and affecting our everyday life profoundly. Moreover, cellular communications have experienced exponential growth and wireless systems that support voice communications have already been deployed with great success. In addition, mobile communication systems and personal communication systems are expected to support a variety of high-speed multimedia services, such as high speed internet access, high-quality video transmission [1].

As reported in [2] a dual-band planar monopole antenna for multiband mobile systems is presented. In [3] a broadband interior antenna of planar monopole type in handsets is proposed. However, these antenna designs [2, 3] are unable to cover the penta-band WWAN (wireless wide area network) operation. Therefore the authors propose in [4] a multiband printed monopole slot antenna for WWAN operation to cover all the five operating bands of GSM850, GSM900, GSM1800, GSM1900 and UMTS for WWAN operation. Also a compact and printed multiband antenna incorporating a conductive wire for WWAN operation is presented in [5]. In [6] a planar hexa-band internal antenna designed for mobile phone applications is proposed that has an area of only 45×12 mm^2 can support the desired hexa-band (WWAN and WLAN) operations. In addition several internal mobile phone antennas capable of covering eight-band WWAN/LTE operation which includes the LTE700/GSM850/900 and the GSM1800/1900/UMTS/LTE2300/2500 has been presented in [7, 8]. Kin-Lu Wong and Wei-Yu Chen propose a small-size printed loop-type antenna integrated with two stacked coupled-feed shorted strip monopoles for eight-band LTE/GSM/UMTS operation in the mobile phone [7].

In this paper, we present a novel planar monopole antenna with a T-shaped coupling feed to cover the desired bands LTE/GSM/UMTS operation in the mobile phone, which includes the LTE700 (698–787 MHz), GSM850 (824–894 MHz), GSM900 (880–960 MHz), GSM1900 (1850–1990 MHz), UMTS (1920–2170 MHz), LTE2300 (2305–2400 MHz) and LTE2500 (2500–2690 MHz) bands. The antenna geometry, design methods and main parameters of the proposed antenna are studied in next sections. The parameters such as return loss, radiation patterns and current distributions for the proposed antenna are discussed.

2   Antenna Design and Configuration

Figure 1 shows the geometry of the proposed planar monopole antenna with a T-shaped coupling feed having a simple structure and comprising a T-shaped driven strip and a coupled radiating structure. Dimensions of the metal pattern of the antenna are given in Fig. 1. In the study, a 1.6-mm thick FR4 substrate with dielectric constant $\varepsilon_r = 4.4$ and loss tangent tan $\delta = 0.02$ of length 120 mm and width 40 mm is used as the system circuit board of the mobile phone. The antenna

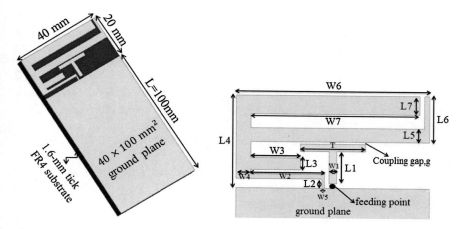

Fig. 1 Geometry of the proposed antenna and dimensions of the metal pattern of the antenna

occupies a compact area of 40×20 mm^2 is printed on the top no-ground portion of the system circuit board with a ground-plane size of 40×100 mm^2, the antenna is fed by a 50 Ohms coaxial cable. Design dimensions of the proposed antenna are listed in Table 1.

The proposed antenna is simulated by using CST Microwave Studio which is based on Finite Integration Technique (FIT). Figure 2 shows the return loss simulation result of proposed antenna. In all simulated results the bandwidth is

Table 1 Design dimensions in mm of the proposed antenna

Parameter	Value	Parameter	Value	Parameter	Value
L3	2.5	W5	0.5	L7	4
L4	20	W6	40	W7	35
L5	3	g	0.5	T	13.5

Fig. 2 Comparison of the simulated return loss for the proposed antenna using CST and HFSS

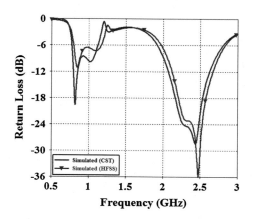

determined from 6 dB (3:1 VSWR) return loss, because the internal antennas of general mobile phones are usually designed based on the bandwidth definition of 6 dB return loss (3:1 VSWR) or even less [9]. It can be seen two wide operating bands, the lower band has a bandwidth of 350 MHz (773–1123 MHz) and the upper band has 880 MHz (1922–2802 MHz).

After this validation into simulation by using CST-MWS we have used another 3D EM simulator, namely, Ansoft High Frequency Structure Simulator (HFSS) which is based on Finite Element Method (FEM). As depicted in Fig. 2, it is clearly observed that simulation results obtained using CST-MWS and HFSS are in good agreement.

Figure 3 shows the comparison of the simulated return losses for the proposed antenna, and two reference antennas (Ref1, Ref2). It is clearly seen that two operating bands are achieved by the proposed antenna.

To study the effects of the parameter T, Fig. 4 plots the simulated return loss as a function of the length T of the coupling strip of the monopole antenna. The results for T varied from 13.5 to 29.5 mm with other parameters fixed, as given in Fig. 1, are presented. Large effects on both the antenna's lower and upper bands are also seen. With the decreased length T (T = 13.5) the lower band can be controlled to cover the desired frequency.

Fig. 3 Simulated return loss for the proposed antenna, *Ref1* and *Ref2*

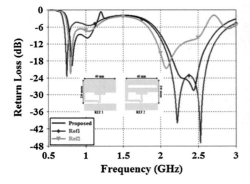

Fig. 4 Simulated return loss as a function of the length T

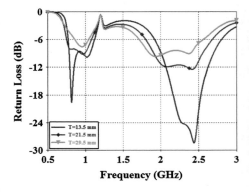

Figure 5 shows the simulated return loss for the coupling gap g varied from 0.5 to 1.5 mm between the T-shaped strip and coupled radiating structure. Large effects of the gap g on the antenna's lower band are seen. When the gap decreases, the lower band can be adjusted to cover the desired frequency. Also some effects of the upper band are seen. The coupling gap in the proposed antenna is selected to be 0.5 mm.

Figure 6 shows the simulated return loss as a function of the length L of the system ground plane. Large effects both in the lower band and upper band can be observed. The length L = 100 mm is the better choice for our antenna design.

The simulated surface current distributions of the proposed antenna at 820, 900, and 2437 MHz are shown in Fig. 7a. Strong excited surface current distributions at 820 and 900 MHz are seen in the radiating structure. While the T-shaped and the system ground plane are more excited at 2437 MHz.

The simulated 3-Dimensional radiation patterns for the proposed antenna has been shown in Fig. 7b, the results for frequencies at 820, 900 and 2437 MHz are studied. Clearly the quasi omni-directional character is more pronounced in low frequency and decreases as the frequency increases. This instability in diagrams is not unique to this antenna. It is noticed in most antennas with radiating more or less complex structures involving slots and meanders.

Fig. 5 Simulated return loss as a function of the coupling gap g

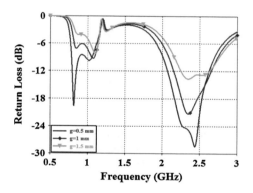

Fig. 6 Simulated return loss as a function of the length L of the main ground

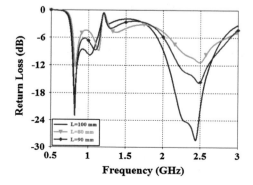

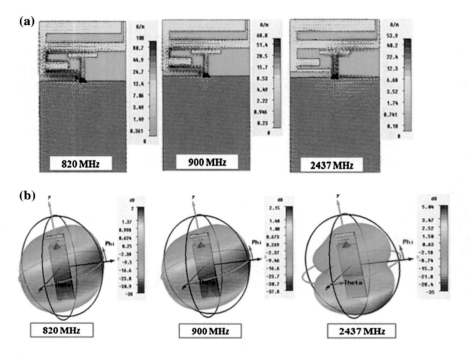

Fig. 7 Simulated (**a**) surface current distributions and (**b**) radiation patterns at 820, 900 and 2437 MHz for the proposed antenna

3 Conclusion

This paper presented a novel compact planar monopole antenna. It is composed of a T-shaped driven strip and a coupled radiating structure. Moreover, the antenna only occupies a small size of 40×20 mm^2 and can be easily printed on one surface of a FR4 substrate, which is used as the system circuit board of the mobile phone. The proposed antenna is validated by using two electromagnetic softwares CST-MWS and Ansoft HFSS. In addition, the simulation results show that the monopole antenna can be suitable for two frequency bands 773–1123 MHz and 1922–2802 MHz. Finally, the obtained simulated results, such as return loss, radiation patterns and current distributions, indicate that the proposed antenna is promising candidate for LTE/GSM/UMTS mobile phones applications.

References

1. Wang, J.: High-speed wireless communications: Ultra-wideband, 3G long-term evolution, and 4G mobile systems. Cambridge University Press, Cambridge (2008)

2. Chen, J.H., Ho, C.J., Wu, C.H., Chen, S.Y., Hsu, P.: Dual-band planar monopole antenna for multiband mobile systems. IEEE Antennas Wirel. Propag. Lett. **7**, 769–772 (2008)
3. Shin, Y.S., Park, S.O., Lee, M.: A broadband interior antenna of planar monopole type in handsets. IEEE Antennas Wirel. Propag. Lett. **4**, 9–12 (2005)
4. Wong, K.L., Lee, L.C.: Multiband printed monopole slot antenna for WWAN operation in the laptop computer. IEEE Trans. Antennas Propag. **57**, 324–330 (2009)
5. Lee, C.T., Su, S.W.: Compact and printed, standalone wireless wide area network antenna incorporating a conductive wire at low cost. Microw. Opt. Technol. Lett. **54**, 109–113 (2012)
6. Sze, J.Y., Wu, Y.F.: A compact planar hexa-band internal antenna for mobile phone. Prog. Electromagn. Res. **107**, 413–425 (2010)
7. Wong, K.L., Chen, W.Y.: Small-size printed loop-type antenna integrated with two stacked coupled-fed shorted strip monopoles for eight-band LTE/GSM/UMTS operation in the mobile. Microwave Opt. Technol. Lett. **52**, 1471–1476 (2010)
8. Lee, C.T., Wong, K.L.: Planar monopole with a coupling feed and an inductive shorting strip for LTE/GSM/UMTS operation in the mobile phone. IEEE Trans. Antennas Propag. **58**, 2479–2483 (2010)
9. Chi, Y.W., Wong, K.L.: Internal compact dual-band printed loop antenna for mobile phone application. IEEE Trans. Antennas Propag. **55**, 1457–1462 (2007)

Theoretical Analysis of SINR Performance for Unsynchronized and Nonlinearly Distorted FBMC Signals

Brahim Elmaroud, Ahmed Faqihi, Mohammed Abbad
and Driss Aboutajdine

Abstract In this paper, we present a theoretical analysis of the joint effect of carrier frequency offset (CFO) and high power amplifier (HPA) nonlinear distortion (NLD) on the signal to interference plus noise ratio (SINR) of filter bank multicarrier (FBMC) systems. A promising class of FBMC modulation, called Cosine Modulated Multitone (CMT), is considered and the analytical SINR is derived in the presence of both HPA NLD and CFO. The simulation results have shown a good agreement with the theoretical analysis.

Keywords Filter bank multicarrier · Cosine modulated multitone · HPA · Nonlinear distortion · Carrier frequency offset · SINR

1 Introduction

Filter bank multicarrier is a class of multicarrier modulations which is taking an increasing attention these last years, thanks to its advantages over conventional multicarrier modulations such as the famous Orthogonal Frequency Division Multiplexing (OFDM). Indeed, FBMC employs a special FIR filter called "prototype filter" to construct a bank of N filters by frequency shifting N copies of the prototype filter (N is the number of sub channels). Furthermore, FBMC provides higher bandwidth efficiency than OFDM since it does not require a guard interval or cyclic prefix extension.

B. Elmaroud (✉) · A. Faqihi · M. Abbad · D. Aboutajdine
LRIT, Associated Unit to CNRST (URAC29), FSR, Mohammed V University,
Rabat, Morocco
e-mail: b.elmaroud@gmail.com

A. Faqihi
ENSIAS, Mohammed V University, Rabat, Morocco
e-mail: faqihi@ensias.ma

© Springer International Publishing Switzerland 2016
A. El Oualkadi et al. (eds.), *Proceedings of the Mediterranean Conference on Information & Communication Technologies 2015*, Lecture Notes in Electrical Engineering 380, DOI 10.1007/978-3-319-30301-7_36

Despite their appealing features, FBMC systems, like any other multicarrier scheme, are very sensitive to timing errors and carrier frequency offsets (CFO). This sensitivity has been illustrated in terms of signal to interference ratio (SIR) in [1, 2], and in terms of signal to noise ratio (SNR) degradation in [3], by deriving accurate and approximate expressions of the inter-carrier interference (ICI) and intersymbol interference (ISI). A simple closed form approximation for SIR over an AWGN channel is proposed in [2] and it is accurate over a relatively broad range of CFO and timing offset. Another real issue for FBMC systems comes from the fact that the transmitted signal is the sum of a large number of independently modulated subcarriers. Thus, it suffers from high Peak-to-Average Power Ratio (PAPR) which make the system very sensitive to nonlinear distortion (NLD) caused by nonlinear devices such as high power amplifiers (HPA). This problem has been largely studied for OFDM systems and the impact of NLD on OFDM signals was presented [4] as well as some solutions to estimate and compensate this NLD [5, 6]. For FBMC systems, the authors in [7] have carried out a theoretical analysis of bit error rate (BER) performance for nonlinearly amplified FBMC/OQAM signals under additive white Gaussian noise (AWGN) and Rayleigh fading channels.

In this paper, we aim to study jointly the effects of synchronization errors and HPA nonlinear distortions on FBMC modulations. A theoretical expression of the signal to interference plus noise ratio (SINR) will be derived for an unsynchronized nonlinearly distorted FBMC signal. The proposed SINR expression is compared to the one obtained for an unsynchronized FBMC signal without considering NLD distortions [1]. It is worth noting that there are limited studies that investigate the problem of joint effects of non-synchronization and HPA nonlinearities. An interesting one is [8], where the authors have analyzed the interference caused by nonlinear power amplifiers together with timing errors for multicarrier OFDM/FBMC transmissions. In this work, we will consider frequency errors instead of timing errors and we will plot the SINR curves as functions of the normalized CFO.

The remainder of this paper is organized as follows: In Sect. 2, the FBMC transceiver is introduced. Section 3 analyses the effect of both NLD and CFO on the studied FBMC signal and provides ICI and ISI expressions. Further, we develop a theoretical analysis of the SINR in Sect. 4. Section 5 includes the evaluation of the obtained SINR expression through simulation results. Finally, Sect. 6 concludes this paper.

2   System Model: CMT Transceiver

The first developments in FBMC systems were introduced by Chang and Saltzberg in the mid 1960s [9]. In particular, Chang proposed a special way to transmit a set of parallel (real valued) pulse amplitude-modulated (PAM) symbols using Vestigial side-band modulation (VSB). This method was called Cosine Modulated Multitone (CMT) [9]. The block diagram of a CMT transceiver is presented in Fig. 1. The

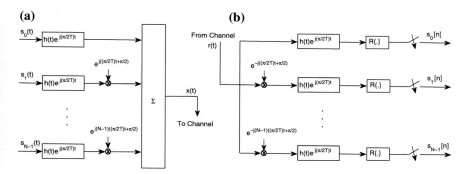

Fig. 1 Block diagram of a CMT transceiver: **a** transmitter; **b** receiver

input signal $s_k(t)$ at the kth subcarrier is the succession of PAM data symbols s_n^k at the rate of $1/T$. It is defined as

$$s_k(t) = \sum_{n=-\infty}^{\infty} s_n^k \delta(t - nT) \qquad (1)$$

According to VSB modulation, each input signal $s_k(t)$ should be passed through a shifted version of the prototype filter $h(t)$ centered at frequency $f = 1/4T$, i.e., $h(t)e^{j(\pi/2T)t}$. The obtained VSB modulated signals are then multiplied by the frequency shifted modulators $e^{jk((\pi/T)t+\pi/2)}$ (with $k = 0,\ldots,N-1$) and summed to form the transmitted CMT signal $x(t)$ given by

$$x(t) = \sum_{n=-\infty}^{\infty} \sum_{k=0}^{N-1} s_n^k h(t - nT) e^{j\frac{\pi}{2T}(t-nT)} \Phi^k(t) \qquad (2)$$

where $\Phi^k(t) = e^{jk((\pi/T)t + \pi/2)}$.

In a distortion-free noiseless channel and according to Fig. 1, the estimated data symbol on the mth subcarrier in the lth time index is given by

$$\hat{s}_l^m = s_l^m \mathcal{R}\left\{ h(t - lT)e^{j\frac{\pi}{2T}(t-lT)} * h(t)e^{j\frac{\pi}{2T}t}\big|_{t=lT} \right\}$$
$$+ \sum_{\substack{n=-\infty \\ (k,n)\neq(m,l)}}^{\infty} \sum_{k=0}^{N-1} s_n^k \mathcal{R}\left\{ h(t - nT)e^{j\frac{\pi}{2T}(t-nT)} \Phi^{k-m}(t) * h(t)e^{j\frac{\pi}{2T}t}\big|_{t=lT} \right\} \qquad (3)$$

where $*$ denote convolution.[1]

---

[1] $(k, n) \neq (m, l)$ is equivalent to $k \neq m \vee n \neq l$.

The first term of the last equality corresponds to the useful signal s_u whereas the second one is the interference term s_i. We have

$$s_u = s_l^m p(t - lT) \cos(\frac{\pi}{2T}(t - lT))|_{t=lT} \tag{4}$$

with $p(t - lT) = \int_{-\infty}^{\infty} h(t - lT - \tau)h(\tau)d\tau$.

To be able to recover the transmitted symbols correctly, $p(t)$ should be a real valued Nyquist filter with zero crossings at the points $t = nT$ for all nonzero integer values of n [9], i.e. $p(nT) = \delta_n$. Hence, the useful signal s_u will be exactly equal to the transmitted signal s_l^m in a distortion-free noiseless channel.

Following a similar procedure, it can be easily shown that the interference term s_i is equal to zero if $p(t)$ is a real valued Nyquist filter, (i.e., $h(t)$ is a symmetric square-root Nyquist filter). See [9] for more details.

3 Joint Effect of HPA NLD and CFO on the CMT Signal

In this section we will show how HPA non linear distortions and carrier frequency offset affect jointly the transmitted CMT signal. Figure 2 presents the system description including NLD and CFO distortion blocks.

3.1 HPA Non Linearity Modeling

A HPA model can be described by its AM/AM and AM/PM conversions which measure the amount of undesired amplitude changes and phase deviations caused by envelope variations of the signal. One of the most commonly used HPA models in the literature is the traveling wave tube amplifier (TWTA) which is defined by the following AM/AM and AM/PM characteristics [10]

$$f_{am}(s_n^k) = \frac{|s_n^k|}{1 + \left(\frac{|s_n^k|}{A}\right)^2}, \quad f_{pm}(s_n^k) = \phi_0 \frac{|s_n^k|^2}{|s_n^k|^2 + A^2} \tag{5}$$

where A is the amplitude saturation of the power amplifier and ϕ controls the maximum phase distortion introduced by this latter.

Fig. 2 System description

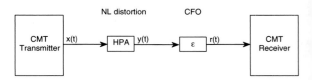

In literature, it is customary to model the non linear distortions, caused by high power amplifiers, using the Bussgang theorem which states that the output signal of the HPA block of Fig. 2 can be written as [4]

$$y(t) = \alpha x(t) + d(t) \tag{6}$$

where $\alpha = |\alpha|e^{\phi_\alpha}$ is a complex factor and $d(t)$ is a zero mean additive noise, which is uncorrelated to $x(t)$.

3.2 Adding CFO

The resulting signal $y(t)$, after the non linearity block, is affected by a carrier frequency offset ϵ, a carrier phase offset ϕ and an AWGN noise $n(t)$. Hence, the received signal can be written as

$$r(t) = y(t)e^{j(2\pi\epsilon t + \phi)} + n(t) = \alpha x(t)e^{j(2\pi\epsilon t + \phi)} + d(t)e^{j(2\pi\epsilon t + \phi)} + n(t) \tag{7}$$

Recall that $x(t)$ is given by Eq. 2.

At the receiver (according to Fig. 1 and following the same derivations as in [1]), the received signal is demodulated, filtered and passed through the $\mathcal{R}\{\cdot\}$ block to produce the estimated data symbols $\hat{s}_n^k$ at each subcarrier index k and time interval n. For given values m and l of k and n respectively, we have

$$\hat{s}_l^m = \mathcal{R}\left\{ \alpha \sum_{n=-\infty}^{\infty} \sum_{k=0}^{N-1} s_n^k h(t - nT)e^{j\frac{\pi}{2T}(t-nT)} e^{j(2\pi\epsilon t + \phi)} \Phi^{k-m}(t) * h(t)e^{j\frac{\pi}{2T}t}\big|_{t=lT} \right\}$$
$$+ \mathcal{R}\left\{ \left(d(t)e^{j(2\pi\epsilon t + \phi)} + n(t) \right) \Phi^{-m}(t) * h(t)e^{j(\pi/2T)t}\big|_{t=lT} \right\} \tag{8}$$

Let us define $\gamma_n^{k-m}(\epsilon, t)$ as:

$$\gamma_n^{k-m}(\epsilon, t) = h(t - nT)e^{j\frac{\pi}{2T}(t-nT)} e^{j(2\pi\epsilon t)} \Phi^{k-m}(t) * h(t)e^{j(\pi/2T)t}\big|_{t=lT} \tag{9}$$

Then Eq. 8 becomes

$$\hat{s}_l^m = \mathcal{R}\left\{ \alpha e^{j\phi} \sum_{n=-\infty}^{\infty} \sum_{k=0}^{N-1} s_n^k \gamma_n^{k-m}(\epsilon, t) \right\} + \hat{d}_l^m + \hat{n}_l^m \tag{10}$$

where $\hat{d}_l^m$ and $\hat{n}_l^m$ are the estimates of the non linear noise d_l^m and the AWGN noise n_l^m, respectively on the mth subcarrier in the lth time index.

By separating the useful signal from the interference signals, we obtain

$$\hat{s}_l^m = \mathcal{R}\left\{\alpha e^{j\phi}\left(s_l^m \gamma_l^0(\epsilon,t) + \sum_{\substack{k=0 \\ k \neq m}}^{N-1} s_l^k \gamma_l^{k-m}(\epsilon,t) + \sum_{\substack{n=-\infty \\ n \neq l}}^{\infty} \sum_{k=0}^{N-1} s_n^k \gamma_n^{k-m}(\epsilon,t)\right)\right\}$$

$$+ \hat{d}_l^m + \hat{n}_l^m = s_l^m \mathcal{R}\{\alpha e^{j\phi} \gamma_l^0(\epsilon,t)\} + \mathcal{ICI} + \mathcal{ISI} + \hat{d}_l^m + \hat{n}_l^m$$

(11)

where $\mathcal{ICI}$ and $\mathcal{ISI}$ are given by

$$\mathcal{ICI} = \sum_{\substack{k=0 \\ k \neq m}}^{N-1} s_l^k \mathcal{R}\{\alpha e^{j\phi} \gamma_l^{k-m}(\epsilon,t)\}, \quad \mathcal{ISI} = \sum_{\substack{n=-\infty \\ n \neq l}}^{\infty} \sum_{k=0}^{N-1} s_n^k \mathcal{R}\{\alpha e^{j\phi} \gamma_n^{k-m}(\epsilon,t)\} \quad (12)$$

4   SINR Analysis

In order to quantify the effect of the HPA NLD, we will ignore the AWGN noise $n(t)$ and consider only the additive noise $d(t)$ due to HPA non linearity.

This approximation can be justified by the fact that we are considering a system working at a very high SNR, i.e., a system transmitting with a high energy. Under these conditions, the effect of HPA NLD is more significant and it was shown that the AWGN noise power is negligible compared to the power of the additive noise caused by NLD [7]. Hence, we have $\sigma_d^2 \gg \sigma_n^2$.

In this case, the signal to interference plus noise ratio is given by

$$SINR(\epsilon) = \frac{P_s}{P_i + \sigma_d^2} \qquad (13)$$

where P_s, P_i and σ_d^2 are the useful signal power, the interference power and the variance of the additive noise d, respectively.

The power of the useful signal is given by

$$P_s = \sigma_s^2 \times \left|\mathcal{R}\{\alpha e^{j\phi} \gamma_l^0(\epsilon,t)\}\right|^2 \qquad (14)$$

Regarding the power of the interference, it is equal to

$$P_i = E\left[\left|\mathcal{ICI} + \mathcal{ISI}\right|^2\right] \qquad (15)$$

with $\mathcal{ICI} + \mathcal{ISI}$ is given by

$$\mathcal{ICI} + \mathcal{ISI} = \sum_{n=-\infty}^{\infty} \sum_{\substack{k=0 \\ (k,n) \neq (m,l)}}^{N-1} s_n^k \mathcal{R}\{\alpha e^{j\phi} \gamma_n^{k-m}(\epsilon, t)\} \tag{16}$$

Thus, the power of the interference becomes

$$P_i = \sigma_s^2 \times \sum_{n=-\infty}^{\infty} \sum_{\substack{k=0 \\ (k,n) \neq (m,l)}}^{N-1} \left| \mathcal{R}\{\alpha e^{j\phi} \gamma_n^{k-m}(\epsilon, t)\} \right|^2 \tag{17}$$

where σ_s^2 is the variance of the symbol s_l^m.

In this paper, we assume that

$$E\left[\left| \mathcal{R}\{ d(t) e^{j(2\pi\epsilon t + \phi)} \Phi^{-m}(t) * h(t) e^{j(\pi/2T)t} |_{t=lT} \} \right|^2 \right] = \sigma_d^2 \tag{18}$$

From Eqs. 14, 17 and 18, we can rewrite the expression of the SINR as

$$SINR(\epsilon) = \frac{\sigma_s^2 \left| \mathcal{R}\{\alpha e^{j\phi} \gamma_l^0(\epsilon, t)\} \right|^2}{\sigma_s^2 \displaystyle\sum_{n=-\infty}^{\infty} \sum_{\substack{k=0 \\ (k,n) \neq (m,l)}}^{N-1} \left| \mathcal{R}\{\alpha e^{j\phi} \gamma_n^{k-m}(\epsilon, t)\} \right|^2 + \sigma_d^2} \tag{19}$$

5   Simulation Results

Considering a noiseless transmission, in this section we will show the effect of non linearity on the SINR of a CMT system. The validity of the proposed theoretical model will also be tested through simulation results. We have considered a CMT system with N = 64 subcarriers transmitting 4-PAM modulated symbols. PHYDYAS prototype filter, obtained by the frequency sampling technique [11], is used. Figure 3 shows the SINR in dB of the studied CMT system as a function of the normalized CFO using expression 19 (i.e., proposed model) and the expression obtained for CMT using the derivations of [1] (ignoring HPA effects).

The simulation curve is obtained using TWTA HPA model with AM/AM and AM/PM distortion given by Eq. 5.

As shown in Fig. 3, the CMT system is highly sensitive to HPA distortions. This sensitivity is illustrated by the degradation of the SINR when HPA effects are considered. We can also observe that the effect of the NLD, due to HPA, begin to take place from a defined value ϵ_0 of the CFO which depends on the number of subcarriers N (we can show that ϵ_0 increases with large values of N). For the

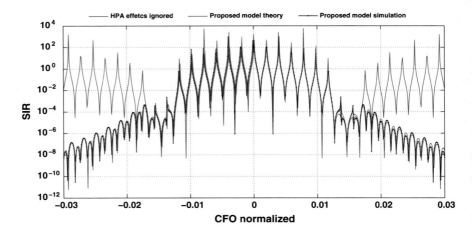

Fig. 3 SINR using the proposed model-theory (Eq. 19) and simulation versus ignoring HPA model [1]

considered system with 64 subcarriers, $\epsilon_0 = 0.017$ kHz. Figure 3 shows a very good agreement between the theoretical expression of the SINR given by Eq. 19 and the simulation results.

On the other hand, we want to show how the HPA distortion affects the SINR of the system. For this purpose, we plot in Fig. 4 the curves of SINR considering two scenarios. In the first one, the additive term $d(t)$ of Eq. 6 is ignored ($d = 0$) and only the parameter α is considered. In the second scenario, α is ignored ($\alpha = 1$) and only the effect of $d(t)$ is evaluated. We can notice from Fig. 4 that the effect of the

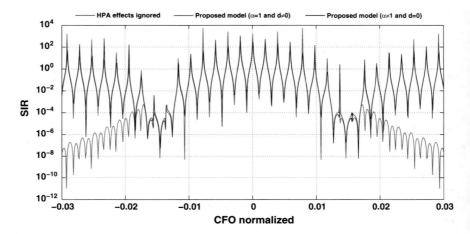

Fig. 4 SINR using the proposed model: effect of α and d

NLD is mainly due to $d(t)$ since the multiplicative parameter α has non-significant impact on the SINR of the system.

6 Conclusion

In this paper, we have studied jointly the impact of carrier frequency offset and HPA nonlinear distortions on FBMC systems. A CMT transceiver was considered and a theoretical expression of the SINR was derived for the corresponding unsynchronized and nonlinearly distorted CMT signal. The proposed model was validated through simulation results. We have shown that the SINR performance is highly sensitive to nonlinear distortions and therefore the effect of HPA can not be ignored for unsynchronized multicarrier systems. It was also shown that the effect of NLD is mainly due to the additive term $d(t)$ and the multiplicative parameter α has non significative impact on the SINR of the system.

References

1. Sourck, H., Wu, Y., Bergmans, J., Sadri, S., Farhang-Boroujeny, B.: Effect of carrier frequency offset on offset QAM multicarrier filter bank systems over frequency-selective channels, pp. 1–6. In: IEEE WCNC, Apr 2010
2. Saeedi-Sourck, H., Wu, Y., Bergmans, J.W., Sadri, S., Farhang-Boroujeny, B.: Sensitivity analysis of offset QAM multicarrier systems to residual carrier frequency and timing offsets. Sig. Process. **91**(7), 1604–1612 (2011)
3. Fusco, T., Petrella, A., Tanda, M.: Sensitivity of multi-user filter-bank multicarrier systems to synchronization errors. In: 3rd International Symposium on Communications, Control and Signal Processing, pp. 393–398 2008, Mar 2008
4. Dardari, D., Tralli, V., Vaccari, A.: A theoretical characterization of nonlinear distortion effects in OFDM systems. IEEE Trans. Comm. **48**(10), 1755–1764 (2000)
5. Tellado, J., Hoo, L., Cioffi, J.: Maximum-likelihood detection of nonlinearly distorted multicarrier symbols by iterative decoding. IEEE Trans. Commun. **51**(2), 218–228 (2003)
6. Dohl, J., Fettweis, G.: Iterative blind estimation of nonlinear channels. In: IEEE ICASSP, pp. 3923–3927, May 2014
7. Bouhadda, H., et al.: Theoretical analysis of ber performance of nonlinearly amplified FBMC/OQAM and OFDM signals. EURASIP J. Adv. Sig. Proc **2014**(1) (2014)
8. Khodjet-Kesba, M., Saber, C., Roviras, D., Medjahdi, Y.: Multicarrier interference evaluation with jointly non-linear amplification and timing errors. In: IEEE 73rd Vehicular Technology Conference (VTC Spring), pp. 1–5, May 2011
9. Farhang-Boroujeny, B., (George) Yuen, C.: Cosine modulated and offset qam filter bank multicarrier techniques: a continuous-time prospect. EURASIP J. Appl. Sig. Process. Article ID 165654, 16 p (2010)
10. Saleh, A.: Frequency-independent and frequency-dependent nonlinear models of TWT amplifiers. IEEE Trans. Commun. **29**(11), 1715–1720 (1981)
11. Bellanger, M.: Specification and design of a prototype filter for filter bank based multicarrier transmission. In: IEEE ICASSP'01, vol 4, pp. 2417–2420 (2001)

Modeling the Migration of Mobile Networks Towards 4G and Beyond

Abdelfatteh Haidine, Abdelhak Aqqal and Hassan Ouahmane

Abstract The deployment of the fourth generation of mobile network (4G) is going round the world, but with different speeds. When considering the Mediterranean area a big difference can be observed, where some countries has started the implementation of 4G, while other countries just started the call to bid for spectrum auctions. In this paper, we discuss the different factors pushing the mobile network operators to decide to invest in 4G infrastructure; by highlighting the challenges and the risks. Then, we model the migration procedure from 2G/3G/3G+ towards a mix infrastructure including the 4G and 4G-beyonds. We model the migration problem as a "shortest path problem" in the state-space. This consists in finding the optimal path, with minimal costs, between start state (current network state) and the target-state at the end of the migration time window. Finally, we focus on algorithms to be applied to solve this shortest-path approach of the migration.

Keywords Long term evolution (LTE) · 4G · Network migration · Shortest path · Combinatorial optimization · CAPEX/OPEX · Migration path

1 Introduction

The explosion of the Internet utilization in the mobile mode pushes the Mobile Network Operators (MNOs) to optimize their infrastructure in order to cover the increasing capacity demand and quality of service requirements. From the technology stand point, there is an answer for any challenge from the users/services side; however, the economical aspect remains one of main hurdles to be well treated

A. Haidine (✉) · A. Aqqal · H. Ouahmane
National School of Applied Sciences—El Jadida, 24000 El Jadida, Morocco
e-mail: a.h.haidine@ieee.org

A. Aqqal
e-mail: aqqal.a@ucd.ac.ma

H. Ouahmane
e-mail: ouahmane.h@ucd.ac.ma

© Springer International Publishing Switzerland 2016
A. El Oualkadi et al. (eds.), *Proceedings of the Mediterranean Conference on Information & Communication Technologies 2015*, Lecture Notes in Electrical Engineering 380, DOI 10.1007/978-3-319-30301-7_37

by the mobile network operators. An optimal design of the economic aspects build the corner-stone in the time when competition is becoming more harder and the reduction of the service prices is the main attracting factor for gaining new clients and keep the fidelity of the existing subscribers. The new reality to which the mobile network operator are facing consists of: increasing number of mobile subscribers, increasing data volume per subscriber, but at the same the revenue are decreasing [10]; some services are not more bringing any gain (like SMS that is losing more field in front of messenger applications; like WhatsApp); circuit-switched telephony losing for VoIP, new applications field (like machine-to-machine communications, Internet of Things–IoT, smart grid, smart cities, etc. There are technologies that can support all these requirements; like Long Term Evolution (LTE), LTE-Advanced (LTE) or the 5th Generation mobile wireless (referred sometimes to as LTE-B); which the roll-out should start by 2020; [9]. However, the MNO has to choose an optimal migration because of two aspects. On one hand, any investment made in new infrastructure must generate new revenues to achieve positive business case. On the other hand, each MNO has to find the optimal time-to-market. In this paper, we propose a model in the transition state space, which can describe any case of the mobile network; built mostly by a set of sites and each site constrains one mobile technology or a mix of technologies. The migration strategy is modeled as a Shortest-Path Problem in the state space to go from the current mobile network state and reach a target state by the end of the migration time window.

The paper is organized as follows: the second section discusses the factors forcing the MNO to invest in the technologies with challenges and risks accompanying such new investment; mostly related to factors that could affects the foreseen and the targeted revenues. These risks are responsible for the gap of the technology deployment in the Mediterranean countries; especially the gaps between North area and the South area. The migration strategy can be mathematically modeled as shortest-path problem; as analyzed and described in third section. The last section discusses the different costs CAPEX and OPEX related to both front-haul and to the backhaul part of the 4G mobile network. Also an overview is given about some metaheuristics that could be used to solve the formulated optimization equations.

2   Migration Challenges

Different factors are pushing a mobile network operator to make investment in new technologies; as it is the case with LTE and even LTE-Advanced (LTE). Some factors could be related to the need of the operator itself and/or the need of his customers; as depicted in Fig. 1. Other factors could be related to the competition and the development of the telecom markets. An example could be the fact when a telecom regulatory put LTE frequencies for an auction. In this case, all MNOs must

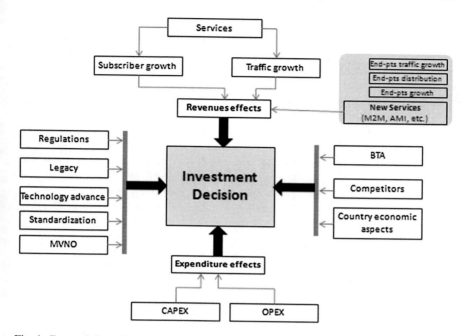

Fig. 1 Factors influencing MNO's decision about new investments

bid in this auction though the use of such frequencies is not foreseen in the near future. This could be better than losing all frequencies to the competitors.

2.1 Challenges and Risks Facing the Migration

The deployment of the modern technology is not only related to the equipment performance; rather also to the required investment together with the expected revenue which should balance the business case. Therefore, the deployment of 4G and beyond is not equally distributed over all countries. When considering the Mediterranean area, we see a large difference between North and South regions. While on the European side of the Mediterranean, e.g. France and Italy play a leading role on deployment of LTE in the world [12]; on the South shore of the Mediterranean, the first LTE in North Africa is just planned to be launched; [3] and in some countries the call for bidding for LTE licenses has just been planned [1].

Different factors are responsible for this gap, such as: (1) LTE was designed to support some more advanced applications and innovations, like Machine-to-machine (M2M) communications, smart grid, and other smart cities applications. These applications are deployed intensively in Europe, but not yet used in North African market; (2) LTE was designed to be able to cover a wide range of spectrum. However, the situation of these frequencies is different from one country to another; (3) In the ITU report about "The ICT Development Index (IDI)" by the end of 2014, it states that "Income inequalities contribute to making broadband unaffordable. Handset-based mobile-broadband prices are affordable for almost all households in the developed world; but unaffordable for some segments of the population in the developing world." [7]; (4) Some competing solutions allowing to the users the access to broadband wireless/mobile Internet, such as Wi-Fi Hotspots that are provided either by Cable-TV and/or DSL operators. In such case, the operator negotiates with his subscribers (residential or enterprise) to allow him to provide WLAN Hotspots to new external subscribers. Examples of this scenario are in [2, 5, 8]; (5) 3rd generation of mobile is still competing with the 4G, because some MNO has opted to upgrade their 3G network to some more recent releases of 3G+ (especially HSPA+ with very attractive throughput allow WCDMA networks upgraded to HSPA and to 7.2, 21 and 42 Mbps; see [4]; (6) the mobile technology migration is depending on the society ICT and economical maturation. So, even by 2020 in some countries a large part of the population in some countries will still be covered by GSM/GPRS-only technology; while in this year already an US MNO is switching off 2G equipment.

2.2 Need for Migration Planning

It is clear that the rollout of the LTE/LTE-A will not happen in an intensive manner to close it in a short time interval. An optimal workflow of the migration is necessary knowing that: (a) the adoption of the LTE is not made by all the countries at the same time and the same intensity. This depends on the maturity of the telecom market; (b) The maturity of the society which should have a real need and the ability to pay more money for LTE and its extras (ICT Index, [7]); and (c) The development of the capacity/bandwidth demand is not equally distributed over the entire country; especially in the countries of the South regions of the Mediterranean. Also, the rural areas are less developed to require more than 2G connections for mainly voice call.

During the design of the migration strategy there are several parameters building the degrees of freedom that affect the decisions of network designer to bring the current network infrastructure towards 4G and beyond. An example we can cite: (1) Which part of spectrum to bid for it; (2) How to use the different portions bought spectrum (knowing that LTE use a wide range of frequencies dispatched over the spectrum; like 800, 1800, 2.4, 2.6, 700 and 800 MHz, [6]; (3) Type of Antennas (mono-, duo- or tri-band and which frequency bands values); (4) Locations for eNB (selection and acquisition) and the type/size of eNB that may fit there; either for indoor or outdoor; (5) Where/when to start the migration and at which speed of deployment; and (6) What to reuse from legacy network (existing 2G/3G/3G+ and their backhauls).

3   Modeling of the Migration

3.1   Migration Tasks and Objectives

The transition from one state to another depends on the migration objective set by the MNO; which has to be optimized by the whole "migration path" going from the current state of the network and reaching a final state at the last time slot of the whole migration time window; as illustrated in Fig. 2.

Different objectives can be set according the priorities fixed by the network owner; such as: (a) LTE must be rolled out only in cells with a given threshold for Average Revenue per User (ARPU) and/or Average Revenue per Minute (ARPM) and/or Average Revenue per Cell Site; (b) LTE must be rolled out only in cells with a given overload; (c) to minimize the whole migration time without constraint (or soft constraints) concerning the migration costs/budgets; (d) LTE to be deployed only in a given percentage of the served area (or served users). This scenario is possible where the regulatory wants to have a certain level of broadband wireless availability in order to attract investors from aboard to invest in some IT offshore parks or duty free zones; (e) Minimize the whole migration costs by extending the transition in the whole time window, with minimum costs at each migration time slot (or the total costs; over the whole migration time). This could be interesting for MNO who wants not to go intensively in the LTE deployment, because they are targeting the time after 2020, but at the same time they have to satisfy some regulatory requirements. The migration of mobile networks towards 4G and beyond is described in Table 1.

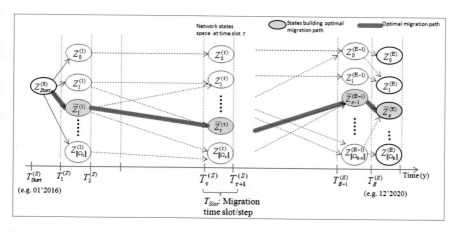

Fig. 2 State space over the migration time window; with a possible migration path

Table 1 Components of network migration as optimization problem

Category	Components
Given	Legacy network (current mobile network infrastructure)
	Candidate new locations (could have limited capacity if shared with other players)
	Users/service/traffic distributions
	Forecasting for the users/services/traffic development during migration
	Costs for CAPEX/OPEX of network elements
	Forecasting for price and costs development
Task	Determine a migration path: (i) going from the current state of the network; (ii) the transition of the network states from a time slot to the next ones; and (iii) arriving to the final state at the last migration slot
	i.e.: Determine which state to migrate to and at which time slot
Objective	Minimize the total migration costs
Constraints	Quality of Services (QoS)
	Requirements from the regulatory
	Capacity of system elements
	Capacity of implementation team/mobile work force management

3.2 Migration Model—Network States and Migration Path

To model and describe the migration, a set of variables and symbols are used. A possible state of the network at the migration time slot τ is represented by Eq. 1. At τ the network can have different possible states as illustrated by Fig. 3.

$$Z_t^{(\tau)} = \left\{ z_0^{(\tau)} \quad z_1^{(\tau)} \quad \cdots \quad z_k^{(\tau)} \quad \cdots \quad z_{\|S\|-1}^{(\tau)} \quad z_{\|S\|}^{(\tau)} \right\} \tag{1}$$

$\tilde{Z}_t^{(\tau)}$ represents a state from time slot τ belonging to the optimal migration path; S builds the set of sites building the geographic area under consideration.

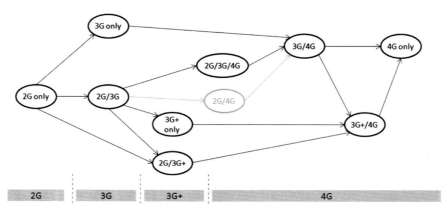

Fig. 3 Different states of network sites and possible transitions in the state space

$z_k^{(\tau)}$ is the state of the site k at time slot τ.

$Z_{Start}^{(0)}$ represents current state of network when starting the design of the migration.

$\tilde{Z}_E^{(E)}$ is the final state of network at the end of the migration time window (i.e. last migration time slot). This state belongs to the optimal migration path.

Ω_τ is the set of states to which network can migrates at time slot τ.

T_{migr} shows size of migration time window constituted by E discrete time slots τ.

The Total costs generated by the migration of site i at migration time slot j is given by Eq. 2; where $C_{i,j}^{(CAPX)}$ and $C_{i,j}^{(OPX)}$ are CAPEX and OPEX generated by the migration of site i at migration time slot j; respectively.

$$C_{i,j} = C_{i,j}^{(CAPX)} + C_{i,j}^{(OPX)} \tag{2}$$

The cost of the migration of a site depends strongly on the starting and the reached state. For example: $C_{i,j;2G \to 3G+}$ covers the costs generated by site i at migration time slot j from 2G only (GSM/GPRS/EDGE) to 3G+ only technology; while $C_{i,j;2G/3G+}$ are costs generated by site i at migration time slot j from 2G only to 3G only technology, which means adding 3G infrastructure to the site.

Before the introduction of the 4G each site can be in of the following states: (a) having 2G only; (b) or both generation 2G/3G; (c) 2G/3G+; (d) 3G only or (e) 3G+ only; as depicted in Fig. 3. We exclude the state 3G/3G+; which means that the site have two variants of third generation, the UMTS (3G) and the HSPA. It is worth to underline that the 3G+ can have three possible variants: (1) High Speed Downlink Packet Access (HSDPA); (2) High Speed Packet Access (HSPA, referred sometimes also as HSDPA/HSUPA); or (3) HSPA+ that is an extended variants of HSPA.

Different objectives can be set for migration planning; such as minimization of the total migration costs. In this case, this optimization objective can be formulated as in Eq. 3. The constraints in Eq. 4 uses the variable indicates if the site i migrates at time slot j. Each site has to migrate only once during the fixed migration window as forced by Eq. 5. Finally, Eq. 6 guarantees that at each tine of the migration the site capacity is larger than the users demand in the site.

$$Minimize \; C_{Migration}^{(Total)} = \sum_{i=0}^{E} \sum_{j=1}^{\|S\|} x_{i,j} C_{i,j} \tag{3}$$

Such that:

$$x_{i,j} = \begin{cases} 1; & migration \\ 0; & otherwise \end{cases} \tag{4}$$

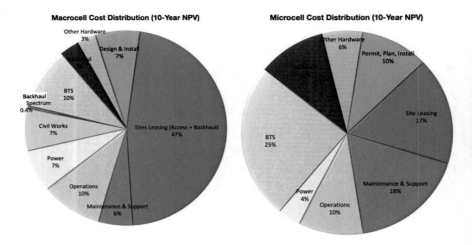

Fig. 4 Costs components for macro-cells and micro-cells (based on [11])

$$\sum_{i=0}^{E} x_{i,j} < 2 \quad \forall j \in S \tag{5}$$

$$D_{i,j} < K_{i,j} \tag{6}$$

3.3 Migration Costs

The deployment costs of 4G networks cover: Cost for radio access part, equipment cost for gateways, cost for transmission link between outer POPs and inner POP (Point of Presence), and cost for handling of traffic behind the gateway (backbone transmission). The different costs components can be categorized as given in Fig. 4.

4 Conclusions

In this paper, we discussed the current situation and challenges facing the mobile network operators, who are forced to look for new technologies that decrease the cost per transmitted Megabyte. The Long Term Evolution (LTE) and LTE-Advanced (LTE-A) bring the technical answers for these challenges; however, the deployment of this technology hide some risks. In spite of that, equipment from different generation will coexist even beyond 2020. Therefore, a migration towards 4G and beyond has to be done in multi-period planning-based strategy. We formulate the migration problem as a search for the shortest migration path in the space state of the network sites. The solution of this problem has to determine in an optimal way which site(s) to be migrated at which migration time slot, so that the total cost are as minimum as possible while respecting some constraints. In future work, optimization algorithms will be applied to some test scenarios in order to check the feasibility of the proposed model. A first interest will be given to some widely used metaheuristics; such evolutionary algorithms and local search, before trying to find some exact algorithms that solve the problem in an acceptable computation time.

References

1. ANRT: Appel d'offre de l'ANRT-Maroc pour l'attribution de licences en vue de l'établissement et l'exploitation de réseaux 4G au Maroc, Nov 2014 (in French)
2. ComCast's Service Homepage: XFINITY® WiFi. http://wifi.comcast.com/
3. Ericsson Press Release: Ericsson to launch North Africa's first LTE and VoLTE network for Algérie Telecom. Ericsson Press Release from 05.05.2014
4. Ericsson: Ericsson mobility report 2014—on the pulse of networked society. Ericsson report, Nov 2014
5. German CableTV (Kabel-Deutschland) WiFi Hotspot Service Homepage (www.hotspot.kabeldeutschland.de/hotspots.html)
6. Hadden, A.: GSA evolution to LTE report. Global Mobile Suppliers Association (GSA) Press release, July 2014
7. International Telecommunication Union (ITU): Measuring the information society report 2014. ITU Report Publications, Dec 2014
8. Lynn, S.: 4 concerns about Comcast's Xfinity Wi-Fi hotspot rollout. PC Magazine (2014)
9. METIS Project: Mobile and wireless communications enablers for the 2020 information society. European R&D Project (EU FP7 ICT-317669-METIS), Project Deliverable D8.3
10. Pietrzyk, P.: Towards 5G—Business Model Innovations. Nutaq Publications (2014)
11. Solheim, A.: Bridging the Gap in LTE Backhaul: Network Architectures and Business Models. DragonWave Corporate Publication (2012)
12. Westwood, S.: Global state of LTE report. OpenSignal Report, Feb 2014

Adaptive Modulation and Coding for the IEEE 802.15.3c High Speed Interface Physical Layer Mode

Mohammed Amine Azza, Moussa El Yahyaoui and Ali El Moussati

Abstract Adaptive modulation and coding (AMC) is a technique that allows the adjustment of modulation and coding based on the quality of the transmission link, to achieve a transmission with higher data rate and better spectral efficiency. The purpose of this paper is to study the influence of AMC on the High Speed Interface Physical layer (HSI PHY) mode of the IEEE 802.15.3c standard. For this we have created a Simulink model representing the schema of the IEEE 802.15.3c transmitter/receiver with an algorithm which allows adaptation in terms of modulation and coding of the transmitter according to the link in accordance with a predefined script.

Keywords IEEE 802.15.3c · Adaptive modulation and coding · OFDM · 60 GHz

1 Introduction

The growth in the use of wireless devices has led to the improvement of the spectral efficiency to allow higher throughputs information. In traditional communication systems, the transmission is designed for a scenario facing variations of the channel, in order to provide an error rate lower than a specific limit. Adaptive character in transmission systems have been developed to solve the problem of variation in the quality of the transmission channel by adjusting the settings to take advantage of available resources by adapting the modulation, coding, transmission rate and other parameters to the conditions on the radio link (e.g. the path loss, the interference due to signals coming from other transmitters, the sensitivity of the receiver, the available transmitter power margin, etc.), to optimize the average spectral efficiency of the link.

M.A. Azza · M.E. Yahyaoui · A.E. Moussati (✉)
Signals Systems and Information Processing, National School of Applied Sciences, Oujda, Morocco
e-mail: aelmoussati@ensa.ump.ma

© Springer International Publishing Switzerland 2016
A. El Oualkadi et al. (eds.), *Proceedings of the Mediterranean Conference on Information & Communication Technologies 2015*, Lecture Notes in Electrical Engineering 380, DOI 10.1007/978-3-319-30301-7_38

365

The AMC technique represents a tool, to increase the spectral efficiency of the radio channels [1], varying in time while maintaining a BER predictable, the order of modulation and Forward Error Correction (FEC) are modified by the coding rate adjustment for variations in the quality of the transmission channel. In the case of high attenuation, for example, a lower Signal-to-Noise Ratio (SNR), the size of the signal constellation is reduced to improve reliability, so a more robust transmission for an effective SNR. Conversely, low attenuation period or high gain, the size of the signal constellation is increased to enable significant modulation rates with a low probability of error, which has an immediate effect on the SNR.

This work falls within the framework of our study of Radio over Fiber (RoF) Systems; its main objective is to develop architecture able to adapt the transmission parameters such as modulation and coding type according to channel quality. For this purpose we have chosen to work on the IEEE 802.15.3c standard to design a model meets all the specifications of the standard and having the advantage of flexibility in terms of modulation and coding. The studied adaptation mechanism is taking into account transparency to the channel type, it allows the adaptation of modulation and FEC code rate regardless of the channel used, radio or optical.

To shed light on the importance of adaptability in terms of modulation and coding, we evaluate the performance of our transmission system before and after the introduction of the AMC mechanism. So, we developed a simulation environment MATLAB/Simulink. First, according to the standard specifications, a model of HSI PHY of IEEE 802.15.3c (TG3c) [2] was designed. Then we went to work on an innovative technique that allows to adapt in terms of modulation and coding with the variations of the transmission channel, while ensuring full transparency against the channel type, to reach our main objective to having a transmission system with high spectral efficiency and stability in terms of transmission quality in the presence of channel variations, and whatever the type of this latter.

2 High Speed Interface Physical Layer HSI PHY of IEEE 802.15.3c Standard

Several reasons justify the attention we have given to the 60 GHz technology, it is clear that this latter offers various advantages over nowadays communication systems. The main reason is the huge and continuous unlicensed bandwidth available; this huge bandwidth provides good solutions in terms of capacity and flexibility. Add to this that the regulation of this band allows a relatively higher transmission power to overcome the greater path loss at the frequency of 60 GHz [3]. Another advantage presented by the 60 GHz technology is the spectral efficiency, and also the ability to support very high data rate applications with a simple modulation.

The interest of 60 GHz radio does not cease to grow especially with the emergence of several international mm-wave standards groups and industry alliances, we find among others, the IEEE 802.15.3c [2] working group (TG3c) which

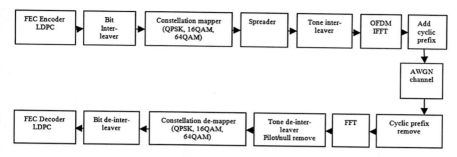

Fig. 1 HSI PHY mode of IEEE 802.15.3c block diagram

was formed to develop an alternative physical layer (PHY) mm-wave for the existing IEEE 802.15.3 WPAN standard 802.15.3-2003, there's also the ECMA TC-48 [4], which began an effort to standardize Medium Access Control (MAC) and PHY for high-speed, last but not least, the Wireless Gigabit Alliance (WiGig) [5] formed in May 2009 to establish a unified specification for 60 GHz wireless technology to create a truly global ecosystem of interoperable products for a wide range of applications (Fig. 1).

As mentioned above, our recent work is based on the IEEE 802.15.3c standard which is the first IEEE wireless standard for data rates over 1 Gb/s. According to the report of TG3c, three PHYs are defined for the mm-Wave PHY, namely single carrier (SC) PHY, high speed interface (HSI) orthogonal frequency division multiplexing (OFDM) PHY and audio video (AV) OFDM PHY. We made the choice to focus our work on second mode HSI PHY, designed for high speed bidirectional data transmission and uses OFDM with an FEC based on LDPC.

The figure above shows the block diagram of the design of HSI PHY mode in Simulink. At first, random data bits are generated and then coded by two LDPC encoders with FEC rates of 1/2, 3/4 and 5/8, the output of LDPC encoder multiplexed to form a single data stream. After the data multiplexer, the bits are interleaved by a bloc inter-leaver. The coded and interleaved bits can be modulated using QPSK, 16QAM and 64QAM modulation. Subcarrier constellations consist of QPSK, 16-QAM, and 64-QAM. Each group of 336 complex numbers are assigned to an OFDM symbol. In each symbol OFDM DC, null, and pilot tones are inserted before the tone inter-leaver. Tone inter-leaver makes sure that neighbouring symbols are not mapped into adjacent subcarriers. The interleaved tones are modulated by OFDM modulator that consists of 512-point IFFT. Finally, a cyclic prefix of 64 tones is inserted to keep the OFDM system from inter-symbol and inter-carrier interferences, and then symbols are transmitted.

At the reception, cyclic prefix is removed from the received signal and then demodulated by OFDM that consist of 512-point FFT. After de-interleaving process, data carriers are identified and QPSK, 16QAM or 64QAM symbols are demodulated by the de-mapper block. The obtained bit stream is de-interleaved and demultiplexed into 2 bit streams to be decoded with 2 LDPC decoders. Different modulation and coding schemes used parameters are given in Table 1.

Table 1 HSI PHY MCS used dependent parameters

MCS index	Data rate (Mb/s)	Modulation scheme	FEC rate
1	1540	QPSK	1/2
2	2310	QPSK	3/4
4	3080	16-QAM	1/2
5	4620	16-QAM	3/4
7	5775	64-QAM	5/8

3 Adaptive Modulation and Coding AMC

As mentioned in the introduction, adaptive modulation and coding is a technique that targets a better use of available resources in a communication system [6], with minimum error probability. Its a technique to adapt and adjust, in real time, the transmission parameters according to the quality of the link (Fig. 2).

At this level, we introduced a channel estimation system (red boxes) with a feedback path between the estimator and the transmitter to which the receiver returns information to the channel state for adapting modulation and coding. The transmitter expects the receiver the channel estimation results to select one of the preset MCSs modes for the next transmission interval. Note here that our adaptive system requires tremendous variations in channel quality, and must also make sure that there's not a huge delay between the estimation of the quality and the actual transmission to ensure effective operation of this system.

The channel estimation metric used is the SNR. For estimating SNR at the reception, we chose a simple method among several available [7]. A real time estimator which is adapted to all types of modulation used is developed. As well known, the SNR is the ratio between the signal and the noise powers.

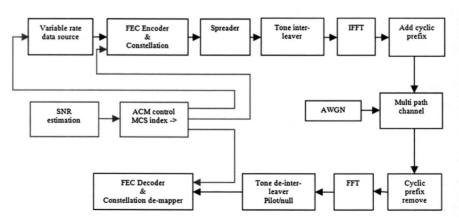

Fig. 2 AMC system and HSI PHY mode of IEEE 802.15.3c block diagram

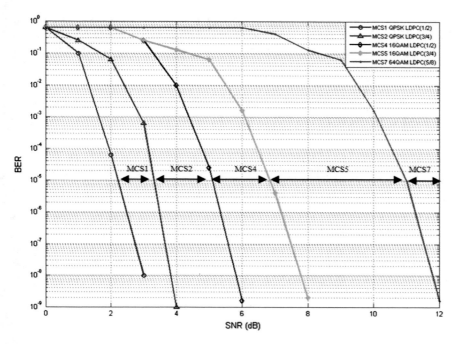

Fig. 3 BER versus SNR thresholds and corresponding to the N = 5 MCS used by the AMC

SNR = S/N, where S represents the average signal power, and N is the average noise power which can be calculated from the noise variance, σ^2.

The choice of the appropriate modulation and coding scheme (MCS) for use at a new transmission, is realized at the AMC control block based on the prediction of the channel status. An SNR threshold ensures a BER below a limit value BER0, which is defined by the system for each method whenever the SNR is above the threshold.

From the BER, SNR thresholds are achieved according to their characteristics for a modulation and coding scheme on an AWGN channel. As described in Fig. 3, this method consists in cutting the SNR range in N + 1 sub regions (N is the number of modulation and coding schema used by the AMC method, in our case N = 5), N + 2 thresholds which we translate as.

Each of N schemes is then used to operate in a particular area of the SNR. When SNR value is within an area, the information associated with the channel status is sent to the AMC control block which then adapts and requires the transmitter a transmission rate, the encoder encoding type and chosen a method of modulating guaranteeing a BER below the threshold BER0. This allows the system to transmit data with high spectral efficiency when the SNR is high and reduced when the SNR is low.

The AMC control block has the function to detect the crossing of decision thresholds for determining the method of modulation and coding to be used in the next frames sent by returning this information not only to the transmitter but also to the receiver.

4 Test and Results

For our tests, we have considered the HSI PHYs MCSs parameters presented in Table 1. The signal through the channel is disturbed by AWGN. Thus, the receiver receives a signal with a time-varying channel signal to noise ratio CSNR thing that affects the total SNR at the reception.

The curves shown in Fig. 3 correspond to different modulation methods and coding permitted by the HIS PHY mode of IEEE 802.15.3c system, MCS1, MCS2, MCS4, MCS5, and MCS7. They were obtained from simulations assuming an accurate knowledge of the coefficients of the AWGN channel. The entire adaptation (switching) threshold are obtained by reading SNR points corresponding to a $BERthreshold = 10^{-5}$.

Figure 3 shows the BER versus SNR curves of 5 encoding and modulation schemes with the margins of use of each scheme mentioned with arrows. The points of switching between MCS are fixed by the BER threshold. From the case where

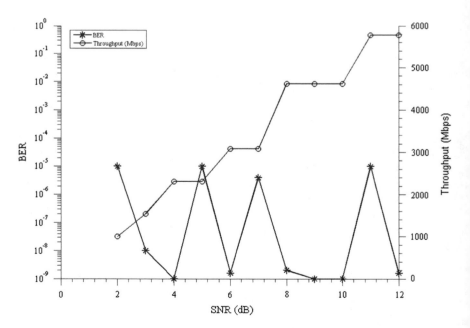

Fig. 4 Curves of BER versus SNR and throughput variation for the AMC method

the SNR is higher (SNR between 10 and 11 dB), we use the modulation and coding scheme MCS7 guaranteeing a BER below the BER threshold, when the estimated SNR at the reception increases, a switch to the next scheme MCS5 is immediately made, and so on until we get to use the first scheme MCS1 in the SNR margin between 2 and 3 dB.

The variation of SNR at the reception due to CSNR variation, introduces a change in the modulation and coding scheme used, which causes a throughput variation as shown in Fig. 4. It can be noticed that with a high SNR levels we can transmit data with fairly high throughput, degradation in the quality of the link causes a change of MCS used to keep an acceptable BER level, but this time with less throughput.

5 Conclusion

In the present paper, a solution of adaptation in terms of modulation and coding for the HSI PHY mode of the IEEE 802.15.3c has been proposed and studied. 5 schemas modulation and coding were used with three different types of modulation (QPSK, 16QAM, 64QAM) and variable FEC rate. The IEEE 802.15.3c model and simulations were made on Matlab/Simulink.

The main conclusion drawn from the results obtained that the adaptive telecommunication systems have an advantage over the conventional system with fixed coding and modulation schemes; this advantage is the spectral efficiency and the stability of the transmission quality against the channel variation.

References

1. Svensson, A.: An introduction to adaptive QAM modulation schemes for known and predicted channels. Proc. IEEE **95**(12), 2322–2336 (2007)
2. IEEE 802.15 TG3c Working Group: Part 15.3: wireless medium access control (MAC) and physical layer (PHY) specifications for high data rate wireless personal area networks, (WPANs). http://www.ieee802.org/15/pub/TG3.html (2009)
3. Yong, S., Xia, P., Valdes-Garcia, A.: 60 GHz technology for Gbps WLAN and WPAN: from theory to practice. Wiley, New York, 296 pp. ISBN 0470747706 (Safari Book)
4. ECMA-378 Standard: High rate 60 GHz PHY, MAC and HDMI PAL. http://www.ecmainternational.org/publications/standards/Ecma-387 (2008)
5. Wireless Gigabit Alliance: http://wirelessgigabitalliance.org/news/wigigalliance-publishes-multi-gigabit-wireless-specification-and-launches-adopter-program/ (2009)
6. Azza, M.A., El Moussati, A., Mekaoui, S., Ghoumid, K.: Spectral management for a cognitive radio application with adaptive modulation and coding. Int. J. Microwave Opt. Technol. **9**(6) (2014)
7. Pauluzzi, D.R., Beaulieu, N.C.: A comparison of SNR estimation techniques in the AWGN channel. In: Proceedings of IEEE Pacific Rim Conference on Communications, Computers and Signal Processing, pp. 36–39 (1995)

The Password Authenticated Connection Establishment Protocol (PACE) Based on Imaginary Quadratic Fields

Taoufik Serraj, Soufiane Mezroui, Moulay Chrif Ismaili
and Abdelmalek Azizi

Abstract Using Buchmann-Williams protocol and relying on the intractability of the discrete logarithm problem in the class group of an imaginary quadratic field, we introduce a new version of Password Authenticated Connection Establishment protocol. This construction (IQF-PACE) can be proved secure in Bellare Pointcheval Rogaway (BPR) model, and provides a variant of the actually elliptic curve instantiation of the PACE protocol to avoid some side channel attacks, particularly, Fault Analysis Attacks (FAAs).

Keywords Imaginary quadratic field · PACE · Security · Smart card

1 Introduction

The security of the most of cryptographic primitives is based on the hardness of some mathematical problems (e.g., factoring, discrete logarithm). Key exchange is an important primitive in cryptographic protocols. To enable two parties, typically a client and a server, to establish a secure and authenticated channel over an insecure network, several Authenticated Key Exchange protocols (AKE) are proposed [1–4] using various authentication means such as: (public or secret) cryptographic keys or short secret keys (password). The latter case is the most important in practice, since both parties generate a strong cryptographic key from a shared human-memorable password (e.g., PIN) without requiring a complex public key infrastructure (PKI), these protocols are called Password-Authenticated Key Exchange (PAKE).

Password Authenticated Connection Establishment (PACE) framework [5, 6] is an example of Password-Authenticated Key Exchange protocols. Actually an elliptic curve based instantiation of PACE is used to secure the communication between machine-readable documents (e.g., e-passport) and reader machines [5–7],

T. Serraj (✉) · S. Mezroui · M.C. Ismaili · A. Azizi
ACSA Laboratory Faculty of Sciences, Mohammed First University,
60000 Oujda, Morocco
e-mail: taoufik.serraj@gmail.com

© Springer International Publishing Switzerland 2016
A. El Oualkadi et al. (eds.), *Proceedings of the Mediterranean Conference on Information & Communication Technologies 2015*, Lecture Notes in Electrical Engineering 380, DOI 10.1007/978-3-319-30301-7_39

373

also PACE may be used in other situations where authenticated key exchange are required (e.g., Near Field Communication Technology).

In the below, the Buchmann-Williams protocol [2] is shown in Sect. 2. In Sect. 3, an Imaginary Quadratic Field instantiation of PACE protocol (IQF-PACE) is described. Finally, the security of the IQF-PACE protocol is discussed, specially, in the presence of faults attacks.

2   A Key Exchange Protocol Based on Imaginary Quadratic Field

This section recalls what is the Buchmann-Williams key exchange system [2].

Let $D<0$ be a square free integer and let $\mathbb{K} = \mathbb{Q}(\sqrt{D})$ be an imaginary quadratic field of discriminant Δ satisfying the conditions of security fixed in [8]. Now, let us take an ideal I of the ring of integers $O_{\mathbb{K}}$ of $\mathbb{K}$. We will denote the positive generator of the ideal $I \cap \mathbb{Z}$ by $L(I)$. The ideal I is called primitive if it is not divided by any principal ideal of $\mathbb{Z}$. Furthermore, it is said to be reduced if it is primitive and there does not exist a nonzero $\beta \in I$ for which $|\beta| < L(I)$.

It is well known that each equivalence class of ideals of $O_{\mathbb{K}}$ contains a reduced ideal. This is a natural result which can be deduced from properties of the fundamental domain of the modular group. Further, there is an algorithm running in polynomial time for finding such a reduced ideal.

Throughout the paper, for any ideal I of $O_{\mathbb{K}}$, we note I_{red} to be the reduced ideal equivalent to I.

Now, we will describe the Buchmann-Williams protocol:

Two users Alice and Bob select a value of Δ so that $|\Delta|$ is large and an ideal I in $O_{\mathbb{K}}$ with sufficiently big order. The values of Δ and the ideal I are public.

1. Alice selects randomly an integer x and computes a reduced ideal J such that J is equivalent to I^x. She sends J to Bob.
2. Bob selects randomly an integer y and computes a reduced ideal M such that M is equivalent to I^y. He sends M to Alice.
3. Alice computes a reduced ideal $(M^x)_{red}$ equivalent to M^x, Bob computes a reduced ideal $(J^y)_{red}$ equivalent to J^y.

Since M^x is equivalent to I^y, and J^y is equivalent to I^{xy}, the reduced ideal computed by Alice and Bob is the same and we can deduce from this $(M^x)_{red} = (J^y)_{red}$. They take as the common secret key $L((M^x))_{red}) = L((J^y)_{red})$.

This protocol is generalized mathematically in [4], and all we can do here by Buchmann-Williams we can do it also by the general methods presented in [4]. The fundamental domain of the modular group is replaced by the Hilbert fundamental domains of some biquadratic number fields, and the algorithm of reduction is replaced by another algorithm due to Stark.

Notice that we will work in this paper with Buchmann-Williams protocol only for simplicity.

3 The Password Authenticated Connection Establishment Protocol Based on Imaginary Quadratic Fields

The Password Authenticated Connection Establishment protocol (PACE) was proposed by the German Federal Office for Information Security (BSI) [5], it is designed to be free of patents. PACE was adopted by the International Civil Aviation Organization (ICAO) [7] to secure the contactless communication between machine-readable travel documents (a chip) and a reader (a terminal).

We define a function F which will be needed in the construction of the protocol.

Let $D < 0$ be a square free integer and let $\mathbb{K} = \mathbb{Q}(\sqrt{D})$ be an imaginary quadratic field.

It is well known that the ring of integers $O_{\mathbb{K}}$ of $\mathbb{K}$ is $\mathbb{Z} + \mathbb{Z}\omega$, where:

$$\omega = \frac{r - 1 + \sqrt{D}}{r} \quad \text{and} \quad r = \begin{cases} 1 & \text{if } D \equiv 2, 3 \,(\mathrm{mod}\,4) \\ 2 & \text{if } D \equiv 1 \,(\mathrm{mod}\,4) \end{cases}. \tag{1}$$

Now, if I is any ideal of $O_{\mathbb{K}}$, then

$$I = \mathbb{Z}a + \mathbb{Z}(b + c\omega), \tag{2}$$

where $a, b, c \in \mathbb{Z}$, $a > 0$, $c > 0$, $c|a$, $c|b$ and $ac|N_{\mathbb{K}/\mathbb{Q}(b + c\omega)}$.

Let div be a function giving as output a divisor $div(t)\,(<t)$ of any random integer $t \in \mathbb{Z}$. The function F maps an integer n from $\mathbb{Z}$ to an element in the class group G of the imaginary quadratic field $\mathbb{K}$, F is defined by:

$$F: \mathbb{Z} \to G$$
$$G \mapsto \mathbb{Z}div\left(\left(n + \frac{r-1}{r}\right)^2 - \frac{D}{r^2}\right) + \mathbb{Z}(n + \omega). \tag{3}$$

3.1 The Protocol Description

We denote a chip by C and a terminal by T, both C and T share a common password π, they also agree on the following parameters and cryptographic primitives:

1. The class group G of an imaginary quadratic field $\mathbb{K}$ of discriminant Δ such that $|\Delta|$ is a prime $\equiv 3$ (mod4) (for security purposes) with class number

h satisfying $h \approx \sqrt{|\Delta|} \approx p$, where p is a prime of sufficiently bigger size [8, 9].
2. A random number generator.
3. A symmetric encryption scheme *ENC*.
4. A hash function *H*.
5. A function *F* mapping an integer to an element in the class group *G*.
6. A message authentication code *MAC*.

The PACE protocol (Fig. 1) comprises four phases: Randomization, Mapping, Key establishment, and key confirmation. It is a hybrid cryptographic protocol that combines symmetric and asymmetric primitives.

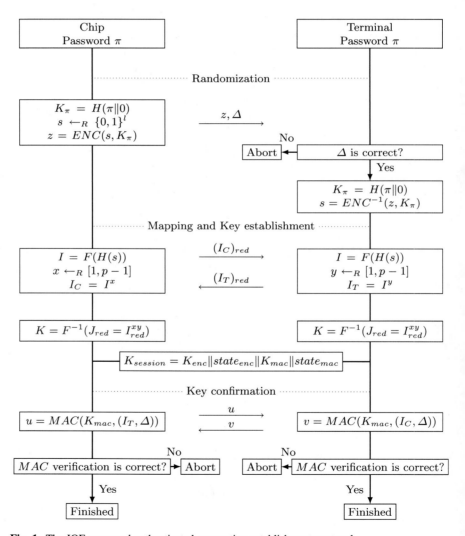

Fig. 1 The IQF-password authenticated connection establishment protocol

In the Randomization phase:

1. C chooses a random nonce s, encrypts it through the cipher ENC using a password-derived key K_π and sends $(z = ENC(s, K_\pi), \Delta)$ to T.
2. T recovers $s = ENC^{-1}(z, K_\pi)$ by decryption z under K_π.

C and T abort the execution if Δ is incorrect.

In the Mapping phase:

1. C and T hold s.
2. They both compute $I = F(H(s))$, where $I \in G$.

In the Key establishment phase:
C and T perform a Diffie-Hellman exchange on the class group G.

1. C selects a random $x \in [1, p-1]$, computes $I_C = I^x$, and sends its reduced ideal to T.
2. T selects a random $y \in [1, p-1]$, computes $I_T = I^y$, and sends its reduced ideal to C.
3. C and T both compute the reduced ideal equivalent to $J = I^{xy}$.
4. C and T derive the key $K = F^{-1}(J_{red})$.

The session key is:

$$K_{session} = K_{enc} \parallel state_{enc} \parallel K_{mac} \parallel state_{mac},$$

where $K_{enc} = H(1 \parallel K)$ and $K_{mac} = H(2 \parallel K)$. Further, $state_{enc}$ and $state_{mac}$ are two sequences identifying the session and the parties C and T.

In the Key Confirmation phase:
C and T confirm knowledge of the session key $K_{session}$ by:

1. C computes $u = MAC(K_{mac}, (I_T, \Delta))$ and sends u to T.
2. T computes $v = MAC(K_{mac}, (I_C, \Delta))$ and sends v to C.

Each party checks the others MAC and they abort the execution in the case of a mismatch.

4 The IQF-PACE Security

The security of IQF-PACE mainly relies on the password, since such password has limited size, an attacker may lead a brute-force attack from suitable dictionary, this attack is called on-line dictionary attack and is unavoidable, but we can limit the use of passwords by blocking the card if a certain number of failed attempts have occurred. In addition, the attacker should not be able to deduce the password of any party in an off-line dictionary attack.

4.1   Security Requirements for IQF-PACE

In [10], a security analysis of the PACE framework is provided by Bender et al., the authors show that the PACE protocol is secure as Password-based Key-Exchange protocol in the real-or-random sense of Abdalla, Fouque and Pointcheval [11] independently of the choice of the function F mapping an integer to an element of the group G if:

1. The encryption scheme and the hash function used in PACE behave ideally.
2. The message authentication scheme is unforgeable against adaptively chosen message attacks.

   In addition, other number theoretic assumptions related to the Diffie-Hellman problem are required. According to Biasse et al. [8], Hamdy et al. [9] and in accordance with actual recommendations of international security organizations (e.g., BSI, NIST, and ANSSI) in order to avoid the following mathematic attacks it suffices:

1. Index-calculus attack, $|\Delta|$ must be large enough ($\approx$1348 bits).
2. Pohlig-Hellman attack, Rho-Pollard attack and Lambda-Pollard attack, the class number $h \approx p$, where p is a large prime number of size at least 687 bits.
3. All key sizes must be at least of 224 bits for the years 2014–2015 and 256 bits for the years 2016–2020.

4.2   Side-Channel Attacks

In the real-world applications, the interaction between the cryptographic primitive and the adversary depends not only on the mathematical description of the primitive, but also on its implementation and the specifics of the physical device on which the primitive is implemented. In this new setting, physical attacks arise from the fact that cryptographic devices can leak some secret information during the execution of the protocol, like Power Consummation Analysis Attacks, Electromagnetic Radiation Attacks, Timing Attacks, Memory Attacks, Fault Analysis Attacks (FAAs) [12], these attacks are called side-channel attacks (SCAs). The first four attacks are passive, i.e., the attacker observes a working cryptosystem without influencing its operations, while Fault Analysis Attacks (FAAs) are active attacks that use faults to influence the operations of the system, the Fault Attacks are serious threats in practice.

   Elliptic curves are widely used in cryptography due to their efficiency for both software and hardware implementations. Unfortunately, elliptic curve cryptosystems are vulnerable to various faults attacks, e.g., Invalid-Curve Fault Attacks [13], the structure of the class group in Imaginary Quadratic Fields does not enable these attacks.

In the randomization phase, the use of a new random nonce s in each protocol execution makes Differential Fault attacks [14] on the symmetric encryption scheme impossible since the best fault attack of [15] needs at least two executions with the same nonce s.

5 Conclusion

In this paper, we propose the use of the class group of an Imaginary Quadratic Fields in PACE protocol, a good choice of IQF-PACE protocol parameters allows us to build a mathematically and physically secure Password-Authenticated Key Exchange protocol, this construction may be a variant of elliptic curve one in the presence of fault attacks.

Acknowledgments This work was supported by Acadamie Hassan 2 under the project "Mathmatiques et Applications: cryptographie" and URAC6-CNRST.

References

1. Bellovin, S.M., Merritt, M.: Encrypted key exchange: Password-based protocols secure against dictionary attacks. In: Proceedings of the IEEE Symposium on Research in Security and Privacy, pp. 72–84. IEEE, Oakland (1992)
2. Buchmann, J., Williams, H.C.: A key-exchange system based on imaginary quadratic fields. J. Cryptol. **1**, 107–118 (1988)
3. Gennaro, R.: Faster and shorter password-authenticated key exchange. In: Canetti, R. (ed.) Theory of Cryptography 2008. LNCS, vol. 4948, pp. 589–606. Springer, Heidelberg (2008)
4. Mezroui, S., Azizi, A., Ziane, M.: A key exchange system based on some bicyclic biquadratic number fields. In: International Conference on Multimedia Computing and Systems (ICMCS), pp. 1260–1264. IEEE Xplore, Marrakech (2014)
5. Federal Office for Information Security (BSI).: Advanced security mechanism for machine readable travel documents extended access control (EAC). Technical report, (BSI-TR-03110) Version 2.05 (2010)
6. Federal Office for Information Security (BSI).: Advanced security mechanisms for machine readable travel documents and eIDAS token. Technical report, (BSI-TR-03110) Version 2.20 (2015)
7. ISO/IEC JTC1 SC17 WG3/TF5 for the International Civil Aviation Organization.: Supplemental access control for machine readable travel documents. Technical report (2010)
8. Biasse, J.F., Jacobson Jr, M.J., Silvester, A.K.: Security estimates for quadratic field based cryptosystems. In: Steinfeld, R., Hawkes, P. (eds.) Information Security and Privacy 2010. LNCS, vol. 6168, pp. 233–247. Springer, Heidelberg (2010)
9. Hamdy, S., Moller, B.: Security of cryptosystems based on class groups of imaginary quadratic orders. In: Okamoto, T. (ed.) Advances in Cryptology ASIACRYPT 2000. LNCS, vol. 1976, pp. 234–247. Springer, Heidelberg (2000)
10. Bender, J., Fischlin, M., Kugler, D.: Security analysis of the PACE key-agreement protocol. In: Samarati, P., Yung, M., Martinelli, F., Ardagna, C.A. (eds.) Information Security 2009. LNCS, vol. 5735, pp. 33–48. Springer, Heidelberg (2009)

11. Abdalla, M., Fouque, P.A., Pointcheval, D.: Password-based authenticated key exchange in the three-party setting. In: Vaudenay, S. (ed.) Public Key Cryptography—PKC 2005. LNCS, vol. 3386, pp. 65–84. Springer, Heidelberg (2005)
12. Boneh, D., DeMillo, R.A., Lipton, R.J.: On the importance of eliminating errors in cryptographic computations. J. Cryptol. **14**, 101–119 (2001)
13. Biehl, I., Meyer, B., Muller, V.: Differential fault attacks on elliptic curve cryptosystems. In: Bellare, M. (ed.) Advances in Cryptology CRYPTO 2000. LNCS, vol. 1880, pp. 131–146. Springer, Heidelberg (2000)
14. Piret, G., Quisquater, J.J.: A differential fault attack technique against SPN structures, with application to the AES and Khazad. In: Walter, C.D., Koc, C.K., Paar, C. (eds.) Cryptographic Hardware and Embedded Systems—CHES 2003. LNCS, vol. 2779, pp. 77–88. Springer, Heidelberg (2003)
15. Tunstall, M., Mukhopadhyay, D., Ali, S.: Differential fault analysis of the advanced encryption standard using a single fault. In: Ardagna, C.A., Zhou, J. (eds.) Information Security Theory and Practice. Security and Privacy of Mobile Devices in Wireless Communication 2011. LNCS, vol. 6633, pp. 224–233. Springer, Heidelberg (2011)

Improving Performance of the Min Sum Algorithm for LDPC Codes

Abdelilah Kadi, Said Najah, Mostafa Mrabti and Samir Belfkih

Abstract In this paper we develop an algorithm for decoding low-density parity-check (LDPC) code that improve performance and reduce latency time. This algorithm is called Variable Factor Appearance Probability Min Sum (VFAP-MS) and is inspired from the Variable Factor Appearance Probability Belief Propagation (VFAP-BP) algorithm. The presented algorithm exploit the existence of short cycles in the code and strategy for reweighting check nodes, and is suitable for wireless communications applications. Simulation results show that the VFAP-MS algorithm outperforms the standard MS described in the literature.

Keywords Short cycle · Tanner graph · VFAP-BP · URW-BP · VFAP-MS

1 Introduction

In 1963 the LDPC codes have been described for the first time in the thesis of Gallager [1], these works were forgotten for 30 years. A commonly accepted reason to explain this omission is the difficulty at the time of designing efficient circuits for

A. Kadi (✉) · M. Mrabti
Information Sciences and Systems Laboratory,
Sidi Mohemed Ben Abdellah University, Fez, Morocco
e-mail: kadi.abdelilah@yahoo.fr

M. Mrabti
e-mail: mostafa.mrabti@yahoo.fr

S. Najah
Intelligent Systems and Applications Laboratory,
Sidi Mohemed Ben Abdellah University, Fez, Morocco
e-mail: najah.lessi@gmail.com

S. Belfkih
Systems Engineering Laboratory, Ibn Tofail University, Marrakesh, Morocco
e-mail: samir.belfkih@hotmail.fr

© Springer International Publishing Switzerland 2016
A. El Oualkadi et al. (eds.), *Proceedings of the Mediterranean Conference on Information & Communication Technologies 2015*, Lecture Notes in Electrical Engineering 380, DOI 10.1007/978-3-319-30301-7_40

processing algorithms. LDPC codes were reintroduced by Mackay and Neal in the year 1996 [2].

Recently, LDPC codes have been widely used, due to their performance in error correction are very close to the Shannon limit.

LDPC codes can be described by a bipartite graph called Tanner graph [3], these are composed of two sets of nodes, namely, variable nodes and check nodes. The signals go from the variable nodes to the check nodes and from the check nodes back to variable nodes. Typically different sets of wires are used for the signals going from variable nodes to check nodes and vice versa.

There are several parameters which may change the performance of an LDPC code, the short cycle or girth is among those parameters, the existence of short cycles in LDPC code can degrade the performance of the code [4, 5]. In this context, the existence of short cycles have a significant impact on the performance of LDPC codes, and require the development of an algorithm that takes into account the number of short cycles in the LDPC code.

Liu and de Lamare [6, 7] have recently proposed (VFAP-BP), which combines the reweighting strategy proposed by Wymeersch et al. in [8], with the knowledge of the short cycles, the results show that the (VFAP-BP) algorithm, outperforms the standard BP and URW-BP [8], and require small number of iterations.

In this paper, we propose to combine between the VFAP method and the MS algorithm, the resulting algorithm is called VFAP-MS. This algorithm also exploits the reweighting strategy [8], and the knowledge of the short cycles obtained by the cycle counting algorithms [9].

The simulation results show that the proposed VFAP-MS algorithm outperforms the MS algorithm in the convergence behavior, and offers a better (BER/SNR) performance than the standard MS particularly when using a small number of iterations.

This paper is organized as follows: Sect. 2 introduces the background in terms of understanding the Min Sum Algorithm. The proposed VFAP-MS algorithm is described in Sect. 3. The simulation and results are discussed in Sect. 4. Finally, Sect. 5 concludes the paper.

2   Literature Review

2.1   Sum Product Algorithms

We assume binary code word $(x_1, x_2, \ldots, x_N)$ is trans-mitted using Binary Phase-Shift Keying (BPSK) modulation. Then the sequence is transmitted over an Additive White Gaussian Noise (AWGN) channel and the received symbol is $(y_1, y_2, \ldots, y_N)$.

We define:

$V_{(i)} = \{j : H_{ij} = 1\}$ as the set of variable nodes which participate in check equation i.
$C_{(j)} = \{i : H_{ij} = 1\}$ Denotes the set of check nodes which participate in the variable node j update.
Also $V_{(i)} \backslash j$ denotes all variable nodes in $V_{(i)}$ except node j.
$C_{(j)} \backslash i$ Denotes all check nodes in $C_{(j)}$ except node i.

Moreover, we define the following variables which are used throughout this paper.
λ_i: Is defined as the information derived from the log-likelihood ratio of received symbol y_i.

$$\lambda_i = \log\left(\frac{P(x_i = 0|y_i)}{P(x_i = 1|y_i)}\right) = 2\frac{y_i}{\sigma^2} \tag{1}$$

where σ^2 is the noise variance.

α_{ij}: Is the message from check node i to variable node j. This is the row processing output.
β_{ij}: Is the message from variable node j to check node i. This is the column processing output.

The Belief propagation algorithm decoding as described by Tinoosh and Bass in [10] can be summarized in the following four steps:

1. *Initialization*
   For each i and j, initialize β_{ij} to the value of the log-likelihood ratio of the received symbol y_j, which is λ_j.
   During each iteration, α β messages are computed and exchanged between variables and check nodes through the graph edges according to the following steps numbered 2–4.
2. *Row processing or check node update*
   Compute α_{ij} messages using β messages coming from all other variable nodes connected to check node C_i, excluding the β information from V_j:

$$\alpha_{ij} = \prod_{j' \in V(i)\backslash j} sign(\beta_{ij'}) \times \phi\left(\sum_{j' \in V(i)\backslash j} \phi(|\beta_{ij'}|)\right) \tag{2}$$

where the non-linear function

$$\phi(x) = -\log\left(\tanh\frac{|x|}{2}\right) \tag{3}$$

The first product term in Equation α is called the parity (sign) update and the second product term is the reliability (magnitude) update.

3. *Column processing or variable node update*

Compute β_{ij} messages using channel information λ_j and incoming α messages from all other check nodes connected to variable node V_j, excluding check node C_i.

$$\beta_{ij} = \lambda_j + \sum_{i' \in C(j) \backslash i} \alpha_{i'j} \tag{4}$$

4. *Syndrome check and early termination*

When column processing is finished, every bit in column j is updated by adding the channel information λ_j and α messages from neighbouring check nodes.

$$z_j = \lambda_j + \sum_{i' \in C_{(j)}} \alpha_{i'j} \tag{5}$$

From the updated vector, an estimated code vector $x = \{x_1, x_2, \ldots, x_N\}$ is calculated by:

$$x_i = \begin{cases} 1 & if \ z_i \leq 0 \\ 0 & if \ z_i > 0 \end{cases} \tag{6}$$

If $H \times x^t = 0$, then x is a valid codeword and therefore the iterative process has converged and decoding stops. Otherwise the decoding repeats from step 2 until a valid codeword is obtained or the number of iterations reaches maximum number, which terminates the decoding process.

2.2 Min Sum Algorithms

The BP algorithm can achieve good performance for LDPC codes, but the magnitude part $\phi(x)$ is too complex for hardware implementation. Otherwise, the Min-sum algorithm makes an ingenious approximation to simplify the calculations of updated messages in check nodes.

The magnitude part of check node update in SP decoding can be simplified by approximating the magnitude computation in the row processing step, with a minimum function. This algorithm is called MinSum (MS) [10], and the row processing output is calculated by:

$$\alpha_{ij \, Min \, Sum} = \prod_{j' \in V(i) \backslash j} sign(\beta_{ij'}) \times \boldsymbol{min}_{j' \in V(i) \backslash j}(|\beta_{ij'}|) \tag{7}$$

All other steps are the same as in SPA decoding.

3 VFAP-MS Algorithm

The proposed VFAP-MS combine between a new strategy for reweighting check node less complicated than the strategy proposed in [8], this strategy consist to multiply the α_{ij} message coming from check node by ρ factors, and method for calculating ρ value proposed in [6, 7].

The different steps of the algorithm are described below:

1. *Initialization*

   The first step in this algorithm consists of calculating a short cycle or girth g in the code and $s_i = (i = 0, 1, \ldots, M-1)$ the number of length-g cycles with respect to every check node $C_i = (i = 0, 1, \ldots, M-1)$, we can make this calculation using the algorithm proposed by Halford [9], and μ_g the average number of length-g cycles passing a check node.

   As described in [6, 7] if $s_i < \mu_g$ the check node C_i is regarded as constructive then $\rho_i = 1$ otherwise this check node is determined as a destructive node and we have $\rho_i = \rho_v$, where, $\rho_i = \frac{2}{\bar{n}_D}$ and is the average connectivity for N variable nodes which is computed as

   $$\bar{n}_D = \frac{1}{\int_0^1 u(x)dx} = \frac{M}{N \int_0^1 v(x)dx} \quad (8)$$

   where $u(x)$ and $v(x)$ are the distributions of variable nodes and check nodes. Then we calculate value of log-likelihood ratio of received symbol y_i.

   $$\lambda_i = \log\left(\frac{P(x_i = 0|y_i)}{P(x_i = 1|y_i)}\right) = 2\frac{y_j}{\sigma^2} \quad (9)$$

2. *Row processing or check node update*

   Compute α_{ij} messages using β messages coming from all other variable nodes connected to check node C_i, excluding the β information from V_j:

   $$\alpha_{ij\,Min\,Sum} = \prod_{j' \in V(i)\backslash j} sign(\beta_{ij'}) \times min_{j' \in V(i)\backslash j}(|\beta_{ij'}|) \quad (10)$$

3. *Column processing or variable node update*

   Compute β_{ij} messages using channel information λ_j and incoming α messages from all other check nodes connected to variable node V_j, excluding check node C_i.

   $$\beta_{ij} = \lambda_j + \sum_{i' \in C(j)\backslash i} \rho_{i'}\alpha_{i'j} \quad (11)$$

4. *Syndrome check and early termination*
   When column processing is finished, every bit in column j is updated by adding the channel information λ_j and α messages from neighboring check nodes.

$$z_j = \lambda_j + \sum_{i \in C_{(j)}} \alpha_{ij} \tag{12}$$

From the updated vector, an estimated code vector $x = \{x_1, x_2, \ldots, x_N\}$ is calculated by:

$$x_i = \begin{cases} 1 & if \ z_i \le 0 \\ 0, & if \ z_i > 0 \end{cases} \tag{13}$$

If $H \times x^t = 0$, then x is a valid codeword and therefore the iterative process has converged and decoding stops. Otherwise the decoding repeats from step 2 until a valid codeword is obtained or the number of iterations reaches maximum number, which terminates the decoding process.

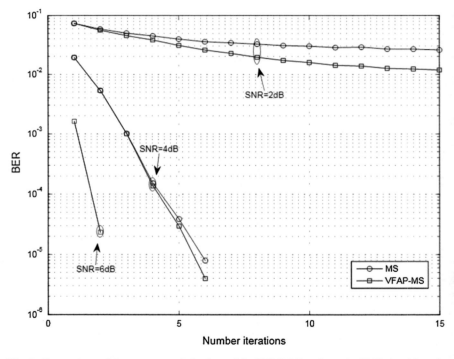

Fig. 1 Comparison of the convergent behaviors of the VFAP-MS and standard MS algorithms for decoding regular LDPC code (252, 504), where SNR equal to 2, 4, 6 dB

4 Simulation and Results

In this part we make comparison between the two algorithm described above the VFAP-MS, and MS, this comparison concerns bit error rate (BER) performance and convergence behaviours.

To test the tow algorithms, we chose progressive edge growth PEG regular $(3, 6)$ with size $(252, 504)$ from Mackay [11].

We have the distribution of variable node $u(x) = x^2$, and $v(x) = x^5$ is the distribution of check node and the average distribution is $n_{D,reg} = 3$ with $k = 4$, by using the algorithm of counting short we find short cycle $g = 8$ and the number of short cycle in the check node 802 and the average of check node is $\mu_g = 3.18$.

The Fig. 1 shows the convergent behaviours of the VFAP-MS, and MS, the results reveals that the proposed VFAP-MS converges faster than the VFAP-BP and BP.

Figure 2 shows that the proposed VFAP-MS outperforms the MS when the maximum of iteration is 20.

The results show that the proposed VFAP-MS is best compared with the MS algorithm in the BER performance and convergent behaviors.

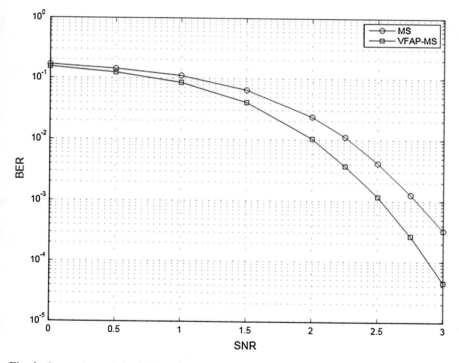

Fig. 2 Comparison of the BER performances of the VFAP-MS, and standard MS algorithms while decoding regular code (252, 504) with a 10 decoding iterations and (k = 4)

5 Conclusion

In this paper, we developed a message passing decoding algorithm that exploits the knowledge of short cycles in the regular LDPC code. The proposed VFAP-MS algorithm outperforms standard MS algorithm in convergence and BER/SNR performance with a small number of iteration. This result will allow us to increase BER performance and reduce latency with small complexity in the hardware implementation.

References

1. Gallager, R.G.: Low-density parity check codes. IRE Trans. Inf. Theory **8**, 21–28 (1962)
2. MacKay, D.J.C., Neal, R.M.: Near Shannon limit performance of low-density parity-check codes. Electron. Lett. **32**, 1645–1646 (1996)
3. Tanner, R.: A recursive approach to low complexity codes. IEEE Trans. Inf. Theory **27**, 533–547 (1981)
4. Sipser, M., Spielman, D.: Expander codes. IEEE Trans. Inf. Theory **42**(6), 1710–1722 (1996)
5. Wiberg, N.: Codes and decoding on general graphs. Linkoping University, Linkoping (1996)
6. Lui, J., de Lamare, R.C.: Low-latency reweighted belief propagation decoding for LDPC codes. IEEE Commun. Lett. (2012)
7. Liu, J., de Lamare, R.C.: Knowledge-aided reweighted belief propagation decoding for regular and irregular LDPC codes with short blocks. In: 9th IEEE International Symposium on Wireless Communications Systems (ISWCS), pp. 984–988, Paris, France, Aug 2012
8. Wymeersch, H., Penna, F., Savic, V.: Uniformly reweighted belief propagation for estimation and detection in wireless networks. IEEE Trans. Wireless Commun. **11**(4), 1587–1595 (2012)
9. Halford, T.R., Chugg, K.M.: An algorithm for counting short cycles in bipartite graph. IEEE Trans. Inf. Theory **52**(1), 287–292 (2006)
10. Mohsenin, T., Baas, B.: Split-row: a reduced complexity, high throughput LDPC decoder architecture. In: Proceedings of ICCD, Oct 2006
11. MacKay, D.J.C.: Encyclopedia of sparse graph codes [EB/OL]. http://www.inference.phy.cam.ac.uk/mackay/codes/data.html

Part VII
Modeling, Identification and Biomedical Signal Processing

Comparative Study of the Reliability of a Moroccan Level Crossing System Using Belief Function Theory and Imprecise Probabilities

Jaouad Boudnaya and Abdelhak Mkhida

Abstract The Levels Crossings (place of crossing of a railway by a road) are one of the most important sources of accidents in the railway sector in Morocco. For this reason, the National Office of the Moroccan Railways (ONCF) has launched a program which aims at removing hundreds of Level Crossings because of their dangerousness. *This* paper presents a comparative study between the belief function theory and the imprecise probabilities so as to evaluate the reliability of the Moroccan Level Crossings system over the time by integrating human factor and components failures data uncertainty.

Keywords Level crossing · Belief function · Imprecise probabilities · Reliability · Human factor · Uncertainty

1 Introduction

The level crossings constitute the major source of the risks of accidents in the railway domain in Morocco. Therefore, it is necessary to approach this problem in order to improve the performances of the railway systems.

Several works related to this problem were presented in the literature. In the paper [1], the authors analyze the functional interactions between the existing level crossing functions and any new technological system in terms of reliability, so as to choose asset owners wishing to upgrade and improve the existing systems reliability. The purpose of the work discussed in [2] is to improve safety of level crossing by analyzing accident/incident data bases and integrating human behaviour

J. Boudnaya (✉) · A. Mkhida
ENSAM Meknes, Laboratory of Mechanics, Mechatronics and Control (L2MC),
Moulay Ismail University, Marjane 2, Al-Mansour Road Agouray, PO Box 15290,
50500 Meknes, Morocco
e-mail: j.boudnaya@gmail.com

A. Mkhida
e-mail: abdelhak.mkhida@gmail.com

© Springer International Publishing Switzerland 2016
A. El Oualkadi et al. (eds.), *Proceedings of the Mediterranean Conference on Information & Communication Technologies 2015*, Lecture Notes in Electrical Engineering 380, DOI 10.1007/978-3-319-30301-7_41

using UML diagrams, in order to bring out the main functions of level crossing protection system which are concerned by different actors of the project. The paper [3] presents a probabilistic method that accounts for the variations of the component design variables of sight distance at level crossings so as to evaluate system reliability. The method is validated using Monte Carlo simulation. The proposed method should result in safer operations at railroad grade crossings. In [4], level crossings are modelled by p-time Petri nets in order to satisfy time specifications defined in safety requirements of railway systems. In [5] the authors propose a global model of the level crossing implying at the same time the rail and road traffic by using stochastic Petri nets. This model is obtained by a progressive integration of the developed elementary models; each of them describes the behaviour of a section. It allows the follow-up and the qualitative and quantitative evaluation of the effect of various factors on the level of the risk.

In this paper, we evaluate the imprecise reliability of the Moroccan level crossing by comparing the results of two approaches: belief function theory and imprecise probabilities by integrating human factor and failure data uncertainty.

2   The Level Crossings in Morocco

Level crossings are crossings at the level of a railway with a highway or pedestrian path. They constitute one of the most important sources of accidents in the railway domain in Morocco. This led early in the railway to choose a radical solution: temporarily prohibiting the road crossing, often physically by barriers. This barrier can be operated manually or automatically.

2.1   Principle of Functioning

The principle of security of the level crossing not guarded is as follows [6], (Fig. 1):

- Rest situation (Level crossing open): the road fires and the bell switched off, and barriers lifted.
- Activation of the system: a device of detection (pedal of announcement) is placed at a distance of the level crossing, when the train attacks this device, the road fires ignite in red and the bell rings (announcement of the train).
- Closure of barriers: after approximately 7 s of the release of the announcement, the barriers begin to fall. The low position of the barriers is reached after 10 s.
- Reopening of the level crossing: when the train arrives at the level crossing (35 s after the announcement), it attacks the device of rearmament (pedal of surrender). After the complete release of the train, the barriers go up, the road fires and the bell stop ringing.

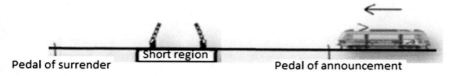

Fig. 1 Principle of functioning of the automated level crossing

2.2 *Prototype of the Moroccan Level Crossing*

Within the framework of the global program of the security of the level crossing of the Moroccan railway, it was decided in July, 2012 to strengthen the safety of the level crossings not guarded and situated on lines with high traffic (approximately 260 level crossings) by a program that extends through 2015. New equipments will be installed on the unguarded level crossing and will allow announcing to the road users the approach of the train. For instance, Fig. 2 represents the first prototype which is put in the level crossing N_3080 situated at km 168 + 088 between Tangier and Sidi Kacem, on May 7th 2013 by a Spanish company [6].

3 The Belief Function Theory

The theory of belief functions (also known as Demspter-Shafer theory and evidence theory) is originated with the work of Dempster in the 1960s [7] and Glenn Shafer in 1976 [8].

It is a quantifying of the credibility attributed to imperfect knowledge.

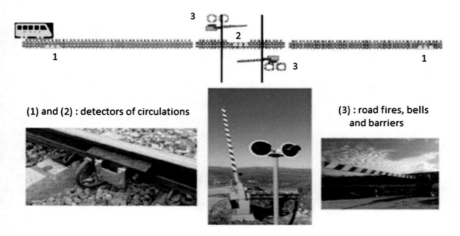

Fig. 2 Prototype of the Moroccan level crossing

Mass function (or belief function). Given a set Ω of N exclusive and exhaustive hypotheses Hi, called frame of discernment. 2^Ω denotes the set of 2^N subsets A_i of Ω. An elementary mass function is defined by:

$$
\begin{aligned}
&m: 2^\Omega \longrightarrow [0, 1] \\
&m(\phi) = 0 \\
&\sum_{2^\Omega} m(A_i) = 1
\end{aligned}
\tag{1}
$$

The focal elements are the subsets of Ω having a non-zero mass.

m (Ω): is the degree of uncertainty or total ignorance.

Credibility function. A credibility function Cr can be defined on the same set by:

$$
\begin{aligned}
&Cr(\phi) = 0 \\
&Cr(\Omega) = 1 \\
&Cr(A_j) = \sum_{A_i \subseteq A_j} m(A_i)
\end{aligned}
\tag{2}
$$

The credibility of Aj is the sum of focal elements A_i that lead A_j.

It measures how the information given by a source support A_j.

Plausibility function. The plausibility function can also be introduced from the elementary mass functions by:

$$
\begin{aligned}
&Pl(A_j) = \sum_{A_i \cap A_j \neq \phi} m(A_i) \\
&\forall\ A \subset \Omega:\ Cr(A) \leq P(A) \leq Pl(A)
\end{aligned}
\tag{3}
$$

This function measures to what extent information given by a source does not contradict Aj.

4   The Imprecise Probabilities

Imprecise probability theory was formalized by Walley [9] in 1991.

Imprecise probability theory is based on the knowledge representation of probability distributions. The imprecise probabilities can be defined as a frame of a probability distribution P(A) by two distributions: (low and high distributions Pmin (A) and Pmax(A)). It respects the following properties:

$$
{}^*0 \leq Pmin(A) \leq Pmax(A) \leq 1
\tag{4}
$$

$$
{}^*Pmin(A) = 1 - Pmax(A)
\tag{5}
$$

$$^*P\text{min}(\phi) = P\text{max}(\phi) = 0 \tag{6}$$

$$^*P\text{min}(\Omega) = P\text{max}(\Omega) = 1 \tag{7}$$

$$^*if \ A \supset B \ then \ P\text{min}(A) \geq P\text{min}(B) \ and \ P\text{max}(A) \geq P\text{max}(B) \tag{8}$$

$$^*P\text{min}(A \cup B) \geq P\text{min}(A) + P\text{min}(B) \ and \ P\text{max}(A \cup B) \leq P\text{max}(A) + P\text{max}(B)$$
$$\tag{9}$$

When Pmin = Pmax it narrows down to a precise probability.

In this work we consider that Rmin and Rmax are respectively lower and upper reliability, they correspond to lower and upper bounds of a family of probability measures P.

The interval [Rmin, Rmax] represents the degree of ignorance or the epistemic uncertainty.

5 Modelling of the Level Crossing

5.1 Description of the System

The Moroccan railway signalling system consists of three parts:

- Rail part: it consists of a material part (train and rail-road) and of a human part (the operator of the train).
- Road part: it contains a material part (vehicle and road) and a human part (the driver of the vehicle).
- Level crossing: it consists of three main parts:

  - Power network and communication network between the components of the railway signalling system.
  - Control part: it consists of Programmable Logic Controller and its program.
  - Operative part: it consists of sensors (Sensor of Ad and Sensor of Surrender) and actuators (the road lights, the alarms and the barriers).

5.2 Reliability Bloc Diagram of the Moroccan Level Crossing

From the description of the railway signalling system, we model the reliability bloc diagram of the Moroccan Level Crossing which is given in Fig. 3. It contains 10 serial components, and 3 parallel configurations each of them contain respectively 2, 4 and 2 elementary components.

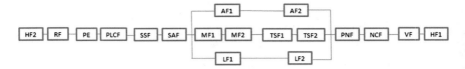

Fig. 3 Reliability bloc diagram of the Moroccan level crossing

The meanings of the basic components are given in the Table 1. We suppose that the reliability of each component follows exponential law with an imprecise failure rates Thus, the reliability of each basic component i at time t is given by

$$\mathbf{R_i(t) = exp(-\lambda_i \cdot t)} \qquad (10)$$

The failure data are taken from the following sources [10–12], as well as the human failure rate is calculated by the following formula assuming that between 80 and 90 % of railway accidents are caused by human error and considering the 118 number of accidents presented in [6] in Morocco during 9 years on 10 busiest lines.

$$\lambda_{HF} = \frac{a \cdot p}{y \cdot r} \qquad (11)$$

where:

a: number of accidents.
p: human error percentage
y: number of years
r: number of railway.

Table 1 Failure rates of components

Symbol	Meaning	Failure rates min and max: λ (h^{-1})
HF	Human failure	[1, 20E−04, 1, 35E−04]
VF	Vehicle failure	[1, 00E−03, 1, 80E−02]
RF	Rail failure	[1, 90259E−06, 2, 85388E−06]
PLCF	Programmable logic controller failure	[0, 000004, 0, 000023]
PE	Program error	[2, 5E−09, 0, 00000005]
NCF	Network communication failure	[0, 0000001, 5, 00E−06]
PNF	Power network failure	[0, 0000001, 5, 00E−06]
SAF	Sensor Ad failure	[0,0001, 0,0004]
SSF	Sensor surrender failure	[0,0001, 0,0004]
AF	Alarm failure	[0,0001, 0, 0008]
LF	Light failure	[0,0001, 0,0008]
MF	Motor failure	[5, 70E−07, 4, 50E−06]
TSF	Transmission system failure	[0, 00004, 0, 00006]

The minimal cuts C and the paths of success P of the Moroccan Level Crossing model are given in the following expressions:

$$C = \{[HF2], [RF], [PE], [PLCF], [SAF], [SSF], [PNF], [NCF], [VF], [HF1], [AF1, LF1, MF1],$$
$$[AF1, LF1, MF2], [AF1, LF1, TSF1], [AF1, LF1, TSF2], [AF1, LF2, MF1], [AF1, LF2, MF2],$$
$$[AF1, LF2, TSF1], [AF1, LF2, TSF2], [AF2, LF1, MF1], [AF2, LF1, MF2], [AF2, LF1, TSF],$$
$$[AF2, LF1, TSF2], [AF2, LF2, MF1], [AF2, LF2, MF2], [AF2, LF2, TSF1], [AF2, LF2, TSF]\}$$

$$(12)$$

$$P = \{[HF2, RF, PE, PLCF, SAF, SSF, PNF, NCF, VF, HF1, AF1, AF2],$$
$$[HF2, RF, PE, PLCF, SAF, SSF, PNF, NCF, VF, HF1, LF1, LF2],$$
$$[HF2, RF, PE, PLCF, SAF, SSF, PNF, NCF, VF, H1, MF1, MF2, TSF1, TSF2]$$

$$(13)$$

Theses expressions are used to evaluate the global reliability of the Moroccan Level Crossing system.

6 Results and Discussions

To evaluate the reliability of the Moroccan level crossing system, we compute upper and lower values of the system reliability at the interval [0, 4000 h] using the two approaches: Belief function and Imprecise Probabilities. The obtained values are shown in Table 2. Then, we plot the curves of upper and lower reliability of the Moroccan Level Crossing as a function of the time at the interval [0, 4000 h] (Fig. 4). As we can see, the two curves of the reliability using the two approaches (Belief function and imprecise probabilities) are almost the same allure, but the

Table 2 Moroccan level crossing reliability by using belief function theory and imprecise probabilities

Approach	Belief function		Imprecise probabilities	
Time (h)	Rmin	Rmax	Rmin	Rmax
0	1	1	1	1
100	0.14769	0.86537	0.04335	0.9048
200	0.0216	0.7491	0	0.819
300	0.00357	0.64815	0	0.741
400	0.00064	0.56049	0	0.67
500	0	0.48558	0	0.607
1000	0	0.23498	0	0.368
2000	0	0.0545	0	0.135
3000	0	0.01256	0	0.05
4000	0	0.00277	0	0.018

Fig. 4 Comparison of
Moroccan level crossing
lower and upper reliability
between the belief function
approach and imprecise
probabilities

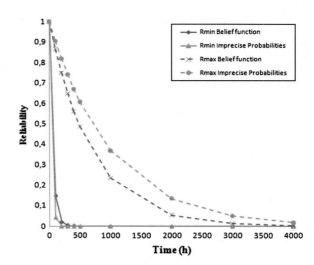

imprecise interval is different between the two previous theories, it is larger in the imprecise probabilities theory than in the belief function theory.

The system lower reliability becomes zero at the time t = 200 h for the imprecise probability approach, but, it becomes zero at the time t = 500 h for the belief function theory. This is due to the fact that no maintenance policies are done on the system in this study.

The area between the two curves represents the epistemic uncertainty of the reliability of the system. It is due to epistemic uncertainties (imprecision) of failure rates of basic components. This area is larger in the imprecise probabilities than in the belief function theory.

7   Conclusion

In this paper, we have proposed a comparative study of the Moroccan Level Crossing reliability using the belief function theory and the imprecise probabilities. It is a methodology to validate the reliability results by integrating human factor and components failures data uncertainty.

In our future works, we will complete our study, by considering dependency between components and using other methods like fuzzy logic analysis.

Acknowledgments In the terms of this paper, we thank the National Office of the Moroccan Railroad (ONCF) as well as the Center of Doctoral Studies of ENSAM MEKNES. We also thank obviously the steering committee of the conference (MediCT 2015), to allow us to share our research with colleagues.

References

1. Silmon, J., Roberts, C.: Using functional analysis to determine the requirements for changes to critical systems: railway level crossing case study. Reliab. Eng. Syst. Safety **95**, 216–225 (2010)
2. Bahloul, K., et al.: Adding technological solutions for safety improvement at level crossings: a functional specification. Proc. Soc. Behav. Sci., 1375–1384 (2012)
3. Easa, S.M.: Reliability-based design of sight distance at railroad grade crossings. Transp. Res. A **28**(1), 1–15 (1994)
4. Dutilleul, S.C., Defossez, F., Bon, P.: Safety requirements and p-time petri nets: a level crossing case study. In: IMACS Multiconference on Computational Engineering in Systems Applications (CESA), Beijing, China (2006)
5. Ghazel, M.: Using stochastic petri nets for level-crossing collision risk. IEEE Trans. Intell. Transp. Syst. **10**(4), 668–677 (2009)
6. Rabat, S.N.: ONCF. securing the railroad crossings at the right of the level crossings: equipment of ungraded level crossings by warning and protection automatic system. Project sheet (2013)
7. Dempster, A.P.: Upper and lower probabilities induced by a multivalued mapping. Ann. Math. Stat. **38**, 325–339 (1967)
8. Shafer, G.: A Mathematical Theory of Evidence. Princeton University Press, New Jersey (1976)
9. Walley, P.: Statistical Reasoning with Imprecise Probabilities, 1st edn. Chapman and Hall, London, pp. 30, 33, 44, 126 (1991)
10. Brissaud, F., et al.: Modelling of failure rates in mechanics: a combination of a Weibull law and a Cox model for the modeling of failure rates over the time and Influence factors. In: 3rd PENTOM Congress (2007)
11. Cabau, E.: Introduction to the conception of safety. Technical document no. 144, Schneider Electric (1999)
12. Houasnia, T.: Weighting of Equipments Failure Rates in Hostile Environments. University of Quebec (1999)

Stabilization of Switched T-S Fuzzy Systems with Additive Time Varying Delays

Fatima Ahmida and El Houssaine Tissir

Abstract This paper is concerned with the problem of robust stabilization continuous switched nonlinear systems with two additive time varying delays via Takagi-Sugeno (TS) fuzzy model approach. The stabilization conditions for nominal system are derived using Lyapunov-Krasoviskii functional approach. Sufficient conditions for the existence of fuzzy state feedback gain are formulated by linear matrix inequalities. The stabilization robustness is then investigated. Finally, a numerical example is given to show the less conservatism of the stability criteria and the effectiveness of the designed method.

Keywords Switched T-S fuzzy control systems · Additive time varying delays · Liner matrix inequality (LMI)

1 Introduction

Switched systems, which provide a unified framework for mathematical modeling of many physical or man-made systems displaying switching features such as power electronics [1], flight control systems [2], network control systems [3]. The stabilization of these systems has been widely studied in recent decades [4, 5] and the references therein. Besides switched systems, it is well known that fuzzy control has attracted increasing attention in the stabilization of nonlinear systems. Nonlinear systems with time-delay constitute basic mathematical models of real phenomena, for instance in biology, mechanics and economics. Control of

F. Ahmida (✉) · E.H. Tissir
Department of Physics, Faculty of Sciences, LESSI, Dhar El Mehraz,
B.P: 1796, 30003 Fès-Atlas, Morocco
e-mail: fatima_ahmida@yahoo.fr

E.H. Tissir
e-mail: elhoussaine.tissir@usmba.ac.ma

© Springer International Publishing Switzerland 2016
A. El Oualkadi et al. (eds.), *Proceedings of the Mediterranean Conference on Information & Communication Technologies 2015*, Lecture Notes in Electrical Engineering 380, DOI 10.1007/978-3-319-30301-7_42

nonlinear systems is a challenging task because there are no systematic mathematical tools that help to find necessary and sufficient conditions guaranteeing the stability performance. Stability and stabilization of fuzzy systems with time-delay has not only important theoretical interest but also practical values. Since time-delays are the main cause of oscillation, divergence or instability, considerable efforts have been made to stability for systems with time delays. The stability of such systems has been addressed by many authors [6–8]. In [9], the stability result and switching law for the considered unforced Takagi-Sugeno time delay fuzzy switched systems are presented but for the case without uncertainties. Reference [10] has studied the robust exponential stability of such systems. It should be pointed out that the stability/stabilization results mentioned are based on systems with one single delay in the state.

Recently in [11] a new model of system with two additive time-varying delay components was proposed. This model has a strong application background in remote control and networked control, e.g., sensors and actuators in a distributed control system will induce network delays.

More recently, the asymptotic stability and robustness performance are investigated in different works for different types of systems [12–14]. The problem of stability for uncertain switched T-S fuzzy systems with two additive time varying delays is discussed in [15] by employing an improved free weighting matrix approach. This motivates us to consider the stabilization of the switched T-S fuzzy system with two additive time varying delays.

In this paper, we study the problems of stabilization and controller design for switched continuous-time systems with two additive time varying delays using T-S fuzzy model. Our aim is to develop improved delay-dependent robust stability criteria over the latest results available from the open literature [9, 15]. Then, we use the Lyapunov Krasovskii method, combined with linear matrix inequalities (LMIs). The state feedback gain and the upper bound of the time delay can be obtained. We provide two illustrative examples to show that the proposed design method.

2 Problem Formulation

Consider a class of switched systems with T-S fuzzy subsystems as follows

$$R^i: \text{If } z_1(t) \text{ is } W_1^i, \quad z_2(t) \text{ is } W_2^i, \dots, \quad z_s(t) \text{ is } W_s^i$$
$$\begin{cases} \dot{x}(t) = \left(A_{ij} + \Delta A_{ij}(t)\right)x(t) + (A_{dij} + \Delta A_{dij}(t))x(t - d_1(t) - d_2(t)) + B_{ij}u(t) \\ \qquad\qquad\qquad x(t) = \phi(t), t \in [-h, 0] \end{cases}$$

$$(1)$$

where $i \in R := \{1, 2, \ldots, r\}$, r is the number of IF-THEN rules, $x(t) \in R^n$ is the state vector, $u(t) \in R^n$ denotes the input vector, $\phi(t) \in ([-h, 0], R^n)$ is a compatible vector-valued initial function, with norm $\|\phi(t)\| = \sup_{\theta \in [-h, 0]} \|\phi(\theta)\|$ $z_1(t), z_2(t), \ldots, z_s(t)$ are the premise variables and $W_l^i, l = 1, \ldots, s$ are the fuzzy sets, the piecewise constant function $\sigma \in F$ with $F = \{1, \ldots, N\}$. $\sigma = j$ means that the jth subsystem is activated. The jth subsystems are described by constant matrices A_{ij}, A_{dij} and B_{ij} are constant matrices with appropriate dimensions, $\Delta A_{ij}(t), \Delta A_{dij}(t)$ are unknown matrices representing parametric uncertainties and are assumed to be of the form:

$$[\Delta A_{ij}(t) \quad \Delta A_{dij}(t)] = H_{ij} F_{ij}(t) [E_{aij} \quad E_{dij}] \tag{2}$$

where E_{aij} and E_{dij} are known constantes matrices with appropriate dimensions, $F_{ij}(t)$ are unknown time-varying matrices with Lebesgue measurable elements bounded by:

$$F_{ij}^T(t) F_{ij}(t) \leq I \tag{3}$$

$d_1(t)$ and $d_2(t)$ represent the two delay components in the state and satisfy the following conditions:

$$0 \leq d_1(t) \leq h_1 \prec \infty, 0 \leq d_2(t) \leq h_2 \prec \infty, h = h_1 + h_2$$
$$0 \leq \dot{d}_1(t) \leq \tau_1 \prec \infty, 0 \leq \dot{d}_2(t) \leq \tau_2 \prec \infty, d(t) = d_1(t) + d_2(t), \tau = \tau_1 + \tau_2 \tag{4}$$

By using the common used center-average defuzzifier, product interference and singleton fuzzifier, the global dynamics of T–S fuzzy system (1) can be inferred as

$$\begin{cases} \dot{x}(t) = \sum_{i=1}^r \mu_i(z(t)) \left[\begin{array}{c} (A_{ij} + \Delta A_{ij}(t))x(t) + (A_{dij} + \Delta A_{dij}(t))x(t - d_1(t) - d_2(t)) \\ + B_{ij}u(t) \end{array} \right] \\ x(t) = \phi(t), t \in [-h, 0] \end{cases} \tag{5}$$

where $\mu_i(z(t)) = w_i(z(t)) / \sum_{i=1}^r w_i(z(t)), w_i(z(t)) = \prod_{l=1}^r W_l^i(z_l(t))$. $W_l^i(z_l(t))$ is the membership value of $z_l(t)$ in W_l^i, some basic properties of $\mu_i(z(t))$ are: $\mu_i(z(t)) \geq 0$, $\sum_{i=1}^r \mu_i(z(t)) = 1$.

The following fuzzy rules for fuzzy-model-based control for the jth subsystem are:

Controller rule k: If $z_1(t)$ is W_1^k, $z_2(t)$ is $W_2^k, \ldots, z_s(t)$ is W_s^k then

$$u(t) = K_{kj}x(t) \tag{6}$$

Hence, the defuzzified output of the controlled rules is given by

$$u(t) = \sum_{k=1}^{r} \mu_k K_{kj} x(t) \tag{7}$$

where $K_{kj} \in R^{n \times m} (k = 1, \ldots, r, j = 1, \ldots, N)$ are to be determined later.

Combining (5) and (7), the resulting closed-loop system can be expressed as follows:

$$\begin{cases} \dot{x}(t) = \sum_{i=1}^{r} \sum_{k=1}^{r} \mu_i(z(t)) \, \mu_k(z(t)) \begin{bmatrix} (A_{ij} + B_{ij} K_{kj} + \Delta A_{ij}(t)) x(t) \\ + (A_{dij} + \Delta A_{dij}(t)) x(t - d_1(t) - d_2(t)) \end{bmatrix} \\ x(t) = \phi(t), t \in [-h, 0] \end{cases} \tag{8}$$

Remark 1 The switching rule is constructed as follows.

Set, $S_j^P = \left\{ x \in R^n; \; \sum_{i=1}^{r} \mu_i(z(t)) \, x^T(t) L_{ij} x(t) < 0 \right\}$

$\overline{S}_1^P = S_1^P \quad \overline{S}_j^P = S_j^P \cup_{k=1}^{j-1} \overline{S}_k^P, \; j = 2, \ldots, N$

[SR]: Step 0: Let x(t) = ϕ(t)

Step 1: $\sigma(x) = argmin_{1 \leq j \leq N} \left\{ x^T(t) [\sum_{i=1}^{r} \mu_i(A_{ij}X + XA_{ij}^T)] x(t) \right\}$

Step 2:   σ(x) = j as long as x(t) ∈ $\overline{S}_j$

Step 3:   if x(t) hits the boundary of $\overline{S}_j$, go to step 1 to determine the next mode.

Theorem 1 *For given scalars* $h_1 > 0$, $h_2 > 0$, $\tau_1 > 0$ *and* $\tau_2 > 0$, *the nominal system of* (8) *is asymptotically stable if there exist matrices*

$X > 0, \overline{Q}_1 > \overline{Q}_2 > 0, \overline{R}_1 > 0, \overline{R}_2 > 0 \; \overline{Z}_1 > 0, \overline{Z}_2 > 0, \overline{Z}_3 > 0, S_{kj}, \; \overline{K} = \begin{bmatrix} \overline{K}_1 \\ \overline{K}_2 \\ \overline{K}_2 \end{bmatrix},$

$\overline{L} = \begin{bmatrix} \overline{L}_1 \\ \overline{L}_2 \\ \overline{L}_2 \end{bmatrix}, \quad \overline{M} = \begin{bmatrix} \overline{M}_1 \\ \overline{M}_2 \\ \overline{M}_2 \end{bmatrix}, \quad \overline{N} = \begin{bmatrix} \overline{N}_1 \\ \overline{N}_2 \\ \overline{N}_2 \end{bmatrix}$ *and* $\overline{W} = \begin{bmatrix} \overline{W}_1 \\ \overline{W}_2 \\ \overline{W}_2 \end{bmatrix}$ *such that the following LMIs hold:*

(i)   *There exist* $0 < \rho_j < 1, \; j = 1 \ldots N$ *such that* $\sum_{j=1}^{N} \rho_j = 1$ *and* $\sum_{j=1}^{N} \rho_j L_{ij} < 0$

$$\tag{9}$$

$$
\text{(ii)} \quad \pi_{ik}^j = \begin{pmatrix} \overline{\pi}_{11} & \overline{\pi}_{12} & \overline{\pi}_{13} & -\overline{M}_1 - \overline{N}_1 + \frac{1}{h}\overline{Z}_3 & -\overline{W}_1 + \frac{1}{h_1}\overline{Z}_1 & \overline{\pi}_{16} \\ * & \overline{\pi}_{22} & \overline{\pi}_{23} & -\overline{M}_2 - \overline{N}_2 & -\overline{W}_2 & 0 \\ * & * & \overline{\pi}_{33} & -\overline{M}_3 - \overline{N}_3 & -\overline{W}_3 & XA_{dij}^T \\ * & * & * & -\overline{R}_1 - \frac{1}{h_2}\overline{Z}_2 - \frac{1}{h}\overline{Z}_3 & \frac{1}{h_2}\overline{Z}_2 & 0 \\ * & * & * & * & -\overline{R}_2 - \frac{1}{h_1}\overline{Z}_1 - \frac{1}{h_2}\overline{Z}_2 & 0 \\ * & * & * & * & * & \overline{\pi}_{66} \end{pmatrix} < 0
$$

$$(10)$$

where

$$L_{ij} = A_{ij}X + XA_{ij}^T$$

$$\overline{\pi}_{11} = B_{ij}S_{kj} + S_{kj}^T B_{ij}^T + \overline{Q}_1 + \overline{R}_1 + \overline{K}_1 + \overline{K}_1^T + \overline{N}_1 + \overline{N}_1^T - \frac{1}{h_1}\overline{Z}_1 - \frac{1}{h}\overline{Z}_3$$

$$\overline{\pi}_{12} = \overline{W}_1 - \overline{K}_1 + \overline{L}_1 + \overline{K}_2^T + \overline{N}_2^T$$

$$\overline{\pi}_{13} = \overline{M}_1 - \overline{L}_1 + \overline{K}_3^T + \overline{N}_3^T + A_{dij}X$$

$$\overline{\pi}_{16} = XA_{ij}^T + S_{kj}^T B_{ij}^T \tag{11}$$

$$\overline{\pi}_{22} = \overline{W}_2 + \overline{W}_2^T + \overline{L}_2 + \overline{L}_2^T - \overline{K}_2 + \overline{K}_2^T - (1 - \tau_1)(\overline{Q}_1 - \overline{Q}_2)$$

$$\overline{\pi}_{23} = -\overline{L}_2 + \overline{M}_2 + \overline{L}_3^T - \overline{K}_3^T + \overline{W}_3^T$$

$$\overline{\pi}_{33} = \overline{M}_3 + \overline{M}_3^T + \overline{L}_3 + \overline{L}_3^T - (1 - \tau)\overline{Q}_2$$

$$\overline{\pi}_{66} = h_1\overline{Z}_1 + h_2\overline{Z}_2 + h\overline{Z}_3 - 2X$$

The feedback control law gains are given by: $K_{kj} = S_{kj}X^{-1}$. *The switching rule is given by* [SR].

Proof The proof is omitted.

Theorem 2 *For given scalars* $h_1 > 0$, $h_2 > 0$, $\tau_1 > 0$ *and* $\tau_2 > 0$, *the system* (8) *is asymptotically stable if there exist matrices* $X > 0, \overline{Q}_1 > \overline{Q}_2 > 0, \overline{R}_1 > 0, \overline{R}_2 > 0,$

$$\overline{Z}_1 > 0, \quad \overline{Z}_2 > 0, \quad \overline{Z}_3 > 0, \quad S_{kj}, \quad \overline{K} = \begin{bmatrix} \overline{K}_1 \\ \overline{K}_2 \\ \overline{K}_2 \end{bmatrix}, \quad \overline{L} = \begin{bmatrix} \overline{L}_1 \\ \overline{L}_2 \\ \overline{L}_2 \end{bmatrix}, \quad \overline{M} = \begin{bmatrix} \overline{M}_1 \\ \overline{M}_2 \\ \overline{M}_2 \end{bmatrix},$$

$$\overline{N} = \begin{bmatrix} \overline{N}_1 \\ \overline{N}_2 \\ \overline{N}_2 \end{bmatrix} \text{ and } \overline{W} = \begin{bmatrix} \overline{W}_1 \\ \overline{W}_2 \\ \overline{W}_2 \end{bmatrix} \text{ such that the following LMIs holds:}$$

(i)   There exist $0 < \rho_j < 1, \ j = 1 \ldots N$ such that $\displaystyle\sum_{j=1}^{N} \rho_j = 1$ and $\displaystyle\sum_{j=1}^{N} \rho_j L_{ij} < 0$

$$(12)$$

(ii)
$$\overline{\pi}_{ik}^{jc} = \begin{pmatrix} \pi_{ik}^{j} & \chi_{ij} \\ * & \mathcal{J} \end{pmatrix} \prec 0 \tag{13}$$

where $\overline{\pi}_{ik}^{j}$ are defined in (11), $\chi_{ij}^{T} = \begin{bmatrix} H_{ij}^{T} & 0 & 0 & 0 & 0 & H_{ij}^{T} \\ E_{aij}X & 0 & E_{dij}X & 0 & 0 & 0 \end{bmatrix}^{T}$

and $\mathcal{J} = \begin{bmatrix} -\varepsilon_1 I & 0 \\ * & -\varepsilon_1 I \end{bmatrix}$

The feedback control low gains are given by: $K_{kj} = S_{kj}X^{-1}$. The switching rule is given by [SR].

3 Numerical Example [16]

Consider the following non switched T-S fuzzy system described by
  Rule 1: If $(x_2/0.5)$ is about 0.5

  Then $\dot{x}(t) = (A_{10} + \Delta A_{10})x(t) + (A_{d1} + \Delta A_{d1})x(t - d_1(t) - d_2(t)) + B_1 u(t)$

  Rule 2: If $(x_2/0.5)$ is about π or $-\pi$

  Then $\dot{x}(t) = (A_{10} + \Delta A_{10})x(t) + (A_{d1} + \Delta A_{d1})x(t - d_1(t) - d_2(t)) + B_1 u(t)$

where

$A_{01} = \begin{bmatrix} 0 & 1 \\ 0.1 & -2 \end{bmatrix}$, $A_{02} = \begin{bmatrix} 0 & 1 \\ 0.1 & -0.5 - 1.5\beta \end{bmatrix}$, $\beta = \dfrac{0.01}{\pi}$; $A_{d1} = A_{d2} = \begin{bmatrix} 0.1 & 0 \\ 0.1 & -0.2 \end{bmatrix}$,

$B_1 = B_2 = \begin{bmatrix} 0 \\ 1 \end{bmatrix}$; $H_i = \begin{bmatrix} -0.03 & 0 \\ 0 & 0.03 \end{bmatrix}$, $E_{ai} = \begin{bmatrix} -0.15 & 0.2 \\ 0 & 0.04 \end{bmatrix}$, $E_{di} = \begin{bmatrix} -0.05 & -0.35 \\ 0.08 & -0.45 \end{bmatrix}$, $i = 1, 2$

By applying Theorem 1 for system without uncertainties, we obtain the stability of the closed loop system for $h_1 + h_2 = 4.91\,e^{15}$. Note that in [16] by tacking $h_1 + h_2 = 7.1254$ it is found that $h_2 = 22$, so that $h = h_1 + h_2 = 29.1254$. In the simulation, we take $h_1 = 30\sin(t)$ and $h_2 = 20\,e^{-t}$. In this case $h_1 + h_2 = 50$, therefore the results of [16] cannot be applied because the feasibility is last.
    By our theorem 1, we obtain the controlled feedback gains: $K_1 = [-0.5192 \quad 0.8253]$, $K_2 = [-0.5192 \quad -0.6699]$. The obtained fuzzy state-feedback controller makes the closed-loop states converge to zero, as is shown in Fig. 1, where the initial condition is assumed to be $x(t) = [-1; 1]$.
    Now, for uncertain systems, similar to [16] we calculate the upper bound $\overline{h}_2$ of $d_2(t)$ when $\overline{h}_1 = 13$. The results are summarized in Table 1.
    We can see from the Table 1 that our method provides less conservative results and smaller feedback gains than that derived in some existing methods [16]. This show the effectiveness of our proposed approach.

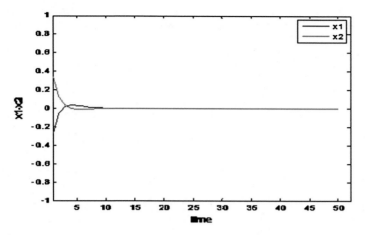

Fig. 1 State response

Table 1 Upper bound $\bar{h}_2$ while $\bar{h}_1$ is fixed and the corresponding gain controller

Methods	Upper bound	K_1	K_2
Theorem 2 [16]	$\bar{h}_1 = 13$ $\bar{h}_2 = 15.4263$	$[-0.928\ -1.1811]$	$[-0.9185\ -1.2058]$
Our Theorem 2	$\bar{h}_1 = 13$ $\bar{h}_2 = 1 \times 10^9$	$[-0.7646\ 0.1954]$	$[-0.7646\ -1.2999]$

4 Conclusion

The stability and control problems have been investigated for switched T-S fuzzy systems with two additive time-varying delays. The LMIs proposed have been obtained by utilizing a Lyapunov Krasovskii functional method, Newton-Leibniz formula. Moreover, the maximum allowable upper delay bound and the feedback controller gain can be obtained simultaneously through solving a set of LMIs. The less conservativeness of the results is shown by a numerical example.

References

1. Tse, C., Bernardo, M.D.: Complex behavior in switching power converters, vol. 90. In: IEEE Proceedings, pp. 768–781 (2002)
2. Pellanda, P.C., Apkarian, P., Tuan, H.D.: Missile autopilot design via a multi-channel LFT/LPV control method. Int. J. Robust Nonlinear Control **12**, 1–20 (2002)

3. Donkers, M.C.F., Heemels, W.P.M.H., Wouw, N.V.D., Hetel, L.: Stability analysis of networked control systems using a switched linear systems approach. IEEE Trans. Autom. Control **56**, 2101–2115 (2011)
4. Zhao, X., Zhang, L., Shi, P., Liu, M.: Stability and stabilization of switched linear systems with mode-dependent average dwell time. IEEE Trans. Autom. Control **57**, 1809–1815 (2012)
5. Tissir, E.H.: Exponential stability and guaranteed cost of switched linear systems with mixed time-varying delays. Int. Sch. Res. Netw. ISRN Appl. Math. **2011**, 14 p (2011)
6. Wu, H.N., Li, H.X.: New approach to delay-dependent stability analysis and stabilization for continuous-time fuzzy systems with time-varying delay. IEEE Trans. Fuzzy Syst. **15**, 482–493 (2007)
7. Liu, F., Wu, M., Yong He, Y., Ryuichi, Y.: New delay-dependent stability criteria for T–S fuzzy systems with time-varying delay. Fuzzy Sets Syst. **161**, 2033–2042 (2010)
8. Peng, C., Tian, Y.C., Tian, E.: Improved delay-dependent robust stabilization conditions of uncertain T–S fuzzy systems with time-varying delay. Fuzzy Sets Syst. **159**, 2713–2729 (2008)
9. Chiou, J.S., Tsai, S.H.: Stability and stabilization of Takagi-Sugeno fuzzy switched system with time-delay. J. Syst. Control Eng. **226**, 615 (2012)
10. Ahmida, F., Tissir, E.H.: Exponential stability of uncertain T-S fuzzy switched systems with time delay. Int. J. Autom. Comput. **10**, 32–38 (2013)
11. Lam, J., Gao, H., Wang, C.: Stability analysis for continuos systems with two additive time-varying delay components. Syst. Control Lett. **56**, 16–24 (2007)
12. Idrissi, S., Tissir, E.H.: Delay dependent robust stability of T-S fuzzy systems with additive time varying delays. Appl. Math. Sci. **6**, 1–12 (2012)
13. Chaibi, N., Tissir, E.H., Hmamed, A.: Delay dependent robust stability of singular systems with additives time varying delays. Int. J. Autom. Comput. **10**, 85–90 (2013)
14. Zabari, A., Tissir, E.H.: Stability analysis of uncertain discrete systems with two additive time delays. Appl. Math. Sci. **18**, 7935–7949 (2014)
15. Ahmida, F., Tissir, E.H.: New stability conditions for switched uncertain T-S fuzzy systems with two additive time varying delays. Int. J. Ecol. Econ. Stat. **34**, 94–103 (2014)
16. Idrissi, S., Tissir, E.H, Boumhidi, I., Chaibi, N.: Delay dependent robust stabilization for uncertain T-S fuzzy systems with additive time varying delays. In: Proceeding of the 8th International conference intelligent systems: Theories and applications, EMI, Rabat Morocco, pp. 65–71, 08–09 May 2013

On Modeling and Fault Tolerance of NanoSatellite Attitude Control System

Meryem Bouras, Hassan Berbia and Tamou Nasser

Abstract The modeling of a satellite system and its orientation in space is a difficult and important step to define before controlling its attitude. Many methods have been developed since the first satellite was built. In this paper we propose a comparative study of the most important attitude representation methods adopted in the literature. We point out the advantages and disadvantages of each of these methods before choosing the one that suits our objective. Then we present, the Simulink blocks for Rotation angles method, which will be used in our attitude model. After that, we will present the implementation of this model in a system on chip, following the proposed architecture.

Keywords Nanosatellite · Attitude representations · Attitude model Fault tolerance · Space environment

1 Introduction

In their orbits, nanosatellites must be oriented around their center of mass, their antennas and their sensors must be pointed at a specific area of the Earth. The attitude control must take into account different external and internal disturbances, acting on the system and affecting its orbit and its attitude or both. To ensure the success and the safety of a nanosatellite mission, an on-board (Attitude Determination and Control System) ADCS is required.

M. Bouras (✉) · H. Berbia · T. Nasser
Embedded and Mobiles Systems Engineering Department, Al Jazari Research Team, ENSIAS, Mohammed V University in Rabat, Avenue Mohammed Ben Abdallah Regragui, Madinat Al Irfane, BP 713, Agdal Rabat, Morocco
e-mail: meryem.bouras@um5s.net.ma

H. Berbia
e-mail: h_berbia@yahoo.com

T. Nasser
e-mail: tamounasser@gmail.com

© Springer International Publishing Switzerland 2016
A. El Oualkadi et al. (eds.), *Proceedings of the Mediterranean Conference on Information & Communication Technologies 2015*, Lecture Notes in Electrical Engineering 380, DOI 10.1007/978-3-319-30301-7_43

409

In spite of the presence of internal and external disturbances, this ADCS provides the ability to maintain a desired orientation in space, carries out a chosen attitude maneuver, and tracks a predefined nominal orbit. This very complex system is crucial to control the attitude, which is a subject of active research [1, 10, 11, 13]. There are various results available in the literature, showing that the desired attitude is often unachieved [14]. This ADCS cannot be optimized without the modeling of the system or its attitude. Different methods of attitude representation have been adopted in the literature; they are more detailed in [7, 12].

In this article we will define and illuminate the advantages and disadvantages of different attitude representations, including the novel method introduced by [3, 4], we will choose, the most appropriate approach for our system, and we will develop Simulink blocks for the chosen method. These blocks will be used in our model, and implemented in a system on chip, with respect to our first developed architecture. In this implementation we will consider the purposes of our project which are real time, the tolerance and correction of the fault caused by the external disturbances.

2   The Attitude

The attitude of a satellite is its rotational orientation in space; it is also the rotation of its defined body coordinate system or (Body Frame) BF with respect to a defined external fixed frame, which may be inertial (Inertial Frame) IF as in [2] or orbital (Reference Frame) RF chosen by us and by [1, 3, 4, 6, 9–11, 13, 14].

Before the development of the ADCS, this attitude should be represented by a mathematical model. Therefore many attitude representation schemes were developed and adopted in the literature.

Three parameters are at the minimum necessary to mathematically model this attitude, with those methods. These representations are the rotation matrix, the Euler angles, the angle-axis Euler, the quaternion, and the rotation angles.

3   Attitude Representation Methods

3.1   The Rotation Matrix

Called the direction cosines matrix, is the transformation matrix C and it is compound of direction cosines. This method is used by [10]. It represents the attitude of the system or the orientation of its body frame relative to a reference frame.

$$C = \begin{bmatrix} C_{11} & C_{12} & C_{13} \\ C_{21} & C_{22} & C_{23} \\ C_{31} & C_{32} & C_{33} \end{bmatrix}. \tag{1}$$

This representation is done in a single rotation, but with nine variables, six of them are redundant. Moreover, it involves the use of trigonometric functions that are slow to calculate on an embedded computer. That makes them only used to perform basic changes.

3.2   Euler Angles

This approach can also be described intuitively by the three Euler angles "α, β, γ" defined by [2–6, 8, 11] in different sequences of three rotations, in the form of the matrix C, with the gimbal lock singularity.

$$C = \begin{bmatrix} c_2 c_3 & c_2 s_3 & -s_2 \\ s_1 s_2 c_3 - c_1 s_3 & s_1 s_2 s_3 + c_1 c_3 & s_1 c_2 \\ c_1 s_2 c_3 + s_1 s_3 & c_1 s_2 s_3 - s_1 c_3 & c_1 c_2 \end{bmatrix}. \tag{2}$$

3.3   The Axis-Angle Euler

The angle-axis Euler cited in [8] is an evolution of the rotational Euler's theorem, which implies that any rotation or sequence of rotations of a rigid body in three-dimensional space is equivalent to a single rotation around the single fixed axis â and the angle φ which describes the magnitude of the rotation around this axis (Fig. 1). This causes the elimination of the singularity gimbal lock.

Fig. 1 The axis-angle Euler

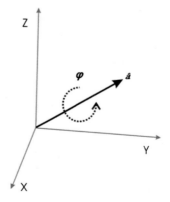

This representation requires four parameters, which is a disadvantage, added to the inherited three previous methods, which is the representation of trigonometric functions; those functions are slow to calculate with an onboard computer.

3.4 Quaternion

Quaternion, sometimes called Euler-Rodrigues parameters, used by [1, 5, 6, 8, 9, 11, 13, 14], is based on the same concept as the axis-angle representation, but with an expression excluding the sine or cosine, which are expensive in terms of calculation.

Quaternion is expressed as four complex numbers with real component η and three imaginary components εx, εy and εz and it is calculated from the angle φ and the axis of rotation $\hat{a} = [ax, ay, az]$ following the expression below:

$$q = \begin{bmatrix} \varepsilon_x \\ \varepsilon_y \\ \varepsilon_z \\ \eta \end{bmatrix} = \begin{bmatrix} a_x \sin(\varphi/2) \\ a_y \sin(\varphi/2) \\ a_z \sin(\varphi/2) \\ \cos(\varphi/2) \end{bmatrix} \quad \text{With } q^T q = \hat{a}^T \hat{a} = 1. \tag{3}$$

Its main advantage is its kinematic equation that is easy to integrate, however it does not give an intuitive idea of the represented rotation.

The rotation matrix is:

$$C = \begin{bmatrix} 1 - 2(\varepsilon_2^2 + \varepsilon_3^2) & 2(\varepsilon_1\varepsilon_2 + \eta\varepsilon_3) & 2(\varepsilon_1\varepsilon_3 - \eta\varepsilon_2) \\ 2(\varepsilon_2\varepsilon_1 - \eta\varepsilon_3) & 1 - 2(\varepsilon_1^2 + \varepsilon_3^2) & 2(\varepsilon_2\varepsilon_3 + \eta\varepsilon_1) \\ 2(\varepsilon_3\varepsilon_1 + \eta\varepsilon_2) & 2(\varepsilon_3\varepsilon_2 - \eta\varepsilon_1) & 1 - 2(\varepsilon_1^2 + \varepsilon_2^2) \end{bmatrix}. \tag{4}$$

3.5 The Rotation Angles

This representation is introduced by [3, 4], to model the unusual or unexpected attitude and eliminate the drawbacks of Euler angles.

These rotation angles are a derivative of the quaternion form in combination with the Euler angles, wherein the sequence of rotations is replaced by a single rotation σ, around the axis E, passing through the fixed point of the lifting unit vector as in Fig. 3, [3, p. 489].

This unit vector is indicated by:

$$\mathbf{e}_\sigma = \mathbf{I}l + \mathbf{J}m + \mathbf{K}n. \tag{5}$$

Thus, the overall angle of rotation can be expressed by the superposition of three simultaneous rotations around the axes of the mobile frame, represented in the form of the following sizes:

$$\xi = \sigma l; \quad \eta = \sigma m; \quad \zeta = \sigma n. \tag{6}$$

These sizes are called rotation angles:

- ξ—Rotation angle around x axis;
- η—Rotation angle around y axis;
- ζ—Rotation angle around z axis.

They retain the advantages and benefits of quaternion and Euler angles, making them directly measurable and with an intuitive physical meaning. They also allow the expression of the kinematic equation in polynomial form, which is an important advantage in the construction of high-speed algorithms and they are easily implemented in a system on chip.

However, they do not have a predefined BLOCK in Matlab/Simulink, as Euler angles or quaternion; In our case, it is a problem, because they will require more work and more time to realize the model.

Their attitude matrix is as follows:

$$A_e = \begin{bmatrix} a\xi^2 + c & a\eta\xi + b\zeta & a\eta\xi + b\zeta \\ a\eta\xi + b\zeta & a\eta^2 + c & a\eta\zeta + b\xi \\ a\zeta\xi + b\eta & a\eta\zeta + b\xi & a\zeta^2 + c \end{bmatrix}. \tag{7}$$

4 Comparison of Representations

Table 1 provides a simple comparison between the different representations cited before.

Table 1 Advantages and disadvantages of attitude representations

Representation	Advantages	Disadvantages
The rotation matrix	• Simple calculation of consecutive rotations • 1 rotation	• 9 parameters
Euler angles	• Intuitive • 3 parameters • Predefined block in Simulink	• Singularity (90°) • 3 rotations
The axis-angle Euler	• No singularity • 1 rotation	• Complex calculations • 4 parameters
Quaternion	• No singularity • 1 rotation • Predefined block in Simulink	• Not intuitive • 4 parameters
The rotation angles	• No singularity • Intuitive • 1 rotation • 3 parameters	• No predefined block in Simulink

5   The Attitude Model Developed Blocks

After the introduction of these representations of the attitude, it is possible to use these methods to model how the orientation of the satellite changes. The mathematical model is in the form of two equations. The dynamic equation, whose detailed demonstration, was presented in [12, 13], and the kinematics equation, where we use the rotation angles method. This method is the best and the most suitable for our model because it is easier to implement in an embedded system.

We have reevaluated the following kinematics equation, based on rotation angles, and developed by [3, 4].

$$\begin{bmatrix} \dot{\xi} & \dot{\eta} & \dot{\zeta} \end{bmatrix}^{\mathrm{T}} = \mathbf{W}_R \begin{bmatrix} p_B & q_B & r_B \end{bmatrix}^{\mathrm{T}}. \tag{8}$$

$$W_R = \begin{bmatrix} f\xi^2 + h & f\eta\xi + \zeta/2 & f\zeta\xi + \eta/2 \\ f\eta\xi + \xi/2 & f\eta^2 + h & f\eta\zeta + \xi/2 \\ f\zeta\xi + \eta/2 & f\eta\zeta + \xi/2 & f\zeta^2 + h \end{bmatrix}. \tag{9}$$

The following Fig. 2 and Fig. 3 depict the blocks that we developed in Matlab/Simulink using the rotation angles method.

This first block (Fig. 2) is a representation of the kinematics equation using 4 parameters. These parameters are (p, q, r), (ξ, η, ζ), h and the single rotation σ to get as result (xidot, etadot, zetadot), which are the first derivative of (ξ, η, ζ).

The purpose of this block (Fig. 3) is to get the rotation angles and the direction cosines matrix (DCM), based on the variables (xidot, etadot, zetadot).

The nanosatellite or satellite model that we will develop with these blocks will be tested and it will be controlled with an active control method, using the magnetorquers and the reaction wheels actuators. However, the paper that introduced the rotation angles representation used them with the passive control method gravity gradient. Finally, the model will be implemented in a system on chip following the architecture in Fig. 4 and this will take into account real time and the tolerance and control of the fault and effects caused by external disturbances.

Fig. 2 xidot, etadot and zetadot

xidot etadot zetadot

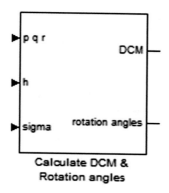

Fig. 3 DCM and rotation angles

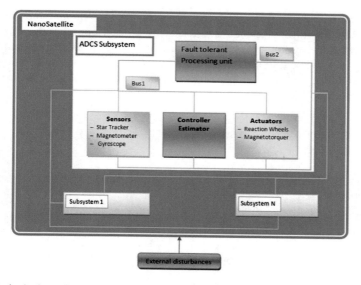

Fig. 4 Attitude determination and control architecture

6 Fault Tolerance and Control

The assessment of disturbance torques is an essential element of rigorous spacecraft attitude control design. The main focus of our research is the tolerance and the control of the fault caused by the external disturbances such as cosmic and solar radiation, geomagnetic field, atmospheric drag and micrometeorites. Those disturbances act on the system and affect its electronic boards, its attitude and its orbit. The ADCS will consider the need to detect and to tolerate the fault and effects caused by those disturbances. The two mostly discussed effects are the Total Ionizing Dose

(TID) and the Single Event Effect (SEE). To minimize the fault and defects caused by these effects, we choose to develop a fault tolerant system with auto-diagnosis. This system will have different levels of redundancy, in the processing unit, in the communication buses, in sensors and actuators.

The functional bloc diagram (Fig. 4) is composed of a fault tolerant processing unit, a fault tolerant estimator controller, redundant buses (Bus1 is the primary and Bus2 is the backup), sensors and actuators. The optimization will be carry out on processing unit level, buses level and the communication between them and the different components of the ADCS system and the other subsystems, as seen in Fig. 4.

7 Conclusion

All in all, after defining the ADCS, the attitude, the different representations methods and the advantages and disadvantages of each one of them. From these methods we chose the best and the most suitable for our purposes: rotation angles and we started to use it in our model. This final model will be tested and it will be implemented in a system on chip, considering our goal, the tolerance and control of fault and defects caused by the external disturbances.

References

1. Alvenes, F.: Satellite attitude control system. Master thesis, Department of Engineering Cybernetics (2012)
2. Bacconi, F.: Spacecraft attitude dynamics and control: Dottorato di Ricerca in Ingegneria Informatica e dell'Automazione, UNIVERSITÀ DEGLI STUDI DI FIRENZE, Dipartimento di Sistemi e Informatica, Anno Accademico (2005–2006)
3. Chelaru, T.V., Barbu, C., Chelaru, A.: Mathematical model for small satellites, using attitude and rotation angles. Int. J. Syst. Appl. Eng. Dev. 5(4) (2011)
4. Chelaru, T.V., Barbu, C.: Mathematical model for sounding rockets, using attitude and rotation angles. Int. J. Appl. Math. Inf. 3(2) (2009)
5. Fossen, T.I.: Mathematical models for control of aircraft and satellites. Centre for Autonomous Marine Operations and Systems, Department of Engineering Cybernetics, Norwegian University of Science and Technology, Apr 2013
6. Gregory, B.S.: Attitude control system design for ion, the Illinois observing nanosatellite. B. S., Marquette University, 2001, Thesis Submitted in partial fulfillment of the requirements for the degree of Master of Science in Electrical Engineering, in the Graduate College of the University of Illinois at Urbana-Champaign, Urbana, Illinois (2004)
7. Groÿekatthöfer, K., Yoon, Z.: Introduction into quaternions for spacecraft attitude representation. Technical University of Berlin, Department of Astronautics and Aeronautics, Berlin, Germany, 31 May 2012
8. Hernando, Y.M.: Conception d'un système de commande autonome pour le simulateur matériel de satellite LABSAT. UNIVERSITÉ DE SHERBROOKE Faculté de GENIE, Département de génie électrique, Mémoire de maîtrise Spécialité: génie électrique, Septembre (2013)

9. Lavet, V.F.: Study of passive and active attitude control systems for the OUFTI nanosatellites. Master thesis, Faculty of Applied Sciences, University of Liège, Academic Year (2009–2010)
10. Menges, B.M., Guadiamos, C.A., Lewis, E.K.: Dynamic modeling of micro-satellite Spartnik's attitude
11. Pukdeboon, C.: Finite-Time Second-Order Sliding Mode Controllers for Spacecraft Attitude Tracking. Hindawi Publishing Corporation Mathematical Problems in Engineering Volume (2013)
12. Shuster, M.D.: A survey of attitude representations. J. Astronaut. Sci. **41**(4), 439–517 (1993)
13. Tregouet, J.F.: Synthèse de correcteurs robustes périodiques à mémoire et application au contrôle d'attitude de satellites par roues à réaction et magnéto-coupleurs. EDSYS: Automatique, Institut Supérieur de l'Aéronautique et de l'Espace (ISAE), Université de Toulouse, soutenue le lundi 3 décembre (2012)
14. Van Nguyen, H.: Estimation d'attitude et diagnostic d'une centrale d'attitude par des outils ensemblistes. Laboratoire Gipsa-lab dans l'Ecole Doctorale Electronique, Electrotechnique, Automatique, Traitement du Signal (EEATS), UNIVERSITÉ DE GRENOBLE, soutenue publiquement le 24 mars (2011)

Difference Spectrum Energy Applied for Fetal QRS Complex Detection

Khalifa Elmansouri, Rachid Latif and Fadel Maoulainine

Abstract The detection of the fetal QRS complex from one composite signal recorded from the abdominal area is a challenging problem for both the biomedical and signal processing communities. In this work, in contrast to the classical approach using Singular Value Decomposition (SVD) based on the analysis of the singular value ratio, we present a new approach using optimal SVD based on difference spectrum energy to detect the QRS complexes of the fetus cardiac activity, also the Undecimated Wavelet Transform (UWT) was used for noise filtering and to determine peaks locations. The result of the R peaks performance measure is calculated for the two approaches. This method is advantageous since it is based on the analysis of one abdominal signal, and it is validated in both the synthetic and real signals.

Keywords Fetal QRS complex · SVD · UWT · Peak detection

1 Introduction

Fetal electrocardiogram (FECG) monitoring enables to evaluate the health and condition of the fetus, and it presents a classical problem in biomedical engineering, that remains a challenging issue. Today, it is based entirely on the fetal heart rate monitoring with the fetal QRS complex is the only information source in early stage diagnostic of fetal health. The fetal signal obtained from the maternal abdomen normally has a low amplitude and an unfavorable signal-to-noise ratio from which the fetal heart rate can hardly be detected.

Besides the inherent low amplitude of FECG signals, compared to the maternal ones, they are severely contaminated by various kinds of noise sources, including

K. Elmansouri (✉) · R. Latif
Signals System and Computer Sciences Laboratory (ESSI), Agadir, Morocco
e-mail: khalifa.elmansouri@edu.uiz.ac.ma

F. Maoulainine
Faculty of Medicine, Team of Child, Health and Development, Marrakech, Morocco

© Springer International Publishing Switzerland 2016
A. El Oualkadi et al. (eds.), *Proceedings of the Mediterranean Conference on Information & Communication Technologies 2015*, Lecture Notes in Electrical Engineering 380, DOI 10.1007/978-3-319-30301-7_44

myographic signals, fetal brain activity, uterine contractions, powerline interference or moving artifacts. Moreover, the frequency range of the FECG signal strongly overlaps with the spectrum of other bio-signals, especially the maternal electrocardiogram, making the separation task more difficult.

Many signal processing techniques have been proposed and detailed in the review of Sameni and Clifford [1] to extract FECG signal. Some of these techniques use multiple leads, such as: neural networks combined with fuzzy logic [2] and blind source separation [3]. Other group of techniques uses a single electrode, such as: nonlinear state-space projection [4] and the classical SVD method [5] in which most of the fetal R-waves are missed in the extracted FECG waveform.

In this paper, we aim to apply difference spectrum energy based on optimal SVD for searching fetal QRS peaks with a high accuracy, also the UWT is used: in preprocessed step for noise filtering and to determine peaks locations after SVD reconstruction. The idea is to extract the fetal peaks from the mother abdominal ECG signal directly without signals separation process. Other contribution of this work is the detection of fetal QRS complex from one abdominal ECG recording without need to have various sensor measurements.

2 Methodology

The proposed method consists of two complementary stages. In the first stage, we try to separate the noise components from the composite signal using UWT and linear expansion of thresholds. In the second, the signal is decomposed using Hankel matrix based on optimal singular value decomposition and difference spectrum energy analysis. The classical approach to which we have compared our work is explained in [5].

2.1 Optimal Singular Value Decomposition

The definition of SVD is: for a matrix $A \in R^{m \times n}$, two orthogonal matrices $U = [u_1, u_2, \ldots, u_m] \in R^{m \times m}$ and $V = [v_1, v_2, \ldots, v_n] \in R^{n \times n}$ are surely existed to meet the following equation [6]:

$$A = USV^{T}. \tag{1}$$

where $S = [\text{diag}(\sigma_1, \sigma_2, \ldots, \sigma_q), O]$ or its transposition, which is decided by $m < n$ or $m > n$, $S \in R^{m \times n}$, while O is zero matrix, $q = \min(m, n)$, and $\sigma_1 \geq \sigma_2 \geq \ldots \geq \sigma_q > 0$. These σ_i (i = 1, 2 ... q) are called the singular values of matrix A.

Consider a discrete signal of length N samples denoted in vector notation as $X = [x(1) \, x(2) \ldots x(N)]$. The Hankel matrix can be created by this signal as follows [7]: if signal length N is even, the columns number of Hankel matrix should be

$n = N/2$ and the rows number $m = N/2 + 1$; if signal length N is odd, the columns number of Hankel matrix should be $n = (N + 1)/2$ and the rows number $m = (N + 1)/2$.

$$A = \begin{bmatrix} x(1) & x(2) & \cdots & x(n) \\ x(2) & x(3) & \cdots & x(n+1) \\ \cdots & & \cdots & \cdots \\ x(N-n+1) & x(N-n+1) & \cdots & x(N) \end{bmatrix}. \qquad (2)$$

where $1 < n < N$, let $m = N - n + 1$, then $A \in R^{m \times n}$

In order to realize the isolation of signal using SVD, the Eq. (1) should be converted to the form of column vectors u_i and v_i

$$A = \sigma_1 u_1 v_1^T + \sigma_2 u_2 v_2^T + \cdots + \sigma_q u_q v_q^T = \sum_{i=1}^{q} A_i. \qquad (3)$$

where $A_i = \sigma_i u_i v_i^T$, $A_i \in R^{m \times n}$

Supposing that $P_{i,1}$ is the first row vector of A_i, and $H_{i,n}$ is a subcolumn in the last column of A_i. According to the creation principle of Hankel matrix, if $P_{i,1}$ and the transposition of $H_{i,n}$ are linked together, then a SVD component signal P_i can be obtained, which can be expressed as the vector form

$$P_i = (P_{i,1}; H_{i,n}^T); \quad P_{i,1} \in R^{1 \times n}; \quad H_{i,n} \in R^{(m-1) \times 1}. \qquad (4)$$

The component signals obtained by SVD can form a simple linear superposition for original signal.

$$X = \sum_{i=1}^{q} P_i. \qquad (5)$$

Let $C_i = u_i v_i^T$ and $A = \sum \sigma_i C_i$. Because the spectra of both fetal and maternal energies overlap, they will contribute equally to each C_i, and the SVD cannot separate energy density concentration in the individual C_i. Then, to separate the densities into smallest regions, the transformed vectors y_i and x_i of orthonormal transformation for the u_i and v_i are created as in [8]; the optimal SVD is as follows:

$$B = \sum^{Q} z_{ik} B_i. \qquad (6)$$

where the singular value z_{ik} is the first Q of the ik-components and have the same function as the σ_i terms in diagonal S for the SVD. Obviously, there are significantly more (m-by-n) of the z_{ik} than the σ_i.

Now, consider the difference of singular value as follows:

$$d_i = z_i - z_{i+1}. \tag{7}$$

The sequence $D = (d_1, d_2 \ldots d_{q-1})$ is called the difference spectrum energy of singular value, which describe the variation status between the two adjacent singular values.

2.2   Undecimated Wavelet Transform

The discrete wavelet transform is not translation invariant and it leads to different wavelet coefficients. The UWT overcomes this and permits to get more comprehensive feature of the analyzed signal. The idea behind UWT is that it does not decimate the signal. Thus, it produces more accurate information for the frequency localization. UWT is detailed in [9].

2.3   Proposed Algorithms for Fetal QRS Complex Detection

The cardiac electrical activity of a fetus is recorded noninvasively from electrodes on the mother's body surface. Such recordings can be expressed as the sum of three signals embedded in the abdominal area: One is a deformed version of the maternal ECG $\hat{x}(t)$, another is the fetal ECG s(t) and a third is additive noise from other sources $\eta(t)$, as in:

$$w(t) = \hat{x}(t) + s(t) + \eta(t). \tag{8}$$

To separate these signals our approach consists of the following steps:

1. De-noising the signal using UWT and linear expansion of thresholds.
2. Create a Hankel matrix A from the original signal (Fig. 1).
3. The Matrix A is decomposed by optimal SVD, so singular value vector is obtained.
4. Searching the point Z where an obvious sudden change occurs in singular value vector [max of difference spectrum (Fig. 2)].
5. The difference spectrum is divided onto three regions (show Fig. 2): the region A between the axis and the point Z, it contains the most dominant maternal components, so the reconstructed signal from A is shown in Fig. 3. The region B between the points Z and Y, it contains both the dominant maternal components and weak fetal components. The region C after the point Y is the region where the fetal and other components are mixed. The value of Y is computed with:

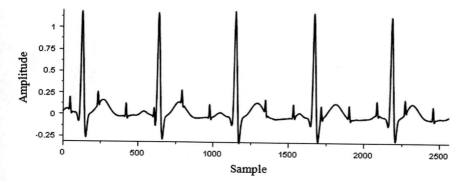

Fig. 1 Synthetic abdominal ECG signal

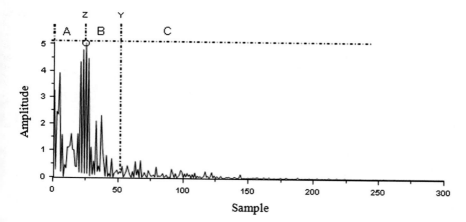

Fig. 2 Difference spectrum energy

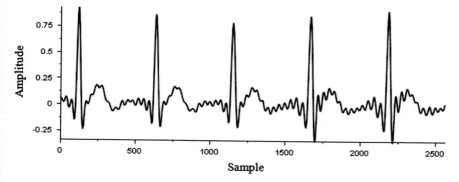

Fig. 3 Reconstructed signal from the region A

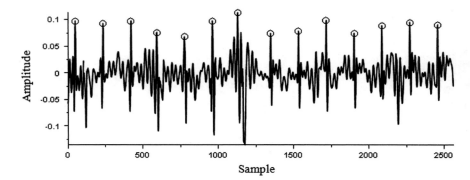

Fig. 4 Reconstructed signal from the region C

$$Y = Z + shift. \tag{9}$$

where shift value is defined with an empirical process.

If shift value is incorrect (Y localized in the region B or in C), then the fetal QRS is not obvious, else the fetal QRS complexes are obvious and detectable.

6. Reconstruction of the component of the region C and peak detection using UWT (Fig. 4).

3   Simulation Results

To investigate the effectiveness of the proposed methodology for detection of fetal QRS complex from a single lead abdominal signal, the process was applied to both synthetic abdominal signals and a real ECG database of abdominal data. The results of our experiments are presented in this section.

For synthetic abdominal signals, the Calculation of the R peaks performance measure is defined as

$$\text{performance} = \frac{\#\text{fetal\_QRS} - (\#\text{misses} + \#\text{false})}{\#\text{fetal\_QRS}} \times 100\,\%. \tag{10}$$

3.1   Synthetic Data

For generating the synthetic ECG signals we use the dynamical model developed by McSharry et al. [10]. Both fetal and maternal ECG signals are synthesized using

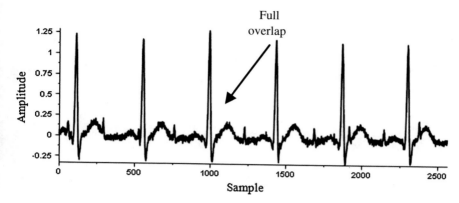

Fig. 5 Synthetic abdominal ECG signal

this model using different parameters to account for different shapes and beat rates of the two signals.

The maternal ECG signal is assumed to have a heartbeat rate of 70 beats/min and the FECG signal is assumed to have a heartbeat rate of 133 beats/min. We have used slightly different parameters for the dynamical model in each case to make the shape of the waveforms different for the maternal ECG and the FECG. Both signals are generated with 512 Hz sampling rate. A more significant challenge is when the maternal and fetal components are fully overlapping. The synthetic abdominal ECG in Fig. 5 presents a full overlap, showing that the fetal beat is totally embedded in the much stronger maternal beat.

After, the signal is processed by UWT using the wavelet db6 through 7 levels and linear expansion of thresholds for denoising purpose. The SVD decomposition was carried out to obtain the difference spectrum energies, and the max of difference spectrum (the point Z) is located in the 24th coordinate.

After several tests, the point Y is located in the 82th coordinate; the reconstructed signal from C is shown in Fig. 6.

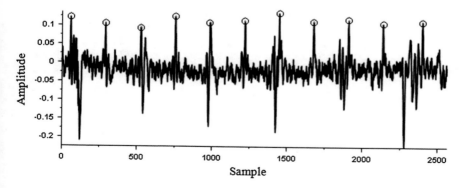

Fig. 6 Reconstructed signal from the region C

Based on prior knowledge of the peak locations in the specific waveform, we find that the performance measure is 100 %.

Other three of maternal abdominal simulation signals with a large range of peaks are processed using the proposed method and the minimum performance measure noted is 99.34 %, in contrast the maximum performance measure noted for classical approach is 64.12%.

3.2 Real Data

We next investigate the application of the proposed methods using data sets from [11]. The data set recording information is:

- Signals recorded in labor, between 38 and 41 weeks of gestation
- Sampling rate: 1 kHz
- Resolution: 16 bits.

The abdominal ECG signal that we selected from the dataset is shown in Fig. 7.

After, the signal is processed by UWT using the wavelet db6 through 7 levels and linear expansion of thresholds for denoising purpose. The SVD decomposition was carried out to obtain the difference spectrum energies. After several tests, the point Y is located in the 16th coordinates; the reconstructed signal from C is shown in Fig. 8, where it can be seen that the fetal peaks is detected.

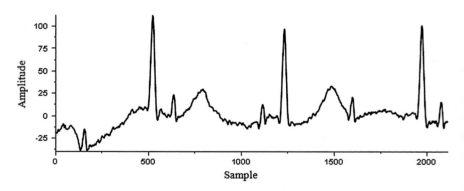

Fig. 7 Real abdominal ECG signal

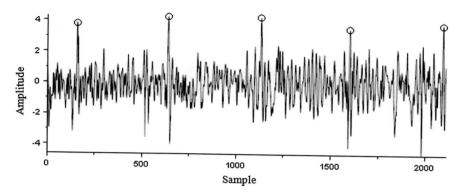

Fig. 8 Reconstructed signal from the region C of real abdominal ECG signal

4 Conclusion

This paper proposed an approach for fetal QRS complex detection from one electrocardiogram signal by using the optimal singular value decomposition. This detection of the fetus cardiac activity is based on the difference spectrum energy which is described by a sequence presenting the variation status between each two adjacent singular values. The UWT is used to get more comprehensive feature of the analyzed signal by eliminating the noise and to produce more accurate information by determining the peaks locations. Experimental results are presented with synthetic and real signals. Promising results have been obtained, such as the minimum performance measure equal to 99.34 %, comparing to classical approach with the maximum performance measure equal to 64.12 %. However, since each method has its own benefits and limitations, our approach requires that the Y parameter be chosen through experience and visual inspection.

References

1. Sameni, R., Clifford, G.D.: A review of fetal ECG signal processing: issues and promising directions. Open Pacing Electrophysiol. Ther. J **3**, 4–20 (2010)
2. Elmansouri, K., Latif, R., Maoulainine, F.: Improvement of fetal electrocardiogram extraction by application of fuzzy adaptive resonance theory to adaptive neural fuzzy system. Int. J. Innov. Appl. Stud. **9**(1), 95–103 (2014)
3. Lathauwer, L., Moor, B., Vandewalle, J.: Fetal electrocardiogram extraction by blind source subspace separation. IEEE Trans. Biomed. Eng. **47**(5), 567–572 (2000)
4. Richter, M., Schreiber, T., Kaplan, D.T.: Fetal ECG extraction with nonlinear state space projections. IEEE Trans. Biomed. Eng. **45**(1), 133–137 (1998)
5. Kanjilal, P.P., Palit, S., Saha, G.: Fetal ECG extraction from single-channel maternal ECG using singular value decomposition. IEEE Trans. Biomed. Eng. **44**(1), 51–59 (1997)
6. Golub, G.H., Van Loan, C.F.: Matrix Computations. John Hopkins University Press, Baltimore (1996)

7. Zhao, X., Ye, B.: Selection of effective singular values using difference spectrum and its application to fault diagnosis of headstock. Mech. Syst. Signal Pr. **25**(5), 1617–1631 (2011)
8. Brenner, M.J.: Non-stationary dynamics data analysis with wavelet-SVD filtering. Mech. Syst. Signal Pr. **17**(4), 765–786 (2003)
9. Elmansouri, K., Latif, R., Maoulainine, F.: Efficient fetal heart rate extraction using undecimated wavelet transform. In: The 2nd World Conference on Complex Systems, pp. 696–701. IEEE Press, Agadir (2014)
10. McSharry, P.E., Clifford, G.D., Tarassenko, L., Smith, L.: A dynamical model for generating synthetic electrocardiogram signals. IEEE Trans. Biomed. Eng. **50**(3), 289–294 (2003)
11. Kotas, M., Jezewski, J., Horoba, K., Matonia, A.: Application of spatio-temporal filtering to fetal electrocardiogram enhancement. Comput. Meth. Prog. Bio. **104**(1), 1–9 (2011)

A Comparison of Wavelet and Steerable Pyramid for Classification of Microcalcification in Digital Mammogram

Khaddouj Taifi, Said Safi, Mohamed Fakir and Rachid Ahdid

Abstract This paper presents a comparative study between wavelet and steerable pyramid transform for microcalcification clusters. Using multiresolution analysis, mammogram images are decomposed into different resolution levels, which are sensitive to different frequency bands, it is important to extract the features in all possible orientations to capture most of the distinguishing information for classification. The experimental results suggest that S-P shows a clear improvement in the classification performance when compared to wavelet (DWT). These multiresolution analysis methods were tested with the referents mammography Base data MIAS Experimental results show that the steerable pyramid method provides a better.

Keywords Microcalcification · Mammography · Detection · Steerable pyramid · Wavelet transform · Homomorphic filtering

1 Introduction

Breast cancer is the first women's cancer. In Morocco, according to the register of the Casablanca region cancers for 2004, breast cancer and cervical cancer are about half of female cancers and 27 % of all cancer. The standardized incidence is 30 for 100,000 women per year. It represents 36.1 % of all female cancers [1, 2].

Many researchers have proposed the powerful tools for analyzing and detection of Microcalcification. Balakumaran et al. [3] presented an algorithm using multiresolution analysis based on the dyadic wavelet transform and microcalcification detection by fuzzy shell clustering. Huddin et al. [4] presented a method to extract features to classify the microcalcification clusters using steerable pyramid decomposition. It is important to extract the features in all possible orientations. Eltoukhy et al. [5, 6], presented an approach for breast cancer diagnosis in digital mammogram

K. Taifi (✉) · S. Safi · M. Fakir · R. Ahdid
Faculty of Science and Technology, Beni Mellal, Morocco
e-mail: Taif_kha@hotmail.fr

© Springer International Publishing Switzerland 2016
A. El Oualkadi et al. (eds.), *Proceedings of the Mediterranean Conference on Information & Communication Technologies 2015*, Lecture Notes in Electrical Engineering 380, DOI 10.1007/978-3-319-30301-7_45

429

using curvelet transform. J.S. Leena Jasmine et al. [7] presented a novel approach for classifying microcalcification in digital mammograms using Nonsubsampled Contourlet Transform (NSCT).

2 Microcalcification Enhancement

In this paper image is enhanced using Normalization and hommomrphic filtering. Finally the enhanced image is obtained with clarity and free from noise.

2.1 Normalization

Normalization is a simple technique that helps in lifting; it is to extend the range of intensities of the pixels of the original image f. The Eq. (1) illustrates this transformation as max is the maximum gray level of image f.

$$g(x) = \frac{\max(f(x)) - f(x)}{\max(f(x))} \tag{1}$$

2.2 Homomorphic Filtering

The geometry of the breast and non-uniformity of lighting. It raises difficulties for analysis. The correction of non-uniform illumination is very important for image analysis. To solve this problem we used the frequency homomorphic filter to keep only the high frequency. Enhancement method described here is shown in Fig. 1 [8].

With $H(u, v) = \dfrac{1}{1 + \left(\dfrac{D(u,v)}{D_0}\right)^{2n}}$; $D(u, v) = \sqrt{u^2 + v^2}; n = 0.05; D_0 = 2$

Figure 2 and 3 shows the result obtained by Normalization and homomorphic filtering.

We see clearly the effect of this filter. Indeed microcalcifications correspond to high frequencies which are accentuated is becoming more visible on the mammographic image.

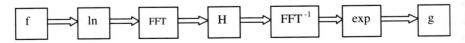

Fig. 1 Homomorphic filtering

Fig. 2 Normalization

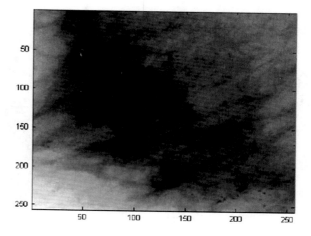

Fig. 3 Homomorphic
filtering

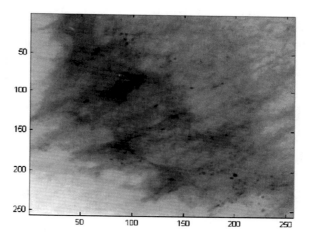

2.3   Wavelet Transform

One of the multi resolution analysis tools that have been widely used in image processing is wavelet analysis. Originally proposed in the form of Mallet's pyramidal algorithm, an image can be successfully decomposed into detail sub-bands at different level of resolutions. This decomposition is known as 2-dimensional (2D) separable discrete wavelets transform (DWT). Our detection method decomposes the original image into sub bands with low-low Approximation (LL), low-high vertical (LH), high-low horizontal (HL), and high-high diagonal (HH) components. Three stages of decomposition are necessary [9, 10].

Figure 4a, shows examples of Wavelet decomposition of images of mammograms at three levels of resolution.

(a) **(b)**

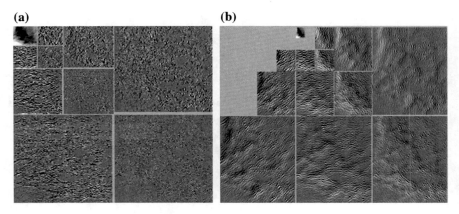

Fig. 4 a Filtering with the filter daub2 with levels 3. b Decomposition images with steerable filters

2.4 Steerable Pyramid

Steerable filters introduced by Freeman and Adelson [11] have the property to provide a decomposition of images as oriented bands can be used to split the features of image into different sub-bands at different levels, with 'approximations' and 'details' [12]. Steerable pyramids were used for various applications such as enhancement, edge detection, texture analysis, noise elimination.

From Fig. 4b, we observe that the number of microcalcifications detected by levels 1, 2 and 3 increased by using S-P.

3 Mammographic Image Classification

In the classification phase are use the nearest Neighbor classifier (KNN) [13].

3.1 Feature Extraction

The statistical measures can be used to identify a texture such as mean μ_i, variance σ_{ij}^2 and entropy e_{ij} of the energy distribution of the multi-resolution coefficients for each sub-band at each decomposition level [14]. Let $I_{ij}(x,y)$ be the image at the specific block j of sub-band i, there feature vector $V_{ij}=\mu_{ij},\sigma_{ij}^2, e_{ij}$.

$$\mu_{ij} = \frac{1}{M \times N}\sum_{X=1}^{M}\sum_{Y=1}^{N}\left|I_{ij}(x,y)\right| \tag{2}$$

$$\sigma_{ij}^2 = \frac{1}{M \times N} \sum_{X=1}^{M} \sum_{Y=1}^{N} \left| I_{ij}(x, y) - \mu_{ij} \right|^2 \tag{3}$$

$$e_{ij} = -\sum_{l=1}^{L} \left((p_l) \times \log(p_l) \right) \tag{4}$$

3.2 Measures for Performance Evaluation

A number of different measures are commonly used to evaluate the performance of the proposed method. These measures including classification accuracy, Precision AC-Accuracy assesses the effectiveness of the classifier [15].

$$Ac = \frac{TP + TN}{TP + TN + FP + FN} \tag{5}$$

3.3 Classification Rates of Normal and Abnormal Image

This paper presents a comparative study between wavelet and steerable pyramid transform to classify microcalcifications into normal or abnormal (Benign or Malignant) cases using multi-orientation and multi-resolution representations (Table 1).

3.4 Classification Rates of Normal and Abnormal Image with Wavelet

We experimented with the aforementioned types of wavelets aiming at incrementing the percentage of energy that corresponds to the horizontal, vertical and diagonal details of the third level of wavelet transform.

From Fig. 5 we presented a comparative study among the orthogonal wavelets, in order to decide on the appropriate family to be used for mammographic image processing.

Table 1 Number of training and testing set for microcalcification

Category	Database	
	No. of training set	No. of testing set
Normal	36	18
Anormal	58	28
(Benign)	3	5
(Malignant)	6	7

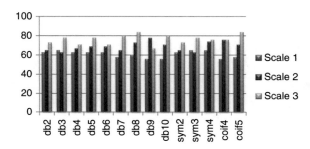

Fig. 5 Results of the experiments of the application of different wavelet families on the mammographic images with microcalcifications. Xaxis indicates the various wavelets used

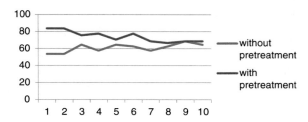

Fig. 6 Classification rate with and without pretreatment

The evaluation criterion for the decision of the wavelet was the degree of similarity between the wavelet and the image profile. This was assessed through the percentage of energy that corresponds to the coefficients of the wavelet transform. Our experimentation showed that the Least Asymmetric Daubechies wavelet with length 8 is the most appropriate, since it accumulates more energy corresponding to the details it can be concluded that the maximum successful classification rate using wavelet obtained by the features extracted at the decomposition level 3.

The effect of pretreatment is very important and particularly remarkable in Fig. 6 classification results. Classification of the performance obtained without pretreatment are significantly lower than those obtained with the pretreatment.

3.5 Classification Rates of Normal and Abnormal Image with Steerable Pyramid

Figure 7 shows the rate of normal and abnormal classification at different scale. The maximum successful classification rate is 82 % with S-P at scale 2 and 3.

Figure 8 shows the rate of normal and abnormal classification at scale 3. The maximum successful classification rate is 82 % with S-P for k=6 and 84 % with DWT for k=1 (Table 2).

Table 2 Classification result with Pretreatment for K=1

K=1					K=6				
S-P	TP	FP	TN	FN	S-P	TP	FP	TN	FN
	22	6	15	3		21	7	17	1
DWT	25	3	14	4	DWT	23	5	13	5

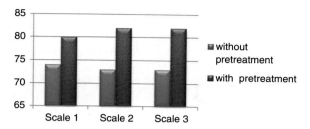

Fig. 7 Classification rate the steerable pyramid with and without pretreatment

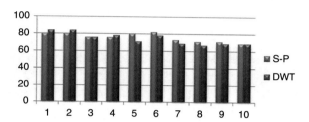

Fig. 8 Shows the rate of normal and abnormal classification at different scale

3.6 Classification Rates of Benign and Malignant

The maximum successful classification rate is 74 % with S-P (Fig. 9) (Table 3).

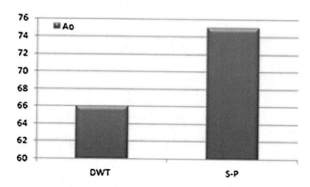

Fig. 9 Shows the rate accuracy of benign or malignant classification at scale 3

Table 3 Classification result with pretreatment and Nearest Neighborhood Classifier (KNN)	*TP*	*FP*	*TN*	*FN*
S-P	7	0	2	3
DWT	6	1	2	3

4 Conclusion

The goal of this work is to detect the MCs in the breast and to classify the tissues by KNN technique. In this paper a new CAD system is introduced, the DWT and S-P transform were used in the feature extraction step. Segmented image contains the suspected region which is given for feature extraction process. The extracted features are classified into normal or abnormal (Benign or Malignant) region using KNN.

In conclusion, the obtained results demonstrate that the S-P features based Nearest Neighborhood classifier is giving higher classification rate.

References

1. Errihani, H.: Caractéristiques psychosociales des patients cancéreux marocains: étude de 1 000 cas recrutés à l'Institut national d'oncologie de Rabat. Rev Francoph Psycho Oncologie Numéro. 2, 80–85 (2005)
2. Fondation Lalla Salma Prévention et traitement "Axe Epidemiologie Situation et Actions" Vol. 2
3. Balakumaran, T., Vennila, ILA., Shankar, C.G.: Detection of microcalcification in mammograms using wavelet transform and fuzzy(IJCSIS). Int. J. Comput. Sci. Inf. Secur. 7(1) (2010)
4. Huddin, A.B., Ng, B.W.H., Abbott, D.: Investigation of multiorientation and multiresolution features for microcalcifications classification in mammograms. In: 7th International Conference on Intelligent Sensors, Sensor Networks and Information Processing, pp. 52–57 (2011)
5. Eltoukhy, M.M., Faye, I., Samir, B.B.: Breast cancer diagnosis in digital mammogram using multiscale curvelet transform. Comput. Med. Imag. Graph. 34, 269–276 (2010). doi:10.1016/j.compmedimag.2009.11.002
6. Eltoukhy, M.M., Faye, I., Samir, B.B.: A comparison of wavelet and curvelet for breast cancer diagnosis in digital mammogram. Comput. Biol. Med. 40, pp. 384–391 (2010)
7. Jasmine, J.S.L., Baskaran, S., Govardhan, A.: A robust approach to classify microcalcification in digital mammograms using contourlet transform and support vector machine. Am. J. Eng. Appl. Sci. 6(1), 57–68, ISSN: 1941-7020 (2013)
8. Delac, K., Grgic, M., Kos, T.: Sub-Image homomorphic filtering technique for improving facial identification under difficult illumination conditions. In: International Conference on Systems, Signals and Image Processing (IWSSIP-06), 21–23 Sept 2006
9. Mallat, S.G.: A theory for multiresolution signal decomposition: The wavelet representation. IEEE Trans. Pattern Anal. Mach. Intell. 11, pp. 674–693 (1989)
10. Vetterli, M., Herley, C.: Wavelets and filter banks: Theory and design. IEEE Trans. Sig. Process. 40, pp. 2207–2232 (1992)
11. Freeman, W.T., Adelson, E.H.: The design and use of steerable. IEEE Trans. Pattern Anal. Mach. Intell. 13(9) (1991)
12. El Aroussi, M., El Hassouni, M., Ghouzali, S., Rziza, M., Aboutajdine, D.: Local appearance based face recognition method using block based steerable pyramid transform. Sig. Process. 91, 38–50 (2011)
13. Boiman, O., Shechtman, E., Irani, M.: In defense of nearest neighbor based image classification. In: IEEE Conference on Computer Vision and Pattern Recognition (VPR) (2008)
14. Li, S.T., Li, Y., Wang, Y.N.: Comparison and fusion of multi resolution features for texture classification. In: International lConference on Machine Learning and Cybernetics. 6, 3684–3684 (2004)
15. Nithya, R., Santhi, B.: Comparative study on feature extraction method for breast cancer classification. J. Theo. Appl. Inf. Technol. 33(2) (2011)

Detection of Sleep Spindles Using Frequency Components Tracking in Raw EEG Data

Imane Zriouil, Fakhita Regragui, El Mehdi Hamzaoui,
M. Majid Himmi and Jamal Mounach

Abstract Sleep spindles are among the hallmarks observed in the electroencephalogram (EEG) that occur during non-rapid eye movement sleep precisely in stage 2. They are transient waveforms of biological and clinical interest. In this paper, we present a method to detect spindles in raw EEG recordings during sleep. The method consists of processing the signal through an adaptive autoregressive model whose features are represented by the zeroes of the model polynomial rather than the prediction coefficients. Tracing in time of the zeros' modulus shows sharp transitions indicating statistical changes due to the occurrence of spindles. The results obtained were compared to those reported by other techniques based on traditional EEG visual reading used by neurophysiologists.

Keywords EEG · Sleep spindle · Adaptive autoregressive model · Zero-tracking

I. Zriouil (✉) · F. Regragui · E.M. Hamzaoui · M.M. Himmi
LIMIARF, Faculty of Sciences Morocco, University Mohammed V,
Rabat, Morocco
e-mail: zriouil.imane@gmail.com

F. Regragui
e-mail: fakhitaregragui@yahoo.fr

E.M. Hamzaoui
e-mail: hamzaouielmehdi@gmail.com

M.M. Himmi
e-mail: himmi.fsr@gmail.com

J. Mounach
Department of Neurophysiology, Military Teaching Hospital Mohammed V,
Rabat, Morocco
e-mail: jmounach@yahoo.fr

© Springer International Publishing Switzerland 2016
A. El Oualkadi et al. (eds.), *Proceedings of the Mediterranean Conference on Information & Communication Technologies 2015*, Lecture Notes in Electrical Engineering 380, DOI 10.1007/978-3-319-30301-7_46

437

1   Introduction

Sleep is an essential requirement in humans. It is a necessity for humans to be healthy and to regenerate their physical and mental physiological state. The role of sleep ensures essential functions, such as growth, brain maturation, development and preservation of cognitive abilities, temperature control, fit many hormonal secretions. In contrast, lack of sleep or poor sleep quality can have a real impact on health and the proper conduct of everyday life.

A normal sleep has 4–5 cycles which succeed and each lasts about 90–110 min. According to American Academy of Sleep Medicine (AASM) [1], a cycle is divided into two phases: slow-wave sleep (stages wake, N1, N2 and N3) and paradoxical sleep. The first is associated with the absence of dreams (NREM "Non Rapid Eye Movements") and the other to the emergence of dreams (REM "Rapid Eye Movements").

During the different stages of sleep, a series of brain waves appears [1]:

- **Stage 1** called somnolence or drowsy sleep; it is the transition between wakefulness and sleep. This stage is characterized by fragmentation of alpha waves (8–13 Hz) to theta waves (4–7 Hz). It only represents about 5 % of the total sleep time.
- **Stage 2** is called light sleep in which brain waves become slower. This stage is marked by the appearance of sleep spindles and K complexes. Stage 2 lasts about 45–50 % of total sleep time.
- **Stage 3** called deep sleep or slow-wave sleep (SWS), it is characterized by delta waves (<3,5 Hz) and rarely with some sleep spindles. It represents around 15–20 % of total sleep time.

The activity of sleep spindles is present throughout the life of a person. A spindle is composed of sinusoidal waves, also known as sigma waves, a short burst of increased brain activity in the region of 11–16 Hz and lasting 0.5–3 s. Their maximum amplitude is obtained when recorded in the central derivations [1]. Spindles are of a biological and clinical interest for their role in cortical development [2] and memory consolidation [3]. The large number of spindles constitutes a sleeping protector against external noise [4]. Neurological disorders can cause changes in the characteristics thereof.

Spindles detection in clinical environment is based on visual reading. This method has several limitations due mainly to the complexity of the recordings, the presence the nonstionaries embedded in the signal and the need to explore long segments of data. Spectral analysis based on the Fourier transform method that assumes time invariance of statistical properties of the signal during short analysis frames is not suitable in the case of EEG recordings since data are very long and requirements of stationarity cannot be satisfied. Abrupt changes in the statistics of the signal are often desirable to detect because they may reveal significant events. For these reasons, the research has been oriented to other techniques using adaptive spectral analysis which consists in tracking frequency components behavior of the signal. Indeed autoregressive modeling applied on visual evoked potentials

(VEP) showed robustness in detecting abnormalities using a few features compared to traditional methods based on the latency measure. The Cepstrum coefficients used to describe the VEP yielded satisfactory results in detecting multiple sclerosis [5] while the zeros of the analysis filter used as the model features combined with a neural network classifier allowed an efficient VEP discrimination [6]. Mobility of the zeros computed from EEG signals was also used for detecting epileptic seizure [7].

In light of the above, this paper proposes spectral analysis techniques to detect spindles based on tracking of frequency changes along EEG recordings during sleep. We processed the signal, through an autoregressive model whose complex zeros calculated adaptively allows detection of the time of occurrence of the spindles. Robustness of the proposed method is assessed by comparing results of the proposed techniques with those obtained traditionally by neurophysiologists using raw EEG data available in DREAMS Sleep Spindles Database [8]. Our method requires a few parameters and works well even in noisy environment.

The rest of the paper is organized as follows. We present the zero-tracking method in Sect. 2. We illustrate the results obtained in Sect. 3 and we conclude the paper in Sect. 4.

2 Adaptive Zero-Tracking Method

Let y represent each EEG recording consisting of a block of samples $y(n)$, $n=0,\ldots,N-1$ where n denotes the time sample and N is the total length of the signal. Spectral analysis of the signal can be performed by processing the signal through an Mth autoregressive model whose polynomial.

$$A(z) = 1 + a_1 z^{-1} + a_2 z^{-2} + \ldots + a_M z^{-M} \qquad (1)$$

A(z) obtained on the basis of the equation:

$$e(n) = y(n) + \sum_{m=1}^{M} a_m y(n - m) \qquad (2)$$

where a is the vector of coefficients a_m, $m=1,\ldots M$ and $e(n)$ is the error of prediction.

A(z) provides an estimate of the spectrum as the inverse of the frequency response $S(\omega) = 1/|A(\omega)|^2$ where ω is the frequency.

Seeking the peaks of the spectrum can alternatively be performed by calculating the roots of A(z) defined such as:

$$A(z) = (1 - z_1 z^{-1})(1 - z_2 z^{-1})\ldots(1 - z_M z^{-1}) \qquad (3)$$

where z_i, $i=1,\ldots$, M are the roots referred herein as the zeroes of the polynomial A (z) of order M. In this way, the spectrum of the signal can be represented in the complex plane by the location of complex zeroes within the unit circle. The value of each zero is represented by a single point. This representation is very interesting especially when dealing with non stationary signals in which the model A(z) becomes time dependent and feature extraction has to be performed adaptively. The set of zeroes of the model are to be calculated at each time instant thus providing a means to track the modulus and the phase of the zeroes versus time. This is very useful for detecting changes of statistical properties during the EEG recording introduced by events such as spindles.

Tracking efficiently these changes requires an efficient adaptation algorithm to estimate accurately at each time instant the values of the zeroes while assuring a fast convergence rate. The Recursive Least Square (RLS) algorithm has been proved to be very suitable even though it is costly for large model order. In addition, to improve tracking of nonstationarities in the signal, the coefficients are estimated such as to minimize the quantity:

$$\varepsilon(i) = \sum_{n=0}^{i} \lambda^{i-n} e(i)^2 \tag{4}$$

Where λ is the forgetting factor that takes a positive value less than 1.

Starting with an arbitrary initialization of the zeroes within the circle of radius 1, the vector of coefficients of prediction a is derived from the set of zeroes through a convolution and the prediction error is computed. The RLS algorithm is then applied to update the set of coefficients which allows adaptation of the set of zeroes according to the adaptation equation:

$$z_m(n) = z_m(n-1) + \Delta z_m(n) \qquad \text{for } m = 1, \ldots, M \tag{5}$$

$\Delta z_m(n)$ is the adaptation term that adjusts the corresponding zero to its optimum value at convergence estimated from the set of coefficients updates $\Delta a_j(n)$:

$$z_m(n) = \sum_{j=1}^{M} \frac{z_m(n)}{a_j(n)} a_j(n) = - \sum_{j=1}^{M} \frac{z_m(n)^{M-j}}{\prod_{i \neq m} (z_m(n) - z_i(n))} a_j(n) \tag{6}$$

Yet, to avoid instability of the model that may occur for large values of the order of the model, Eq. (5) was slightly modified by introducing a weighting scalar to reduce the amount of the correcting term such that: $z_m(n) = z_m(n-1) + \alpha \Delta z_m(n)$, for m=1,...,M. In our application $\alpha = 0.5$ was found suitable to guarantee stability of the model forcing all the zeros to converge within the unit circle even those whose magnitude exceed 1 during transient state [6].

3   Materials and Methods

For testing our method, we used data available in DREAMS Sleep Spindles Database constituted of raw EEG signals recorded during sleep for 6 subjects [8]. We used in this study signals of 30 min length, recorded from the C3-A1 channel and sampled at the sampling frequency of 100 Hz, where spindles were annotated by two experts for sleep spindles scoring.

3.1   *The proposed method*

We processed raw EEG recordings through the adaptive zero-tracking method of order M in which complex zeroes were extracted at each time instant. The modulus and the phase of each zero were used to track EEG frequency components. This can be observed using the locations of the zeroes of the model in the complex plane when convergence is achieved. In addition, the time behavior of the modulus of the complex zeroes plotted versus time were used to detect any changes in statistical properties of the EEG. In this work, best fit of the model was found for low order model (3–5). Furthermore, appropriate detection of spindles in the EEG reflected as abrupt jumps observed in the trajectories of the zeroes were obtained for values assigned to the forgetting factor between 0.997 and 0.999.

Figure 1 shows results obtained for an analysis frame of 12 s. The frame was divided into 3 segments; the one in the middle only includes the spindle. Locations of the zeroes at convergence for the corresponding segments plotted in the complex

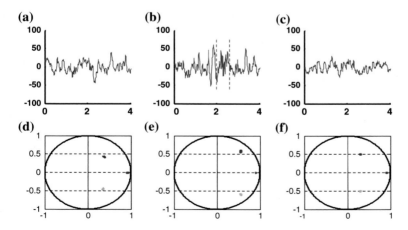

Fig. 1 EEG frames of analysis on top and the corresponding locations of the zeroes in the complex plane computed from a third-order model in the interval of convergence for 3 analysis frames: EEG without spindle EEG (*left*), EEG with spindle (*middle*); EEG without spindle EEG (*right*); λ = 0.999

plane show a pair of complex zero and a real zero. This latter indicating a very low frequency is similarly found in the 3 representations (Fig. 1d–f). Whereas the location of the pair of zeroes is pushed towards the unit circle with the modulus increase from 0.58 to 0.82 when the spindle appears as illustrated in the representation in the middle of Fig. 1e, reflecting the sinusoidal nature of the spindle.

To test capabilities of our method for detecting successive spindles in EEG analysis frame, we analyzed a recording of duration 30 s containing 4 spindles identified (interval between dotted lines) by an expert as shown in Fig. 2a. The plot of the corresponding tracing in time of the modulus of the pair of zeroes shows first a transient state lasting about 2 s needed for the algorithm to converge. Once convergence is achieved, four abrupt jumps appear where the spindles occur as shows in Fig. 2b. While the algorithm attempts to converge after each jump which explain the presence of different regions of convergence, the modulus of the zeroes is always less than 1 indicating that the model remains stable.

The proposed method provides a tool that detects accurately the time of occurrence of the spindle on the basis of a few parameters. Applied on several other EEG recordings, similar results were obtained. The robustness of the method is revealed through its capabilities in even detecting spindles close to each other.

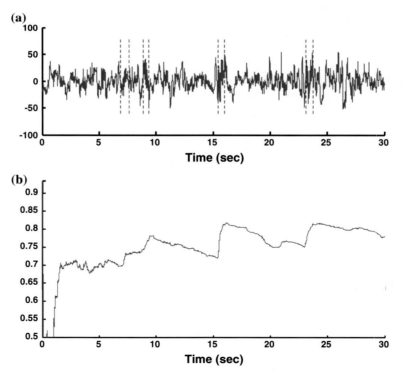

Fig. 2 EEG recording of length 30 s during sleep including 4 spindles and the corresponding modulus of the complex zeroes' behavior versus time. **a** Sleep EEG, **b** modulus of the zeros

4   Conclusions

In this paper, we present a method to detect spindles in raw EEG recordings during sleep. The method consists of processing the signal through an adaptive autoregressive model whose features are represented by the zeroes of the model polynomial as alternative to frequency components estimation in a nonstationary environment. Tracing in time of the zeroes modulus shows abilities in detecting sharp transitions indicating statistical changes due to the occurrence of spindles. The method is stable, it requires a low order model and no preprocessing is needed. New development to characterize the spindle are in progress.

Real time implementation of the method would help neurophysiologists in their interpretation and significantly save time in exploration of long data recordings as well.

References

1. Iber, C., Ancoli-Israel, S., Chesson, A., Quan, SF.: For the American Academy of Sleep Medicine: The AASM Manual for the Scoring of Sleep and Associated Events: Rules, Terminology and Technical Specifications. American Academy of Sleep Medicine, Westchester (2007)
2. Khazipov, R., Sirota, A., Leinekugel, X., Holmes, G.L., Ben-Ari, Y., Buzsaki, G.: Early motor activity drives spindle bursts in the developing somatosensory cortex. Nature **432**, 758–761 (2004)
3. Fogel, S.M., Smith, C.T.: The function of the sleep spindle: A physiological index of intelligence and a mechanism for sleep-dependent memory consolidation. Neurosci. Biobehav. Rev. **35**, 1154–1165 (2011)
4. Dang-Vu, T.T., McKinney, S.M., Buxton, O.M., Solet, J.M., Ellenbogen, J.M.: Spontaneous brain rhythms predict sleep stability in the face of noise. Curr. Biol. **20**, R626–R627 (2010)
5. Drissi, H., Regragui, F., Bennouna, M., Antoine, J.P.: Wavelet transform analysis of visual evoked potentials: Some preliminary results. ITBM-RBM **21**, 84–91 (2000)
6. Mghari, A., Himmi, M.M., Amaloud, A., Regragui, F.: Visual evoked potentials discrimination based on adaptive zero-tracking neural network. Comput. Biol. Med. **36**, 408–418 (2006)
7. Akay, M.: Biomedical Signal Processing. Academic Press, San Diego New York (1994)
8. University of MONS–TCTS laboratory: The DREAMS sleep spindles database. http://www.tcts.fpms.ac.be/_devuyst/Databases/DatabaseSpindles/ (2014)

Mammograms Multiple Watermarking Scheme Based on Visual Cryptography

Meryem Benyoussef, Samira Mabtoul, Mohamed El Marraki and Driss Aboutajdine

Abstract Telemedicine applications require exposing medical data over open networks, which leads to illegal exploitations. However, safety and confidentiality are required for medical images, because critical judgment is done on those images. Therefore, we propose in this paper, a medical images multiple watermarking scheme. Considering the high sensitivity of medical images, the watermark is embedded without modifying the original host image. Based on visual cryptography concept, difference expansion technique and dominant blocks extraction, this scheme achieves copyright protection by a robust watermark and image authentication by a fragile watermark. The experimental results show that the proposed method can withstand several image processing attacks, and can easily prove the integrity of the image.

Keywords Robust watermarking · Fragile watermarking · Visual cryptography · Difference expansion · Wavelet transform · Copyright protection · Image authentication

1 Introduction

Digital image watermarking techniques have been developed to provide security of image in digital form. It is realized by embedding a signature, called also "the watermark pattern", into the original image. The watermark pattern can be embedded either in the spatial domain by directly modifying the pixel intensity of the original image, or in the transform domain such as DCT, DFT, DWT.

There is a lot of medical image watermarking techniques described in the literature that we can classify in two main categories. The first one consists in using

M. Benyoussef (✉) · S. Mabtoul · M.E. Marraki · D. Aboutajdine
Faculty of Science, Department of Physics, LRIT Associated Unit
to the CNRST-URAC N 29, Mohammed V University, Rabat, Morocco
e-mail: benyoussef.meryem@gmail.com

© Springer International Publishing Switzerland 2016
A. El Oualkadi et al. (eds.), *Proceedings of the Mediterranean Conference on Information & Communication Technologies 2015*, Lecture Notes in Electrical Engineering 380, DOI 10.1007/978-3-319-30301-7_47

classical watermarking methods while minimizing the distortion [1, 2]; while the second category corresponds to reversible watermarking [3, 4], which mean that once the embedded content is read, the watermark can be removed from the image allowing retrieval of the original image, however the main drawback of this category is the low embedding capacity compared with the traditional digital watermarking techniques.

The conventional watermarking systems suffer from the tradeoff between the conflicting requirements of capacity, transparency, and robustness. Zero-watermarking has emerged as a new paradigm of watermarking which eliminates the imperceptibility issues due to watermark embedding. This approach does not embed a watermark into the host image physically, whereas it is logically embedded. With the apparition of the concept of Visual Cryptography (VC), first introduced by Moni Noar and Shamir [5], a new rang of zero watermarking techniques based on this concept have been deployed. VC, described as a secret sharing scheme of digital images, involves breaking up the image into n shares using a codebook. Those shares are binary images usually presented in transparencies; so that each participant can hold a transparency (share). The act of decryption is to simply stack shares and view the secret image that appears on the stacked shares. In accordance with cryptography, the security of a cryptosystem does not reside in the algorithm, but resides in the secret key; that is, the security will maintain well even if the algorithm has been published. The first method of image watermarking based on VC was proposed by Hwang in 2000 [6]. Afterwards, others related works have been proposed [7–9]. The watermarking methods based on VC consists, in the embedding process, to create a binary matrix B based on image features. The matrix B is used to generate the secret share from the watermark pattern using a VC codebook. In the extraction process, the same process is repeated to generate the public share. Finally, this later is superimposes on the secret share to recover the ownership label; and since the security characteristics of VC, the watermark pattern is difficult to detect or recover from the marked image in an illegal way.

In this paper, we propose a multiple watermarking scheme for medical image. This scheme achieves copyright protection by a robust watermarking and image authentication by a fragile watermarking. Our method is based on VC concept, FS-DWT transform and difference expansion technique.

The rest of this paper is organized as follows: in Sect. 2, we explain the concept of FS-DWT transform and we describe the robust watermarking algorithm, while the fragile watermarking algorithm is given in Sect. 3. The experimental results and some comparisons are shown in Sect. 4. Finally, Sect. 5 concludes this paper.

2 Robust Watermarking

Our watermarking scheme is divided in two parts, the first one is a copyright protection performed by a robust watermarking. In the embedding process, a cover image, a watermark pattern and a secret key are given as inputs, to create a secret share generated by VC codebook and a secret embedding map generated using some features of FSDWT transform. These later are used in the embedding process of the second part to perform a fragile watermarking. In the extraction process the inputs are the outputs of the same process of the fragile watermarking part.

2.1 FSDWT Transform

Faber-Schauder DWT (FSDWT) has the same construction principle and some identical properties as the S. Mallat wavelet transform. The main difference is the fact that the FSDWT basis is not orthogonal. This transformation is also well adapted to contour detection as it eliminates constant and linear correlation of smooth regions [10]. It uses only the first neighbouring coefficients and gives more precise edge detection than higher order spline wavelets. The main interest of FSDWT is that this wavelet transform is obtained by lifting scheme formulation with only arithmetic operation and no boundary treatment.

In our image watermarking method, we have to extract some image features to construct the secret share. However, medical images are very similar between patients. By using statistical features of the FS-DWT coefficients, wa can create a blocks dominants map consisting of local features such as contours or edges which are unique to each image, and therefore, can act as a signature of the image [11].

2.2 Robust Watermarking Algorithm

Embedding Process:
Inputs: Host Image I ($m \times n$), Watermark Image W ($r \times c$), Secret Key S
Outputs: Secret Share **SS** ($r \times 2c$), Embedding map **EM** ($rc \times 2$)
Step 1:   Perform the FS-DWT transform on I, and find all the dominant blocks
Step 2:   Use S as a seed to select random $r \times c$ dominant blocks. Construct a dominant blocks map **EM**, containing the positions of the selected dominant blocks
Step 3:   Construct a feature image F, which contain averages of the selected dominant blocks. Let F_{avg} be the average of F
Step 4:   Construct a binary matrix B:

$$B(x, y) = \begin{cases} 1, & \text{if } F(x, y) \geq LL_{avg} \\ 0, & \text{if } F(x, y) < LL_{avg} \end{cases} \quad (1)$$

Step 5:   Use the bits in matrix B to select columns in Table 1 for generating the secret share **SS**

Table 1 Codebook used to generate public and secret shares

Pixel	□		■	
Matrix B	0	1	0	1
Public Share	■□	□■	■□	□■
Secret Share	■□	□■	□■	■□
Public Share ⊕ Secret Share	■□	□■	■	■

Extracting Process:

Inputs: Attacked image I′ ($m \times n$), Secret Share **SS** ($r \times 2c$), Embedding map **EM** ($rc \times 2$)

Outputs: Watermark Image W′ ($r \times 2c$)

Step 1:   Perform the FS-DWT transform on I′

Step 2:   For each location **EM**(x, y), select a 77 size sub-image area centered at this location. Construct a feature image F containing averages of these blocks. Let F_{avg} be the average of F

Step 3:   Construct a binary matrix B using Eq. (1)

Step 4:   Use the bits in matrix B to select columns in Table 2 for generating a public share **PS**

Step 5:   Perform logical OR operation on **PS** and **SS** to extract the watermark

Step 6:   Apply the reduction process on the extracted watermark

Reduction Process

Due to the proposed codebook that we used to generate the two shares, the extracted watermark has a size of ($r \times 2c$) compared to the original one. To retrieve the original size and to mitigate the noise effect caused by the watermark extraction, we use a post-process called "reduction process" that can reduce the redundancy data caused by VSS scheme; that is, a block data with two pixels located in each group will be transferred into a corresponding pixel which means that if the block is composed of a black and white pixel or two white pixels then it will be replaced with a white pixel, but if it is composed of a two black pixels then it will be transferred into a black pixel.

Table 2 Visual comparison of extracted watermarks after attacks

Attacked Images	Our Results	[8]'s Results	Attacked Images	Our Results	[8]'s Results
Cropping 25%			Cropping 50%		
Compression 10%			Compression 50%		
Compression 80%			Rotation 3		
Scale 50%			Median Filter (3x3)		
Salt & Pepper Noise 20%			Blurring		

3 Fragile Watermarking

The second part of the watermarking algorithm is a fragile watermarking that detects any manipulation made to a digital image to guarantee the content integrity. We use the image's hash to test the integrity of the image; the hash is inserted, in a reversible manner using difference expansion (DE) technique, into the secret embedding map to generate a secret matrix.

3.1 Difference Expansion

DE is a reversible data-embedding technique, introduced by Tian [12], which involves pairing the pixels of the host image and transforming them into a low-pass image containing the integer averages and a high-pass image containing the pixel differences.

3.2 Fragile Watermarking Algorithm

Embedding Process

Inputs: Host Image I, Secret Key S, Secret Share **SS**, Embedding map **EM**
Outputs: Secret Matrix **SM**
Step 1:   Calculate the hash message (H) of I using MD5 algorithm
Step 2:   Concatenat **SS** with H
Step 3:   Embed the bits into **EM** using DE technique to form **SM**, the process ends when all bits are embedded
Step 4:   Use S as a seed to encrypt **SM**

Extracting Process

Inputs: Attacked image I' ($m \times n$), Secret Key S, Secret Matrix **SM**
Outputs: Secret Share **SS**, Embedding map **EM**
Step 1:   Use S as a seed to decrypt **SM**
Step 2:   Extract H and **SS** from **SM** using DE inverse to retreave **EM**
Step 3:   Calculate the hash (H') of I' using MD5 algorithm, compare it with H to test the integrity of I'

4 Simulation Results

In this section, we present some experimental results concerning the proposed method. Our test mammograms are from MIAS Database of size 1024×1024 pixels, and the watermark is a binary image of size of 100×100 pixels.

4.1 Robustness and Integrity Tests

To test the robustness of the algorithm to attacks, our test images are subjected to several common attacks. A visual comparison between our results and results of a

Table 3 Integrity test

Attacks	PSNR (dB)	NC (%)	Extracted hash	Integrity
Without	Inf	100.00	2da72b1b90989308e7c2717b4e6fe60f	Yes
Cropping 0.1 %	69.07	100.00	3d23ca8f5e95e5ecf5df63181fd7072c	No
Cropping 0.3 %	83.74	100.00	db45d2861f384e6cf8310f6e862d6bf4	No
Sharpening	49.52	99.45	f911b9c8d44d7bd3e85c5cfa2b146116	No
Median Filter 3 × 3	61.78	99.93	53d29dc1e879f93aac895e125e71b3ef	No
Salt& pepper 0.01 %	87.00	99.94	85c06dd2f4a020aa7386650b7c56b12f	No
Rotation 0.1	55.27	99.15	aa077c1b88e3c042c6363d2ca978fc8d	No
Scale 50 %	55.32	99.87	ef813c2504867455527c8d31aacbd9ba	No
JPEG 5	45.38	97.44	c581d1fcb816049d21d7aa29fb064834	No
Crop 10 lines around	60.29	99.86	d6cf275ebdca0f19788628e86dc7e2ca	No

zero watermarking based also on VC [8], which outperform many other approaches, is shown in Table 2; from this table we can see the advantage of our method.

We give in Table 3 the PSNR values of the attacked images and the corresponding NC values of the extracted watermark from each one. We can see that even with a very high NC values, when the watermark is extracted completely, we can prove that the image is not authentique using the fragile extracted watermark.

4.2 Reliability Test

In these simulations, we test the reliability of the proposed method with different input images. Table 4 shows the images used in the embedding and extracting schemes and the extracted watermark using our method and [8] method. The results show that our method is very reliable.

4.3 Capacity Comparison

DE is one of the outstanding reversible embedding techniques, because of its high embedding capacity compared to other reversible schemes. But regarding the high sensitivity of medical images, the capacity is still low to preserve the image features. However, since the embedding in our scheme, is done on the secret embedding map and not into the image itself, we will not consider the underflow and overflow problems; Thus we can use a multiple-layer embedding without worrying about the degradation of the visual quality, which will increase the

Table 4 Reliability test with different input images

Image used in the embedding Process	Image used in the extracting process	Extracted watermark using our method	Extracted watermark using [8] method

Table 5 Hiding capacity comparison

Scheme	Embedding technique	Hiding capacity (bpp)
Zain [1]	Spatial LSB	0.44
Giakoimaki [2]	Frequency DWT	0.08
Guo [3]	DE	0.75
Al-Qershy [4]	Modifed DE	0.5
Our scheme	Multiple layer DE	1

embedding capacity. Table 5 shows a hiding capacity comparison with other schemes.

5 Conclusion

In this work we proposed a multiple watermarking scheme for medical images. The purpose of this method is copyright protection and Image Authentication. In this scheme, we combined VC, DE and FS-DWT transform, to benefit from the many characteristics of this later, from the security of the VC concept and its high protection of medical image content and also from the reversibility and high embedding capacity of DE technique.

The proposed method has been tested in mammograms images from MIAS database, and the experimental results show that this method is reliable and can withstand several image processing attacks such as cropping, filtering and compression etc.

References

1. Zain, J.M., Fauzi, A.R.M.: Medical image watermarking with tamper detection and recovery. In: Proceedings of the 28th IEEE EMBS Annual International Conference (2006)
2. Giakoumaki, A., Pavlopoulos, S., Koutsouris, D.: Multiple image watermarking applied to health information management. IEEE Trans. Inf. Technol. Biomed. 10(4), 722–732 (2006)
3. Guo, X., Zhuang, T.-G.: A region-based lossless watermarking scheme for enhancing security of medical data. J. Digit. Imaging, 1–12 (2007)
4. Osamah, M.: Al-Qershi and Bee Ee Khoo: Authentication and data hiding using a hybrid ROI-based watermarking scheme for DICOM images. J. Digit. Imaging 24(1), 114–125 (2011)
5. Naor, N., Shamir, A.: Visual cryptography. Advances in Cryptology: Eurocrypt'94, pp. 1–12 (1995)
6. Hwang, R.J.: A digital copyright protection scheme based on visual cryptography. Tamkang J. Sci. Eng. 3(3), 97–106 (2001)
7. Abusitta, A.H.: A visual cryptography based digital image copyright protection. J. Inf. Secur. 3, 96–104 (2012)
8. Surekha, B., Swamy, G.N.: Sensitive digital image watermarking for copyright protection. Int. J. Netw. Secur. 15(1), 95–103 (2013)
9. Benyoussef, M., Mabtoul, S., El Marraki, M., Aboutajdine, D.: Blind invisible watermarking technique in DT-CWT domain using visual cryptography. ICIAP, Part I, LNCS 8156, 813–822 (2013)
10. Douzi, H., Mammass, D., Nouboud, F.: Faber-Schauder wavelet transformation application to edge detection and image characterization. J. Math. Imaging Vis. 14(2), 91–102 (2001)
11. El hajji, M., Douzi, H., Harba, R.: Watermarking based on the density coefficients of Faber-Schauder wavelets. ICISP, LNCS. pp. 455–462 (2008)
12. Tian, J.: Reversible data embedding using a difference expansion. IEEE Trans. Circuits Syst. Video Technol. 13(8), 890–896 (2003)

Comparative Analysis of Classification, Clustering and Regression Techniques to Explore Men's Fertility

Anwar Rhemimet, Said Raghay and Omar Bencharef

Abstract Data Mining aims is to extract maximum of knowledge automatically or semi-automatically from huge databases using interactive exploration tools. In this article, we focus on the exploration of medical data to study the case of analyzed sperm samples according to the criteria of the WHO (World Health Organization) using a powerful analysis tool for exploring the results of classification algorithms, clustering and regression. For this research, 100 volunteers provide a semen sample and they were also asked to complete a validated questionnaire on their lifestyle and health status. Sperm concentration is also linked to socio-demographic and environmental factors.

Keywords Data mining · Classification · Regression · Clustering · Sperm fertility

1 Introduction

Semen analysis is the most important consideration in assessing male fertility. From the main examination, semen analysis accurately measures parameters such as sperm count, sperm motility, the size and shape of the sperm, semen volume, and the dosage of certain substances normally present.

Following the publication of a meta-analysis on the prediction of sperm quality with artificial intelligence methods (David Gil et al.) [1], there is a debate about a possible decline in sperm quality. In the last two decades, many studies have shown a decrease in sperm parameters [2–4].

Our approach is to study and predict the results of semen analysis from questionnaire responses before the biologic exam of information using socio-demographic data, environmental factors, health status, and individuals' life habits.

A. Rhemimet · S. Raghay
Faculty of Science and Technology, Cadi Ayyad University, Marrakech, Morocco

O. Bencharef (✉)
Higher School of Technology-Essaouira, Cadi Ayyad University, Marrakech, Morocco
e-mail: bencharef98@gmail.com

© Springer International Publishing Switzerland 2016
A. El Oualkadi et al. (eds.), *Proceedings of the Mediterranean Conference on Information & Communication Technologies 2015*, Lecture Notes in Electrical Engineering 380, DOI 10.1007/978-3-319-30301-7_48

The goal is to be convenient for the early diagnosis of patients suffering from disorders or during the selection of candidate's sperm donors. As an example, this research can be done automatically, by implementing Data Mining algorithms (DM) such as classification, clustering and regression.

Indeed, comparative studies have been made on methods of data minig on different areas to make the choice on the most suitable algorithms: SS Baskar et al. [5] and Vrushali Bhuyar [6] made a comparative study to soil classification Naive Bayes, JRip and J48, and found that the J48 algorithm is the best method to use for this event. While D Ramesh et al. [7] used multiple linear regression to predict the yield of rice. On the other hand, Suman et al. [8] combines the data using the K-means Clustering on soil dataset then he applied the linear regression to classify the clusters.

In this paper, we propose to use the Oracle Data Miner exploration tool to make efficient data analysis for our case study.

2   Background

Algorithms used as test example will be described below for the two types of learning. Indeed, there are two types: the supervised data mining, which is used primarily for data classification and unsupervised data mining is used in research associations or groups of individuals.

2.1   Supervised Learning

Supervised learning mainly concerns data classification methods (we know the input and the output is to be determined) and regression (output is known and you want to find the entrance).

- **Generalized Linear Models (GLM)**: GLM implements logistic regression for classification of binary targets and linear regression for continuous targets. GLM classification supports confidence bounds for prediction probabilities. GLM regression supports confidence bounds for predictions [9].
- **Support Vector Machine (SVM)**: Distinct versions of SVM use different kernel functions to handle different types of data sets. Linear and Gaussian (nonlinear) kernels are supported. SVM classification attempts to separate the target classes with the widest possible margin. In addition, SVM regression tries to find a continuous function such as the maximum number of data points that lies within an epsilon-wide tube around it [10].
- **Decision Tree (DT)**: Decision trees extract predictive information in the form of human-understandable rules. The rules are if-then-else expressions; they explain the decisions that lead to the prediction [11].

- **Naive bayes (NB)**: Naive Bayes makes predictions using Bayes' Theorem, which derives the probability of a prediction from the underlying evidence, as observed in the data [12].

2.2 Unsupervised Learning

In unsupervised learning situations all variables are treated in the same way, there is no distinction between explanatory and dependent variables. However, in contrast to the name undirected data mining there is still some target to achieve. This target might be as general as data reduction or more specific as clustering. The dividing line between supervised learning and unsupervised learning is the same that distinguishes discriminant analysis from cluster analysis. Supervised learning requires that the target variable is well defined and that a sufficient number of its values are given. For unsupervised learning typically either the target variable is unknown or has only been recorded for too small number of cases.

- **k-Means**: k-Means is a distance-based clustering algorithm that partitions the data into a predetermined number of clusters. Each cluster has a centroid (center of gravity). Cases (individuals within the population) that are in a cluster are close to the centroid [13].
- **O-Cluster:** O-Cluster creates a hierarchical, grid-based clustering model. The algorithm creates clusters that define dense areas in the attribute space. A sensitivity parameter defines the baseline density level [14].

3 Results

3.1 Database

The database was obtained following a study conducted by the World Health Organization in 2010 and the information was organized by the Biotechnology Department at the University of Alicante.

To perform this study, volunteer's students from the University of Alicante were recruited as a young population and healthy males. It was previously established that the use of volunteers are not introduced by selection through male fertility studies [15], and [1] we were interested in the impact of environmental factors and the lifestyle in the semen parameters. It was decided to not study population with known or suspected reproductive disorders (e.g. fertility clinic patients). In addition, It is also important to note that the study population of this work mainly represents the characteristics of candidates becoming semen donors (i.e. young male,

Table 1 The table below presents all relevant attributes information about the dataset

Season in which the analysis was performed	(1) winter, (2) spring, (3) Summer, (4) fall. (-1, -0.33, 0.33, 1)
Age at the time of analysis	18–36 (0, 1)
Childish diseases	(1) yes, (2) no. (0, 1)
Accident or serious trauma	(1) yes, (2) no. (0, 1)
Surgical intervention	(1) yes, (2) no. (0, 1)
High fevers in the last year	(1) less than three months ago, (2) more than three months ago, (3) no. (-1, 0, 1)
Frequency of alcohol consumption	(1) several times a day, (2) every day, (3) several times a week, (4) once a week, (5) hardly ever or never (0, 1)
Smoking habit	(1) never, (2) occasional (3) daily. (-1, 0, 1)
Number of hours spent sitting per day	ene-16 (0, 1)
Output: diagnosis	normal (N), altered (O)

university students) [16]. 100 volunteers between 18 and 36 years participated in the study.

Table 1 shows a description and the varying range of values used in the study and also the standardized values after treatment in the different fields of database entry. The values were converted to the input data in a range of standardization based rules:

(a) Numerical variables such as age are normalized into the interval (0–1).
(b) The variables with only two independent attributes are prearranged with binary values (0, 1).
(c) The variables with three independent attributes, such as "Vaccines received", "High fevers in the last year" and "Smoking habit" are prearranged using the ternary values (-1, 0, 1).
(d) The variables with four independent attributes, such as "Season in which the analysis was performed" are prearranged using the four different and equal distance values (-1, -0.33, 0.33, 1).

3.2   Classification

A classification model is tested by applying it to test data with known target values and comparing the predicted values with known values.

The test data must be compatible with the data used to build the model and should be prepared in the same manner as construction data. Generally, compile and test data come from the same set of historical data. A percentage of records are used to build the model; the remaining records are used to test the model. The test

Table 2 The table presents the results of the classification methods test

	Performance of matrix			Lift		Profit
	Average accuracy	Averall accuracy	Cost	Lift cumulative %	Gain cumuative %	General profit
Decision Tree	43.33	61.36	46.86	0.97	44.44	−7.07
Naïve Bayes	50	88.63	–	1	45.45	−6.16

parameters are used to assess the accuracy with the model predicts and the known values. If the model works well and meets the needs of the business, it can be applied to new data to predict the future [17].

Oracle Data Miner provides four algorithms to solve classification problems; the nature of the data determines which method will provide the best business solution, so in this part we will test the best model for each algorithm and choose the best of them for deployment.

In this section, we will study the two algorithms "*Decision Tree* and *Naive bayes*" because the following methods "*Generalized Linear Models (GLM) and Support Vector Machine (SVM)*" concern the classification and regression at the same time so we will test them in the last chapter to evaluate the regression.

After testing the actions of classification algorithms, the table (Table 2) below presents a comparison of the results:

Concerning the performance matrix, the overall accuracy refers to the percentage of correct predictions made by the model when compared with the actual classifications in the test data. While the average accuracy is the measure of the reproducibility of an observation, to what extent a result can be correctly determined independently of a reference to the true value.

After this detailed experimentation, the integrated model "Naïve Bayes" in the previous sections is determined to be the best solution to the problem studied regarding the results and execution time. Therefore, the model is ready to be applied for the whole population, or for new data.

3.3 Clustering

Clustering is used to identify distinct segments of the population and explain the common characteristics of a cluster member, and also to determine what distinguishes the members of a cluster from members of another cluster. ODM provides two clustering algorithms, improved k-means and O-cluster.

The gravity center is the most typical case in a cluster. This is a prototype that does not necessarily describe a given case assigned to the cluster. The attribute values for the gravity center are the average of numerical attributes and categorical attributes mode.

Table 3 The table represents the comparison of the test results of the confidence estimate for the example of the top ranked attributes

Confidence (%)	Childish deseases	Age	Frequency alcohol	High fevers
K-means	100	83.33	66.67	50
O-cluster	50	25	40	33.34

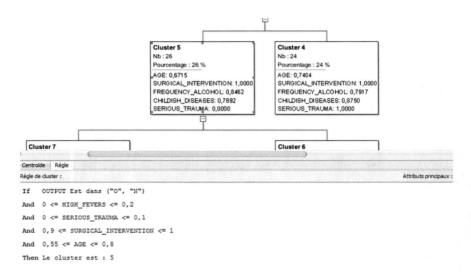

Fig. 1 Tree construction with estimating rule for the k-means algorithm (Cluster 5)

For each cluster in the tree, we can see the number of cases in each group and the percentage of all cases for this cluster. The Cluster Rules tab contains a set of rules that define the cluster in an IF/THEN statement format (Table 3 and Fig. 1).

The version ODM k-means goes beyond classical implementation by defining hierarchical parent-child relationship clusters.

3.4 Regression

To illustrate this feature, we will use the same criteria for the classification problem, but the difference is that we will select as target the continuous attribute: AGE. We will test the following algorithms: Generalized Linear Models (GLM) and Support Vector Machine (SVM).

After testing the actions of both regression algorithms, the table (Table 4) below presents a comparison of the results:

Table 4 The table presents the results of the test to the regression methods

	Predictive confidence	Mean absolute deviation	Mean square deviation	Expected average value	Average real value
GLM	12.34	0.08	0.1	0.68	0.66
SVM	15.08	0.08	0.097	0.68	0.66

SVM distinguishes between small mistakes and errors; the difference is defined by the value of epsilon. The algorithm will calculate and optimize the internal epsilon value.

SVM is determined to be the best solution to the regression problem as SVM is more powerful to determine errors and more efficient for the running time.

3.5 Interpretation

As far as we know, this is the first time that the methods in question were compared accurately on the ODM tool to solve the problem of the relationship between the lifestyle and the quality of sperm. On the data provided, it was found that the most important element linked to sperm concentration is "Age". This result corresponds to previous research [18] indicating that aging increases the risk of decreased sperm concentration.

Other important factors that are related to temperature may affect the quality of the sperm, and we cite time sitting, high fever during the last year and the season when the semen sample is taken. Following the results explaining that the sperm samples obtained in winter are probably better than the semen samples obtained in other seasons, it is shown that the increase of the temperature of the environment may cause an elevation of testicular temperature.

Regarding factors that are least important in sperm concentration, as "Smoking", "Alcohol consumption", "Surgical intervention" and "Childish diseases" they remains, in our study, with no impact if they stay in the normal range. Men with low quality of sperm and who wish to have children should stop smoking and reduce alcohol consumption. All in all, a larger sample size can surely increase the confidence in the observed studies.

4 Conclusion

In this paper, we proposed an analysis of data on sperm concentration using different algorithms and classification technique, clustering and regression. It was based on the most accurately predicted attribute using the powerful data mining tool Oracle Data Miner (ODM) and that gave better results by analyzing the fertility experience. For the classification, Naïve Bayes model was determined to be the best solution to the problem studied following the results and execution time. About clustering, it was found that the ODM version of k-means going beyond the conventional implementation by defining a hierarchical parent-child relationship clusters. And finally regarding the regression method, Support Vector Machine algorithm (SVM) is more effective to determine errors and also more efficient for execution time. Indeed, these algorithms are recommended for this case study and the models are ready to be applied to the general population, or for new data.

References

1. Gil, D., Girela, J.L., De Juan, J., Gomez-Torres, M.J., Johnsson, M.: Predicting seminal quality with artificial intelligence methods. Expert Syst. Appl. **39**, 12564–12573 (2012)
2. Inhorn, M.C.: Global infertility and the globalization of new reproductive technologies: illustrations from Egypt. Soc. Sci. Med. **56**, 1837–1851 (2003)
3. Lutz, W., O'Neill, B.C., Scherbov, S.: Demographics. Europe's population at a turning point. Science **299**, 1991–1992 (2003)
4. Grant, J., Hoorens, S., Sivadasan, S., Loo, M.V., Davanzo, J., Hale, L., Butz, W.: Trends in European fertility: should Europe try to increase its fertility rate… or just manage the consequences? Int. J. Androl. **29**, 17–24 (2006)
5. Baskar, S.S., Arockiam, L., Charles, S.: Applying data mining techniques on soil fertility prediction. Int. J. Comput. Appl. Technol. Res. **2**(6), 660–662 (2013)
6. Bhuyar, V.: Comparative analysis of classification techniques on soil data to predict fertility rate for Aurangabad district. Int. J. Emerg. Trends Technol. Comput. Sci. **3**(2), 200–203 (2014)
7. Ramesh, D., Vishnu Vardhan, B.: Region specific crop yield analysis: a data mining approach. UACEE Int. J. Adv. Comput. Sci. Appl. (IJCSIA) **3**(2) (2013). ISSN 2250–3765
8. Suman, B.B.N.: Soil classification and fertilizer recommendation using WEKA. IJCSMS Int. J. Comput. Sci. Manag. Stud. **13**(5) (2013)
9. Senn, S.: A conversation with John Nelder. Stat. Sci. **18**(1), 118–131 (2003). doi:10.1214/ss/1056397489
10. Cortes, C., Vapnik, V.: Support-vector networks. Mach. Learn. **20**(3), 273 (1995). doi:10.1007/BF00994018
11. Quinlan, J.R.: Simplifying decision trees. Int. J. Man-Mach. Stud. **27**(3), 221 (1987). doi:10.1016/S0020-7373(87)80053-6
12. Webb, G.I., Boughton, J., Wang, Z.: Not so naive Bayes: aggregating one-dependence estimators. Mach. Learn. **58**(1), 5–24 (2005). doi:10.1007/s10994-005-4258-6 (Springer)
13. Coates, A., Ng, A.Y.: Learning feature representations with k-means. In: Montavon, G., Orr, G.B., Müller K.-R. (eds.) Neural Networks: Tricks of the Trade. Springer, Heidelberg (2012)
14. Zhao, D., Tang, X.: Cyclizing clusters via zeta function of a graph. In: Advances in Neural Information Processing Systems (2008)
15. Eustache, F., Auger, J., Cabrol, D., Jouannet, P.: Are volunteers delivering semen samples in fertility studies a biased population? Hum. Reprod. **19**, 2831–2837 (2004)
16. Thorn, P., Katzorke, T., Daniels, K.: Semen donors in Germany: a study exploring motivations and attitudes. Hum. Reprod. **23**, 2415–2420 (2008)
17. Dogan, N., Tanrikulu, Z.: A comparative analysis of classification algorithms in data mining for accuracy, speed and robustness. Inf. Technol. Manag. **14**(2), 105–124 (2013)
18. Kidd, S.A., Eskenazi, B., Wyrobek, A.J.: Effects of male age on semen quality and fertility: a review of the literature. Fertil. Steril. **75**, 237–248 (2001) (Algorithms 2014, 7 417)

The Dynamics of a Population of Healthy People, Pre-diabetics and Diabetics with and without Complications with Optimal Control

Wiam Boutayeb, Mohamed E.N. Lamlili, Abdesslam Boutayeb and Mohamed Derouich

Abstract In this study we propose an extension of the mathematical model developed by Derouich et al. [1] by considering the dynamics of healthy people. Moreover an optimal control approach is proposed in order to reduce the burden of pre-diabetes and diabetes with its complications. Our model shows that the number of pre-diabetics and diabetics with and without complications can be limited by a control and hence the overall burden of diabetes can be reduced. The dynamics of a population of healthy, pre-diabetics and diabetics in presence and absence of complications is studied in a period of 10 years with and without optimal control.

Keywords Diabetes · Mathematical model · Simulation · Optimal control

1 Introduction

In 2013 the International Diabetes Federation (IDF) assumed that the number of adults having diabetes has reached 382 million people worldwide (8.3 %), and by the end of 2013, diabetes will have caused more than five million deaths [2].

In order to prevent adults from this disease, efficient and optimal strategies are needed to reduce the burden of diabetes and its complications.

During the last decades, a large number of publications were devoted to mathematical modelling for diabetes as indicated by recent reviews [2–7]. Based on

W. Boutayeb (✉) · M.E.N. Lamlili · A. Boutayeb
Department of Mathematics, Faculty of Sciences, URAC04,
Boulevard Mohamed VI, BP: 717, Oujda, Morocco
e-mail: wiam.boutayeb@gmail.com

M. Derouich
National School of Applied Sciences, Univerisity Mohamed Premier,
Boulevard Mohamed VI, BP: 717, Oujda, Morocco

© Springer International Publishing Switzerland 2016
A. El Oualkadi et al. (eds.), *Proceedings of the Mediterranean Conference on Information & Communication Technologies 2015*, Lecture Notes in Electrical Engineering 380, DOI 10.1007/978-3-319-30301-7_49

previous mathematical models for the burden of diabetes and its complications [1, 8], our study aims to use an optimal control strategy acting on the reduction of pre-diabetes and consequently minimizing the number of diabetics in presence and absence of complications.

2   The Mathematical Model

To formulate the model, we consider an adult population of size N composed by healthy people, pre-diabetics and diabetics with and without complications.

Let $P = P(t)$, $E = E(t)$, $D = D(t)$ and $C = C(t)$ be respectively the numbers of healthy, pre-diabetics and diabetics with and without complications.

This population evolves continuously in the time interval $[0; T]$, at any time $t \in [0; T]$, we have: $N = N(t) = P(t) + E(t) + D(t) + C(t)$. Figure 1 illustrates the dynamics of an adult population with related settings.

Following the description and illustration above, the proposed model can be written as:

$$\begin{cases} \frac{dP}{dt} = & n - (I_1 + I_2 + I_3 + \mu)P + \gamma_1 E \\ \frac{dE}{dt} = & I_1 P - (\gamma_1 + \beta_1 + \beta_3 + \mu)E \\ \frac{dD}{dt} = & I_2 P + \beta_1 E + \gamma_2 C - (\beta_2 + \mu)D \\ \frac{dC}{dt} = & I_3 P + \beta_2 D + \beta_3 E - (\gamma_2 + \mu + \delta)C \end{cases} \quad (1)$$

where:

- n: denotes the incidence of adult population
- I_1: denotes the rate of healthy persons to become pre-diabetic,
- I_2: denotes the rate of healthy persons to become diabetic,
- I_3: denotes the rate of healthy persons developing complications,
- μ: natural mortality rate,

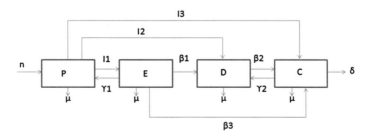

Fig. 1 The dynamics of healthy pre-diabetic and diabetic people with and without complications

- β_1: the probability of a pre-diabetic person to become diabetic,
- β_2: the probability of a diabetic person developing a complications,
- β_3: the probability of a pre-diabetic person developing a complication,
- γ_1: rate at which a pre-diabetic person becomes healthy,
- γ_2: rate at which complications are cured,
- δ: mortality rate due to complications,

The behaviour of this model is studied in a period of 10 years with and without optimal control.

3 The Optimal Control: Existence and Characterization

The optimal control approach is used to reduce the number of pre-diabetics in order to prevent adults from diabetes and its complications.

For the optimal control problem, we consider the control variable $u = u(t)$ to be the percentage of healthy people being prevented from pre-diabetes per unit of time.

The controlled model based on system (1), is written as follows:

$$
\begin{cases}
\frac{dP}{dt} = & n - I_1(1-u)P - (I_2 + I_3 + \mu)P + \gamma_1 E \\
\frac{dE}{dt} = & I_1(1-u)P - (\gamma_1 + \beta_1 + \beta_3 + \mu)E \\
\frac{dD}{dt} = & I_2 P + \beta_1 E + \gamma_2 C - (\beta_2 + \mu)D \\
\frac{dC}{dt} = & I_3 P + \beta_2 D - (\gamma_2 + \mu + \delta)C + \beta_3 E
\end{cases}
\tag{2}
$$

- The problem is then to minimize the objective functional defined as:

$$
\mathcal{J}(u) = \int_0^T \left(E(t) + Au^2(t) \right) dt.
\tag{3}
$$

where A is a positive weight that balances the size of the terms.
U is the control set defined as:

$$
U = \{u/u \text{ is measurable, } 0 \le u(t) \le 1, \ t \in [0, T]\}
$$

- The optimal control $u^* \in U$ satisfies:

$$
\mathcal{J}(u^*) = \min_{u \in U} \mathcal{J}(u).
\tag{4}
$$

3.1 Existence and Positivity of Solutions

Theorem 1 *The set* $\Omega = \{(P, E, D, C) \in \mathbb{R}^4/0 \leq P, E, D, C \leq \frac{n}{\mu}\}$ *is positively invariant under system* (2).

Proof We assume that there exists $t^* > 0$ such that $P(t^*) = 0$, other variables are positive and $P(t) > 0$ for $t \in [0, t^*[$. We have:

$$\frac{dP}{dt} = n - (I_1(1 - u) + I_2 + I_3 + \mu)P + \gamma_1 E \tag{5}$$

$$\frac{dP e^{(I_1 + I_2 + I_3 + \mu)t}}{dt} = e^{(I_1 + I_2 + I_3 + \mu)t}(n + uI_1 P + \gamma_1 E) \tag{6}$$

After integrating the Eq. (6), we obtain:

$$P(t^*) = e^{-(I_1 + I_2 + I_3 + \mu)t^*}\left[P(0) + \int_0^{t^*} e^{-(I_1 + I_2 + I_3 + \mu)t^*}(n + uI_1 P + \gamma_1 E)dt\right] \tag{7}$$

therefore $P(t^*) > 0$ which contradicts $P(t^*) = 0$, thus: $P(t) > 0 \,\forall\, t \in [0, T]$ Similarly, we prove that $E(t) > 0$, $D(t) > 0$ and $C(t) > 0$.

Also, one assumes that:

$$\frac{dN}{dt} = n - \mu N - \delta C \leq n - \mu N$$

$$\Rightarrow N(t) \leq \frac{n}{\mu} - \left(\frac{n}{\mu} - N(0)\right)e^{-\mu t}$$

$$\Rightarrow N(t) \leq \frac{n}{\mu}$$

Theorem 2 *The controlled system* (2) *that satisfies a given initial condition.* $(P(0), E(0), D(0), C(0)) \in \Omega$ *has a unique solution.*

Proof Let

$$X = \begin{pmatrix} P(t) \\ E(t) \\ D(t) \\ C(t) \end{pmatrix}, \quad \varphi(X) = X_t = \begin{pmatrix} \frac{dP(t)}{dt} \\ \frac{dE(t)}{dt} \\ \frac{dD(t)}{dt} \\ \frac{dC(t)}{dt} \end{pmatrix}$$

We can write the system (2) as: $\varphi(X) = X_t = AX + B$ (8).
Where

$$A = \begin{pmatrix} -(\mu + I_1(1 - u(t) + I_2 + I_3) & \gamma_1 & 0 & 0 \\ I_1(1 - u(t)) & -(\mu + \beta_1 + \beta_3 + \gamma_1) & 0 & 0 \\ I_2 & \beta_1 & -(\mu + \beta_2) & \gamma_2 \\ I_3 & \beta_2 & \beta_3 & (\gamma_2 + \mu + \delta) \end{pmatrix}$$

and

$$B = \begin{pmatrix} n \\ 0 \\ 0 \\ 0 \end{pmatrix}$$

then,

$$\|\varphi(X_1) - \varphi(X_2)\| \le \|A\| \cdot \|(X_1 - X_2)\|. \tag{8}$$

Thus, it follows that the function φ is uniformly Lipschitz continuous. So from the definition of the control $u(t)$ and the restriction on $P(t) \ge 0$, $E(t) \ge 0$, $D(t) \ge 0$ and $C(t) \ge 0$, we conclude that a solution of the system (2) exists.

3.2   Existence of an Optimal Control

Theorem 3 *Consider the control problem with system* (2). *There exists an optimal control* $u^* \in U$ *such that*

$$\mathcal{J}(u^*) = \min_{u \in U} \mathcal{J}(u). \tag{9}$$

Proof Using a result by Fleming and Rishel [9], we can prove the existence of the optimal control checking the following points:

- The Theorems 2 and 3 indicate that the set of controls and corresponding state variables is not empty
- $\mathcal{J}(u) = \int_0^T (E(t) + Au^2(t))dt$ is convex in u.
- The control space $U = \{u/u \text{ is measurable, } 0 \le u(t) \le 1, t \in [0, T]\}$ is convex and closed by definition.
- All the right hand sides of equations of system (2) are continuous, bounded above by a sum of bounded control and state, and can be written as a linear function of u with coefficients depending on time and state.
- The integrand in the objective functional, $(E(t) + Au^2(t))$ is clearly convex in U.
- There exist constants $\alpha_1, \alpha_2 > 0$, and $\alpha > 1$ such that $E(t) + Au^2(t)$ satisfies $E(t) + Au^2(t) \ge \alpha_1 + \alpha_2|u|^\alpha$

The state variables being bounded, let $\alpha_1 = \frac{1}{2}\inf_{t\in[0,T]} E(t)$, $\alpha_2 = A$ and $\alpha = 2$ then it follows that: $E(t) + Au^2(t) \geq \alpha_1 + \alpha_2|u|^2$.

Then from Fleming and Rishel [9] we conclude that there exists an optimal control.

3.3   Characterization of the Optimal Control

The necessary conditions for the optimal control arise from the Pontryagin's maximum principle [10].

The Hamiltonian is defined as follows:

$$H = E + Au^2 + \lambda_1 \frac{dP}{dt} + \lambda_2 \frac{dE}{dt} + \lambda_3 \frac{dD}{dt} + \lambda_4 \frac{dC}{dt}$$

Given an optimal control u^* and solutions P^*, E^*, D^* and C^* of the system (2), there exist adjoint variables λ_1, λ_2, λ_3 and λ_4 satisfying:

$$\begin{cases} \lambda_1' = & -\frac{dH}{dP} = \lambda_1 I_1(1-u) + \lambda_1(I_2 + I_3 + \mu) - \lambda_2 I_1(1-u) - \lambda_3 I_2 - \lambda_4 I_3 \\ \lambda_2' = & -\frac{dH}{dE} = -1 + \lambda_2(\beta_1 + \beta_3 + \gamma_1 + \mu) - \lambda_3\beta_1 - \lambda_4\beta_3 \\ \lambda_3' = & -\frac{dH}{dD} = \lambda_3(\beta_2 + \mu) - \lambda_4\beta_2 \\ \lambda_4' = & -\frac{dH}{dC} = -\lambda_3\gamma_2 + \lambda_4(\gamma_2 + \mu + \delta) \end{cases}$$

The optimal control u^* is obtained from the optimally condition:

$$\frac{dH}{du} = 0 \Rightarrow 2Au + \lambda_1(I_1 P) - \lambda_2(I_1 P) = 0 \Rightarrow u^* = \frac{1}{2A}\left[-PI_1(\lambda_1 - \lambda_2)\right]$$

4   Numerical Discretization

Following Gumel et al. [11], we use a Gauss-Seidel-like implicit finite-difference method. The time interval $[t_0, T]$ is discretized with a step h (time step size) such that $t_i = t_0 + ih$, $i = 0, 1, \cdots, m-1$ and $t_m = T$.

At each point t_i we will note

$P_i = P(t_i), E_i = E(t_i), D_i = D(t_i)$ and $C_i = C(t_i)$,

$\lambda_1^i = \lambda_1(t_i)$, $\lambda_2^i = \lambda_2(t_i)$, $\lambda_3^i = \lambda_3(t_i)$, $\lambda_4^i = \lambda_4(t_i)$,

$u_i = u(t_i)$

For the approximation of the derivative we use simultaneously forward difference for $\frac{dP(t)}{dt}$, $\frac{dE(t)}{dt}$, $\frac{dD(t)}{dt}$ and $\frac{dC(t)}{dt}$ and backward difference for $\frac{d\lambda_1(t)}{dt}$, $\frac{d\lambda_2(t)}{dt}$, $\frac{d\lambda_3(t)}{dt}$ and $\frac{d\lambda_4(t)}{dt}$.

So the derivatives $\frac{dP(t)}{dt}$, $\frac{dE(t)}{dt}$, $\frac{dD(t)}{dt}$ and $\frac{dC(t)}{dt}$ are approached by the following finite differences:

$$\left.\begin{array}{l}
\dfrac{dP_{i+1}}{dt} \approx \dfrac{P_{i+1} - P_i}{h} \\[2mm]
\dfrac{dE_{i+1}}{dt} \approx \dfrac{E_{i+1} - E_i}{h} \\[2mm]
\dfrac{dD_{i+1}}{dt} \approx \dfrac{D_{i+1} - D_i}{h} \\[2mm]
\dfrac{dC_{i+1}}{dt} \approx \dfrac{C_{i+1} - C_i}{h}
\end{array}\right\} \quad for \ \ i = 0, \ldots, m-1.$$

Similarly, $\frac{d\lambda_1(t)}{dt}$, $\frac{d\lambda_2(t)}{dt}$, $\frac{d\lambda_3(t)}{dt}$ and $\frac{d\lambda_4(t)}{dt}$ are approached by finite differences. Hence the problem is given by the following numerical scheme:

$$P(0) = P_0, \quad E(0) = E_0, \quad D(0) = D_0, \quad C(0) = C_0,$$
$$u(0) = 0, \quad \lambda_1^m = 0, \quad \lambda_2^m = 0, \quad \lambda_3^m = 0, \quad \lambda_4^m = 0.$$

for $i = 0, \ldots, m-1$ we have:

$$\left\{\begin{array}{l}
P_{i+1} = \dfrac{P_i + hn + h\gamma_1 E_i}{1 + hI_1(1 - u_i) + hI_2 + hI_3 + h\mu} \\[4mm]
E_{i+1} = \dfrac{E_i + hI_1(1 - u_i)P_{i+1}}{1 + h\beta_1 + h\beta_3 + h\gamma_1 + h\mu} \\[4mm]
D_{i+1} = \dfrac{D_i + hI_2 P_{i+1} + h\beta_1 E_{i+1} + h\gamma_2 C_i}{1 + h\beta_2 + h\mu} \\[4mm]
C_{i+1} = \dfrac{C_i + hI_3 P_{i+1} + h\beta_2 D_{i+1} + h\beta_3 E_{i+1}}{1 + h\gamma_2 + h\mu + h\delta} \\[4mm]
\lambda_1^{m-i-1} = \dfrac{\lambda_1^{m-i} + hI_1(1 - u_i)\lambda_2^{m-i} + hI_2\lambda_3^{m-i} + hI_2\lambda_4^{m-i}}{1 + hI_1(1 - u_i) + hI_2 + hI_3 + h\mu} \\[4mm]
\lambda_2^{m-i-1} = \dfrac{h + \lambda_2^{m-i} + h\beta_1\lambda_3^{m-i} + h\beta_3\lambda_4^{m-i} + h\gamma_1\lambda_1^{m-i-1}}{1 + h\beta_1 + h\beta_3 + h\gamma_1 + h\mu} \\[4mm]
\lambda_3^{m-i-1} = \dfrac{\lambda_3^{m-i} + h\lambda_4^{m-i}\beta_2}{1 + h\beta_2 + h\mu} \\[4mm]
\lambda_4^{m-i-1} = \dfrac{h\gamma_2\lambda_3^{m-i-1} + \lambda_4^{m-i}}{1 + h\gamma_2 + h\delta + h\mu}
\end{array}\right.$$

$$M^{i+1} = \frac{1}{2A} \left[I_1 P_{i+1} (\lambda_2^{m-i-1} - \lambda_1^{m-i-1}) \right]$$

$$u_{i+1} = \min(1, \max(0, M^{i+1}))$$

5   Results

A simulation is carried out on an interval time of T = 10 years, subdivised into *m* subintervals of step *h* so that:

$$m = 1000; \quad h = 0.01; \quad T = mh = 10 \text{ years};$$

The following parameters values are based on [3, 9–11]:

$$n = 6{,}000{,}000; \quad I_1 = 0.2; \quad I_2 = 0.3; \quad I_3 = 0.1; \quad \beta_1 = 0.5; \quad \beta_2 = 0.5; \quad \beta_3 = 0.5;$$
$$\mu = 0.02; \quad \gamma = 0.08; \quad \delta = 0.05; \quad A = 1{,}000{,}000;$$

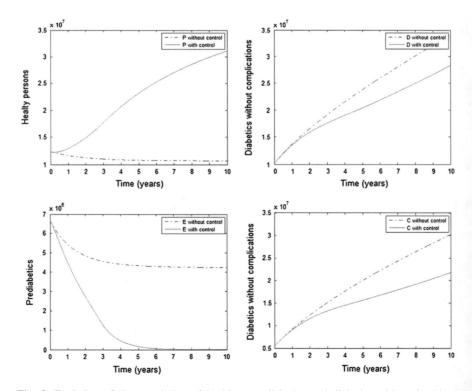

Fig. 2 Evolution of the population of healthy, pre-diabetics and diabetics with and without complications in 10 years

Our model assumes that the optimal control strategy will lead to an increase in healthy people (from 1.07×10^7 to 3.10×10^7), and a decrease of pre-diabetics (from 4.25×10^6 to 1.1×10^3) and diabetics with (from 3×10^7 to 2.14×10^7) and without complications (from 3.4×10^7 to 2.8×10^7) (see Fig. 2).

6   Conclusion

In a period of 10 years our optimal control strategy ensures an efficient reduction of the people evolving from healthy state to pre-diabetes state. Consequently, the number of pre-diabetics and diabetics with and without complications will decrease by 99 %, 28.7 % and 17.7 % respectively, while the healthy population will increase by 280 %. In real life our approach can be accomplished by making people aware of healthy diet, physical activity, smoking reduction and metabolic risks like overweight/obesity and hypertension.

References

1. Derouich, M., Boutayeb, A., Boutayeb, W., Lamlili, M.: Optimal control approach to the dynamics of a population of diabetics. Appl. Math. Sci. 56, 2773–2782 (2014)
2. International Diabetes Federation.: IDF Diabetes Atlas, 6th edn. DF, Bressels, Belgium (2013)
3. Boutayeb, A., Chetouani, A.: A critical review of mathematical models and data used in diabetology. BioMed. Eng. Online 5, 43 (2006)
4. Makroglou, A.: Mathematical models and software tools for the glucose-insulin regulatory system and diabetes: an overview. Appl. Numer. Math. 56, 559–573 (2006)
5. Li, J., Johnson, J.D.: Mathematical models of subcutaneous injection of insulin analogues: a mini-review. Discrete Contin. Dyn. Syst. Ser. B 12, 401414 (2009)
6. Bergman, R.N., Finegood, D.T., Ader, M.: Assessment of insulin sensitivity in vivo. Endocr. Rev. 6(1), 45–86 (1985)
7. Ajmera, I., Swat, S., Laibe, C., Le Novére, N., Chelliah, V.: The impact of mathematical modeling on the understanding of diabetes and related complications: a review. Phamacometrics Syst. Pharmacol. 2(e54), 1–14 (2013)
8. Boutayeb, A., Twizell, E.H., Achouyab, K., Chetouani, A.: A mathematical model for the burden of diabetes and its complications. BioMed. Eng. Online 3, 20 (2004)
9. Fleming, W.H., Rishel, R.W.: Deterministic and Stochastic Optimal Control. Springer Verlag, New York (1975)
10. Pontryagin, L.S., Boltyanskii, V.G., Gamkrelidze, RV. and Mishchenko, EF.: The Mathematical Theory of Optimal Process, vol. 4. Gordon and Breach Science Publishers (1986)
11. Gumel, A.B., Shivakumar, P.N., Sahai, B.M.: A mathematical model for the dynamics of HIV-1 during the typical course of infection. In: Proceeding of the 3rd World Congress of Nonlinear Analysts, vol. 47, pp. 2073–2083 (2011)

Part VIII
Photovoltaic Cell & Systems

Tb^{3+}/Yb^{3+} Activated Silica-Hafnia Glass and Glass Ceramics to Improve the Efficiency of Photovoltaic Solar Cells

S. Belmokhtar, A. Bouajaj, M. Britel, S. Normani, C. Armellini, M. Ferrari, B. Boulard, F. Enrichi, F. Belluomo and A. Di Stefano

Abstract A down-conversion layer placed on the front side of silicon solar cells waveguides has the potential to cute one high-energy photon into two low energy photons. This paper examines the Tb^{3+}/Yb^{3+} energy transfer efficiency in a 70SiO$_2$–30HfO$_2$ glass and glass-ceramics waveguide in order to convert absorbed photons at 488 nm in photons at 980 nm. The evaluation of the transfer efficiency between Tb^{3+} and Yb^{3+} is obtained by comparing the luminescence decay of Tb with and without Yb co-doping ions. A transfer efficiency of 25 % obtained with glass-ceramic sample and 6 % with glass sample proving that glass-ceramic can be a viable system to fulfil our requirements.

Keywords Down-conversion · Rare earths · Glass-ceramic · Energy transfer · Photovoltaic solar cells

S. Belmokhtar (✉) · A. Bouajaj · M. Britel
Laboratory of Innovative Technologies, LTI, ENSA–Tangier,
University Abdelmalek Essaâdi, Tangier, Morocco
e-mail: belmoukhtar@ensat.ac.ma

S. Normani · C. Armellini · M. Ferrari
CNR-IFN, Istituto di Fotonica e Nanotecnologie, CSMFO Lab,
Via allaCascata 56/C, Povo 38123, Trento, Italy

B. Boulard
Institut des Molécules et Matériaux du Mans, UMR 6283, Equipe Fluorures,
Université du Maine, Av. Olivier Messiaen, 72085 Le Mans cedex 09, France

F. Enrichi
Veneto Nanotech, Laboratorio LANN, C.so Stati Uniti 4, 35127 Padua, Italy

F. Belluomo · A. Di Stefano
Meridionale Impianti SpA, Via Senatore Simonetta 26/D, 20867 Caponago (MB), Italy

© Springer International Publishing Switzerland 2016
A. El Oualkadi et al. (eds.), *Proceedings of the Mediterranean Conference on Information & Communication Technologies 2015*, Lecture Notes in Electrical Engineering 380, DOI 10.1007/978-3-319-30301-7_50

1 Introduction

Recent years have seen the development of large research on renewable energy and in particular photovoltaic. Indeed, improving the efficiency of solar cells is now an ecological challenge; this efficiency is theoretically about 30 % for cells based silicon and is strongly limited by several factors [1, 2]. On one hand, a fraction of solar spectrum (about 18 %) is not exploitable by the photovoltaic cells because the corresponding photons have low energy. On the other hand, 30 % of the solar spectrum is made up of high energy ultraviolet photons where part of the energy is converted into heat and wasted [3].

The sector of multi junctions devices, called "third-generation" allows to access to the best result in the field of photovoltaic energy. This technology relies on the use of several cells of different bandgaps, each cell is optimized for a different portion of the solar spectrum [4, 5]. However, this technique is very complicated and expensive to be economically competitive. In order to improve the efficiency of photovoltaic cells, a research track is to limit the thermalization developing layers of photonic conversion that aim to convert a UV photon in several visible or near IR photons whose energy is fully exploited by the solar cells [6–8]. A more simple and interesting approach is the modification of the solar spectrum by the application of luminescent down-converting and up-converting coatings [9].

In down-conversion, multiple low-energy photons are generated to exploit the energy of one incident high-energy photon. In up-conversion, two or more incoming photons generate at least one photon with a higher energy than the incoming photons. We focus on a down-conversion process using cooperative energy transfer between a Tb^{3+} ion and two Yb^{3+} ions permits to cut high energy photon at wavelength shorter than 488 nm into two low energy photons around 980 nm [10].

In this work we investigated the $70SiO_2$–$30HfO_2$ glass ceramic planar waveguides co-activated by Tb^{3+}/Yb^{3+} ions fabricated by sol gel route. Sol gel-derived silica-hafnia is a reliable and flexible system that has proved to be suitable for rare earth doping and fabrication of glass ceramic planar waveguides with excellent optical and spectroscopic properties for photonic applications [11]. Our system is tested to achieve the target of high-performance and low-cost devices.

The Tb^{3+}: $^5D_4$ energy level corresponds at about twice the energy of the Yb^{3+}: $^2F_{5/2}$ energy level. The Yb^{3+} ions don't present an energy level above the $^2F_{5/2}$ level up to the UV region. The cooperative energy transfer between a Tb^{3+} ion and two Yb^{3+} ions can be the main relaxation route to achieve the NIR luminescence of the Yb^{3+}. Therefore two NIR photons are emitted by Yb^{3+} ions after the absorption of a single photon by a Tb^{3+} ion. To evaluate the transfer efficiency, we used decay curve analysis.

2 Experimental

A series of glass ceramic 70SiO$_2$–30HfO$_2$ planar waveguides samples co-doped by fixed 0.5 mol% Tb^{3+} and different mol% Yb^{3+} were prepared by sol–gel route using the dip-coating technique.

The starting solution, obtained by mixing tetraethylorthosilicate (TEOS), ethanol, deionized water and hydrochloric acid as a catalyst, was pre-hydrolyzed for 1 h at 65 °C. The molar ratio of TEOS:HCl:H$_2$O was 1:0.01:2. An ethanolic colloidal suspension was prepared using as a precursor HfOCl$_2$ and then added to the TEOS solutions, with a Si/Hf molar ratio of 70/30. The quantity of ethanol was adjusted for each solution in order to obtain a final total [Si+Hf] concentration of 0.448 mol/l. Terbium and ytterbium were added as Tb(NO$_3$)$_3$;5H$_2$O and Yb(NO$_3$)$_3$;5H$_2$O. The final mixture was left at room temperature under stirring for 16 h. The obtained sol was filtered with a 0.2 µm Millipore filter. Silica–hafnia films were deposited on cleaned pure SiO$_2$ substrates by dip-coating, with a dipping rate of 40 mm/min. Before further coating, each layer was annealed in air for 50 s at 900 °C. After a 10 dipping cycle, the film was heated for 2 min at 900 °C. Final films, obtained after 30 dips, were stabilized by a treatment for 5 min in air at 900 °C. As a result of the procedure, transparent and crack-free films were obtained. An additional heat treatment was performed in air at a temperature of 1000 °C for 30 min in order to nucleate nanocrystals inside the film 70SiO$_2$–30HfO$_2$ glass ceramic planar waveguides doped with rare earth ions were thus produced. Table 1 gives the compositional and optical parameters of the obtained silica-hafnia planar waveguides.

The thickness of the waveguides and the refractive index at 632.8 and 543.5 nm were obtained by a m-lines apparatus (Metricon, mod2010) based on the prism coupling technique, using a Gadolinium Gallium Garnet (GGG) prism, with the setup reported in [12].

XRD measurements were carried out at room temperature an X'Pert PRO diffractometer (Panalytical). A Cu anode (with Kα1, 2 lines) was used as a radiation source. Owing to the small thickness of the investigated waveguides, the grazing incidence x-ray diffraction (GIXRD) geometry was employed. XRD spectra were collected in continuous scan mode in the 2θ range 10°–100°, with a scanning step of 0.1° and counting time of 60 s. The transmittance spectra were recorded in an UV near-infrared spectrophotometer. Photoluminescence spectroscopy was performed by far-field excitation using the 476 nm line of an Ar+ion laser as excitation source.

Table 1 Rare earth concentration, relative index and layer thickness of the sample

Sample label	Terbium concentration in mol%	Ytterbium concentration in mol%	n@543.5 nm TE[±0.001]	n@632.8 nmTE [±0.001]	Layer thickness [±0.2 µm]
AR1	0.5	0	1.621	1.616	1.0
A1	0.5	1	1.626	1.621	1.0
A2	0.5	2	1.631	1.626	1.0
A3	0.5	3	1.633	1.628	1.1

The luminescence spectrum in the region of the transition $^2F_{5/2} \rightarrow {}^{11}F_{7/2}$ of Yb^{3+} ion was analyzed by a single grating monochromator with a resolution of 2 nm and detected using a Si/InGaAs two-color photodiode and standard lock-in technique. Luminescence decay measurements of the $^5D_4$ state of Tb^{3+} ion were performed after excitation with the third harmonic of a pulsed Nd-YAG laser. The visible emission was collected by a double monochromator with a resolution of $5\ cm^{-1}$ and the signal was analyzed by a photon-counting system. Decay curves were obtained recording the signal by a multichannel analyzer Stanford SR430. More information about the experimental setups can be found in [13].

3 Results and Discussion

Figure 1 presents the XRD peaks of silica-hafnia waveguides after heat treatment at 900 °C (glass, Fig. 1a) and 1000 °C (glass ceramic, Fig. 1b) for the waveguide doped 0.5 mol% Tb^{3+} and 0 mol% Yb^{3+} and silica-hafnia waveguides doped 0.5 mol% Tb^{3+} and 3 mol% Yb^{3+} treated at 1000 °C (glass ceramic, Fig. 1c).

All the XRD spectra show that there is an amorphous phase within the waveguides, which is clearly presented by the hump centred at $2\theta \approx 21°$. This non-crystalline phase is contributed to the amorphous structure of silica.

Figure 3a shows that the thin film treated at 900 °C are completely amorphous by against in Fig. 3b we see clearly the bragg diffraction peaks and we note that the crystallization occurs after heat treatment at 1000 °C for 30 min.

Diffraction peaks become more intensive by increasing the concentration of rare earth, showing that in silica-hafnia glass ceramic, the rare earth ions are embedded in hafnia nanocrystals. Similar result have founded by Rocca et al. in Er^{3+} doped $70SiO_2$–$30HfO_2$ planar waveguide [14].

Fig. 1 Observed XRD patterns of $70SiO_2$–$30HfO_2$ waveguides: **a** 0.5 mol% Tb^{3+}: 0 mol% Yb^{3+} treated at 900 °C. **b** AR1. **c** A3

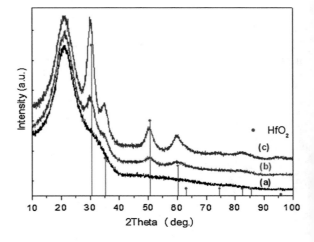

From the comparison between XRD data and the ICSD database, we attribute the crystalline phase in waveguides to the metastable tetragonal hafnium oxide (t-HfO$_2$), The diffraction peaks calculated from the matched ICSD card (No 85322) are shown (vertical lines). The ICSD card belongs to the isostructural metastable t-HfO$_2$ phase [14].

Figure 2 shows the optical transmission spectra of the investigation 70SiO$_2$–30HfO$_2$ waveguides doped 0.5 % Tb^{3+} treated at 900 °C (glass) and that treated at 1000 °C (glass ceramic). The spectra reveals that, despite the difference in refractive index between the v-SiO slabs (n = 1.46 at 632.8 nm) as substrates and the layer glass ceramic (∼1.612 at 632.8 nm) the transmittance of the glass ceramic reaches as high as 90 % in the visible infrared range, which is due to the much smaller size of the precipitated crystals than the wavelength of the visible light [15].

Figure 3 shows the photoluminescence spectra of the Tb^{3+} → Yb^{3+} obtained under the 476 nm excitation for different samples. The intense emission band

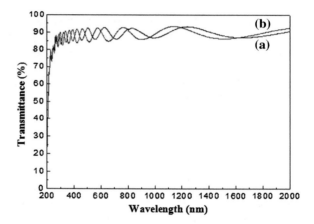

Fig. 2 Optical transmission spectra for 70SiO$_2$–30HfO$_2$ waveguides with 0.5 mol % Tb^{3+}: 0 mol% Yb^{3+} **a** treated at 900 °C. **b** treated at 1000 °C

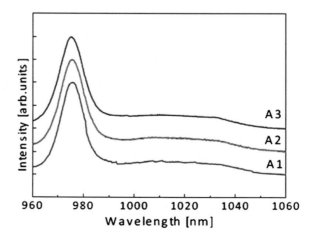

Fig. 3 Room temperature photoluminescence spectra of the $^2F_{5/2}$ → $^2F_{7/2}$ transition of Yb^{3+} ions after excitation at 476 nm for the three samples: **a** A1; **b** A2; **c** A3. Each spectrum was normalized to the maximum of the luminescence intensity

centered at 977 nm, with a shoulder at 1027 nm, is attributed to the $^2F_{5/2} \rightarrow {}^2F_{7/2}$ transition of Yb^{3+} ions. The emission of the Yb^{3+} ions after excitation in the blue region is an indication of the presence of an efficient energy transfer from Tb^{3+} to Yb^{3+} and so of an effective down-conversion. For application of Tb^{3+} and Yb^{3+} co-doped $70SiO_2-30HfO_2$ glass ceramic layer as down converter, high external quantum efficiency is required. However, it is not possible to evaluate the conversion efficiency only on the base of the photoluminescence spectra. Assessment of the conversion efficiency is obtained from the estimation of the energy transfer efficiency between terbium and ytterbium. The evaluation of the energy transfer efficiency between Tb^{3+} and Yb^{3+} can be obtained by comparing the luminescence decay of terbium with and without ytterbium co-doping ions.

In Fig. 4 the decay curves of the $Tb^{3+}{}^5D_4 \rightarrow {}^7F_5$ emission at 543.5 nm are plotted for the different samples. Nearly single exponential luminescence decay is observed for the sample without Yb^{3+}. The fast luminescence decay observed for the co-doped samples is attributed to the energy transfer from the Tb^{3+}: $^5D_4$ to the Yb^{3+}: $^2F_{5/2}$ [16].

The not exponential behavior of the decay can be explained by different distributions of Yb^{3+} ions around the Tb^{3+} ions, which lead to different energy transfer rates for the different Tb^{3+} ions. The energy transfer efficiency η_{Tb-Yb} can be obtained experimentally by dividing the integrated intensity of the decay curves of the $Tb^{3+} \rightarrow Yb^{3+}$ co-doped glass ceramics by the integrated intensity of the Tb^{3+} single doped curve [16]:

$$\eta_{Tb-Yb} = 1 - \frac{\int I_{Tb-Yb} dt}{\int I_{Tb} dt} \tag{1}$$

The effective quantum efficiency can be defined by the ratio between the number of emitted photons and the number of photons absorbed by the material. In our case, a perfect down-conversion system would have an effective quantum efficiency value of 200 %, corresponding to the emission of two photons for one absorbed.

Fig. 4 Decay curves of the luminescence from the $^5D_4$ metastable state of Tb^{3+} ions for the second samples series under excitation at 355 nm

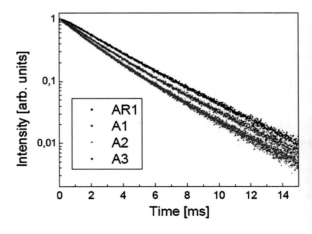

Table 2 Transfer efficiency and effective quantum efficiency as function of Yb^{3+} molar concentration 70SiO$_2$-30HfO$_2$ *glass waveguides* where Tb^{3+} content fixed at 0.5 mol%

Composition (Yb concentration in mol%) (%)	1	2	3
Transfer efficiency (glass) (%)	2	4	6
Effective quantum efficiency (glass) (%)	102	104	106

Table 3 Transfer efficiency and effective quantum efficiency as function of Yb^{3+} molar concentration 70SiO$_2$–30HfO$_2$ *glass ceramic waveguides* where Tb^{3+} content fixed at 0.5 mol%

Composition (Yb concentration in mol%) (%)	1	2	3
Transfer efficiency (glass ceramic) (%)	14	24	25
Effective quantum efficiency (glass ceramic) (%)	114	124	125

The relation between the transfer efficiency and the effective quantum efficiency is linear [16] and is defined as:

$$\eta_{EQE} = \eta_{Tb-r}(1 - \eta_{Tb-Yb}) + 2\eta_{Tb-Yb} \qquad (2)$$

where, the quantum efficiency for Tb^{3+} ions η_{Tb-r} is set equal to 1. The evaluated values of energy transfer efficiency and effective quantum efficiency for the different samples treated at 900 °C (glass) are reported in Table 2.

The evaluated values of energy transfer efficiency and effective quantum efficiency for the different samples treated at 1000 °C (glass ceramics) are reported in Table 3.

By comparing the values of the energy transfer efficiency obtained in glass ceramic and in glass samples, we find that this values is about 24–25 % in the glass ceramic samples, but does not exceed 4–6 % in the glass samples.

These results are attributed to the fact that hafnia nanocrystals are characterized by a cut off frequency of about 700 cm^{-1}, so that non-radiative transition rates are strongly reduced, thus increasing the luminescent quantum yield of the rare-earth ions.

4 Conclusion

70SiO$_2$–30HfO$_2$ glass ceramics co-doped Tb^{3+}/Yb^{3+} with Tb^{3+} content kept constant at 0.5 mol% and increasing Yb^{3+} molar concentration were prepared by sol-gel method and dip coating processing. XRD spectra show that the crystallization occurs after heat treatment at 1000 °C for 30 min and the rare earth ions are embedded in hafnia nanocrystals. Near-infrared emission at 980 nm assigned to the $^2F_{5/2} \rightarrow \,^2F_{7/2}$ transition of the Yb^{3+} ions was observed upon excitation at 476 nm. The energy transfer efficiencies were estimated from the decay curves of the $^5D_4$ metastable state of the Tb^{3+} ion. The Tb–Yb energy transfer efficiency increases with the increase of the molar ratio Yb/Tb and the value found in glass ceramic samples is more significant (25 %) compared to that found in the glass samples (6 %). Among the two systems investigated in this work the glass ceramics is best candidates for down-conversion process, they combine the optical properties of glasses with the spectroscopic properties of the crystals activated by luminescent species.

Acknowledgments The research activity was performed in the framework of the CNR-CNRST joint project (2014–2015).

References

1. Kamat, P.: J. Phys. Chem. C **111**, 2834 (2007)
2. Miles, R.W., Zoppi, G., Forbes, I.: Mater. Today **11**(10), 20 (2007)
3. Shockley, W., Queisser, H.J.: Detailed balance limit of efficiency of pen junction solar cells. J. Appl. Phys. **32**(3), 510e9 (1961)
4. Green, M.A.: Photovoltaic principles. Physica E: Low-Dimensional Syst Nanostruct. **14**(1e2), 11e7 (2002)
5. Dimroth, F.: High-efficiency solar cells from IIIeV compound semiconductors. Phys. Status Solidi (c) **3**(3), 373e9 (2006)
6. Lim, M.J., Ko, Y.N., Kang, Y.C., Junq, K.Y.: RSC Adv. **4**, 10039 (2014)
7. Liu, M., Lu, Y., Xie, Z.B., Chow, G.M.: Sol. Energy Mater. Sol. Cells **95**, 800 (2011)
8. He, W., Atabaev, T.S., Kim, H.K., Hwang, Y.H.: J. Phys. Chem. C **117**, 17894 (2013)
9. Lian, H., Hou, Z., Shang, M., Geng, D., Zhang, Y., Lin, J.: Energy **57**, 270–283 (2013)
10. Bouajaj, A., Belmokhtar, S., Britel, M., Normani, S., Armellini, C., Enrichi, F., Ferrari, M., Boulard, B., Belluomo, F., Di Stefano, A.: Rare earths and metal nanoparticles in silicate glass-ceramics to improve the efficiency of photovoltaic solar cells. In: Proceeding of IEEE, pp. 457–462 (2014)
11. Alombert-Goget, G., Armellini, C., Berneschi, S., Chiappini, A., Chiasera, A., Ferrari, M., Guddala, S., Moser, E., Pelli, S., Rao, D.N., Righini, G.C.: Tb^{3+}/Yb^{3+} co-activated Silica-Hafnia glass ceramic waveguides. Opt. Mater. **33**, 227–230 (2010)
12. Ronchin, S., Chiasera, A., Montagna, M., Rolli, R., Tosello, C., Pelli, S., Righini, G.C., Gonçalves, R.R., Ribeiro, S.J.L., De Bernardi, C., Pozzi, F., Duverger, C., Belli, R., Ferrari, M.: Erbium-activated silica-titania planar waveguides prepared by rf-sputtering. SPIE 4282, 31 (2001)
13. Alombert-Goget, G., et al.: Er^{3+}—activated photonic structures fabricated by sol-gel and rf-sputtering techniques. In: Proceedings of SPIE, vol. 7366, 73660E-1–73660E-15 (2009)
14. Afify, N.D., Dalba1, G., Rocca, F.: XRD and EXAFS studies on the structure of Er^{3+}-doped $SiO_2–HfO_2$ glass-ceramicwaveguides: Er^{3+}-activated HfO_2 nanocrystals. J. Phys. D: Appl. Phys. **42** 115416 (11 pp) (2009)
15. Hendy, S.: Light scattering in transparent glass ceramics. Appl. Phys. Lett. **81**(7), 1171–1173 (2002)
16. Vergeer, P., Vlugt, T.J.H., Kox, M.H.F., Den Hertog, M.I., van der Eerden, J.P.J.M., Meijerink, A.: Quantum cutting by cooperative energy transfer in YbxY1 − xPO4:Tb^{3+}. Phys. Rev. B 71, 014119-1 – 014119-11 (2005)

Stability Analysis of a Boost Converter with an MPPT Controller for Photovoltaic Applications

A. El Aroudi, M. Zhioua, M. Al-Numay, R. Garraoui and K. Al-Hosani

Abstract This paper presents an analysis of the nonlinear dynamics of a current-fed boost converter for PV applications with an MPPT algorithm. First, the model of the converter is derived. Simulation results using this model show that the system can exhibit period doubling bifurcation and subharmonic oscillation when some design parameters are varied. Then, the discrete-time modeling approach and Floquet theory is used to perform a stability analysis of the system periodic orbits, to explain the mechanism of losing their stability and to locate the stability limits in the parameter space. Based on the study, a design of the system free from the undesired subharmonic oscillation can be done. The theoretical results are validated by means of numerical simulations from the circuit-level switched model.

Keywords DC-DC converters · PV applications · MPPT · Instability

1 Introduction

Due to oil product price increasing and the exhaustion of conventional energy resources due to increasingly growing demand and the associated environmental pollution problems, there has been a progressive interest in renewable energy

A. El Aroudi (✉)
Universitat Rovira I Virgili, Tarragona, Spain
e-mail: abdelali.elaroudi@urv.cat

M. Zhioua
University of Tunis El Manar, Tunis, Tunisia

M. Al-Numay
King Saud University, Riyadh, Kingdom of Saudi Arabia

R. Garraoui
University of Gabes, Gabes, Tunisia

K. Al-Hosani
Petroleum Institute (PI), Abu Dhabi, United Arab Emirates

© Springer International Publishing Switzerland 2016
A. El Oualkadi et al. (eds.), *Proceedings of the Mediterranean Conference on Information & Communication Technologies 2015*, Lecture Notes in Electrical Engineering 380, DOI 10.1007/978-3-319-30301-7_51

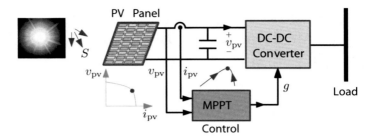

Fig. 1 Schematic diagram of a PV system

sources and this issue is becoming a hot topic of research studies. A promising candidate which is currently attracting much attention is solar energy, as it is both green and inexhaustible. A Maximum Power Point Tracking (MPPT) controller is employed in such a way that the photovoltaic (PV) system is forced to deliver the maximum power. In fact, PV generators have a nonlinear characteristic with a maximum power point (MPP) depending on the temperature and irradiation. The goal of the MPPT system is to ensure that the PV module always operates at its MPP regardless of the temperature, insolation and load variations. The output of this controller is fed-back to a switching converter so that the panel delivers a maximum of energy. Generally speaking, a PV system is composed of three fundamental elements as shown in Fig. 1. The first element consists of the source of energy (PV panels), the second one is the DC-DC converter. The third element is the control system carrying out the regulation of some variables of interest with the aim to extract the MPP from the PV panels. This task is possible thanks to the use of DC-DC converters as an interface which allows the adaptation of the energy flow between the PV module and the load. There are many possibilities to perform this interfacing between the PV source and the load. Impedance matching is one attractive solution to impose an optimum conductance in the input port of the system [1, 2].

The operation of DC-DC converters is mainly based on the switching between different configurations [3]. The switching aspect of such systems together with the feedback action makes them complex and may exhibit a variety of nonlinear behaviors such as bifurcations, quasi-periodicity and chaotic regimes [3, 4]. A literature review reveals that most of the previous works on nonlinear phenomena in switching converters consider voltage-fed converters with resistive loads. An exception is in [5] where the authors analyzed the dynamics of boost converter supplied with a PV source. However, in that paper the MPPT control was not addressed and the input capacitance was not considered. This paper aims to analyze the bifurcation behavior of a current-fed boost converter for photovoltaic applications with a constant voltage as a load. To improve the stability of the system at the slow-time scale a Loss Free Resistor (LFR) behavior is imposed by controlling the input current to be proportional to the input voltage [1]. The proportionality factor is a conductance g which is supposed to be provided by an MPPT controller.

For simplicity, the dynamics of the MPPT algorithm is neglected because it is usually much slower than the converter dynamics. However its output is not considered constant but time periodic. Because the period of the MPPT controller is usually much larger than the switching period a quasi-static approach can be used. The study is done through a discrete-time model which is obtained by using a systematic approach [3] consisting of integrating exactly the system linear configurations, obtaining the local maps during each phase and composing them to obtain the complete map. The stability of the system is then addressed by using Floquet theory and the monodromy matrix [6]. The rest of the paper is organized as follows. Section 2 provides a description of the current-fed boost converter. Simulation results from this model are performed in Sect. 3 showing an unstable behavior in the form of subharmonic oscillation. Continuous-time modeling of the system is performed in Sect. 4. The monodromy matrix is obtained using discrete-time model in Sect. 5 from Floquet theory and Filippov method to perform a stability analysis of the system periodic orbits and to explain the mechanism of losing their stability. Finally, conclusions of this work are presented in the last section.

2 The Boost Converter with a PV Source and Loaded by a Constant Voltage

Figure 2 shows a simplified schematic diagram of the system considered in this study. It consists of current-fed switching DC-DC boost converter structure used to interface the PV array with a storage battery. The first stage is the PV panel which presents a nonlinear current-voltage $i(v)$ characteristic. The second stage is a boost converter, which will receive the relatively low PV panel voltage and step-up it to an appropriate higher level for the storage battery. In practice, an output LC filter is employed to reduce the switching ripple to provide a smooth current i_{bat} to the output. However, it can be demonstrated that this filter apart from smoothing the current injected to the battery, its effect on the fast time-scale behavior of the system is negligible.

The boost converter is operated as an LFR imposed by a suitable control strategy making the input current and the input voltage proportional with the aim to improve the stability at the slow scale [1]. Namely, the control of the current-fed boost converter is as follows. The input inductor current is controlled by a Proportional-Integral (PI) controller to set its average value to a desired reference i_{ref} which is made proportional to the input voltage $i_{ref} = g v_{pv}$. The comparator compares the output of the PI controller, the control voltage v_{con}, with a sawtooth ramp T—periodic signal v_{ramp} and the result of the comparison generates the ON/OFF driving signal for the switch by a set-reset flip-flop in such a way that the switch is ON at the start of each ramp period and it is OFF whenever the control voltage v_{con} crosses the ramp signal v_{ramp}.

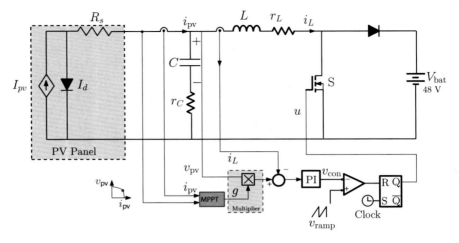

Fig. 2 A boost converter supplied from a PV panel feeding a constant voltage load

3 Behavior of the PV System from Numerical Simulations

Time-domain responses from a circuit-level model based on PSIM have been obtained. The conductance g is considered as a periodic time varying signal whose frequency is much lower than the switching frequency in such a way that quasi-static approach can be used. In most of MPPT controllers such as Perturb and Observe (P&O or ES), g has a triangular shape. In this paper we consider that g varies according to the following triangular signal to emulate the output from the MPPT controller

$$g(t) = \begin{cases} G(1 + \frac{t}{2\Delta t}) & \text{if} \quad 0 \leq t \leq \frac{\Delta t}{2} \\ G(1 + \frac{\Delta t - t}{2\Delta t}) & \text{if} \quad \frac{\Delta t}{2} \leq t \leq \Delta t \end{cases} \qquad (1)$$

where Δt is the MPPT controller period or step-size. The fixed circuit parameters are shown in Table 1. The Equivalent Series Resistance (ESR) r_C of the capacitor has been neglected and that of the inductor was selected as $r_L = 0.1\,\Omega$. The PSIM solar module has an open circuit voltage around 21 V based on the data of BP585 [7]. Note that the input voltage depends on the weather conditions and can be regulated to a value between 0 and the open circuit voltage with a nominal MPP value of 17 V approximatively which changes very little with irradiance.

Table 1 The used parameter values for this study

L	C	τ	κ_p	f_s	Δt	G
200 μH	10 μF	0.001 s	0.9	50 kHz	0.01 s	0.19 S

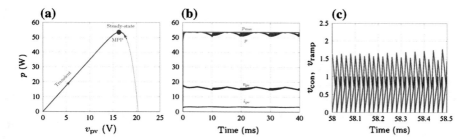

Fig. 3 Waveforms of the system power supplied by a PV panel with maximum power of 54 W and detailed view of the control signal

Figure 3 shows the start-up response and the steady-state behavior of the system obtained from PSIM. It can be observed that the MPP is reached in steady-state as it is clearly observed in Fig. 3a, b. However, the detailed view in Fig. 3c shows that the system behavior in steady-state is not periodic and that subharmonic oscillations are taking place at the fast scale within some switching cycles because the system T —periodic orbit losses its stability by a period doubling bifurcation during these cycles. In particular, the conductance g is a slowly time-varying parameter for performing the MPPT control. Therefore in a real implementation, strict limits on its variation must be imposed. However, several actions that may be taken for the system to be in periodic regime may be in conflict with the operation of the system at the MPP. For instance limiting the dynamic range of the conductance g will limit the power delivered by the PV panel. While this can be desirable when the production of energy is larger than its consumption under light load conditions, it may not be the case under full load operation. Other choice for improving the stability while maintaining the system at the MPP operation is selecting a slope m_a of the ramp modulator which avoid instability under all operating conditions. Before embarking on the appropriate design of the system to avoid undesired behaviors, a careful modeling approach will be used of the complete system.

4 Modeling of the PV System

Consider that the converter is operating in continuous conduction mode (CCM). Therefore, it switches between two different configurations corresponding to the ON and the OFF switch states respectively. The ON phase takes place when the switch S is closed and the diode D is blocked; the OFF phase occurs when the switch S is open and the diode D is conducting.

In accordance with Fig. 2, a continuous-time mathematical model describing the boost converter connected to the PV panel may be written as follows

$$\frac{dv_C}{dt} = \frac{1}{C}(i_{pv} - i_L) \tag{2}$$

$$\frac{di_L}{dt} = \frac{1}{L}v_C + \frac{-r_L - r_C}{L}i_L + \frac{r_C}{L}i_{pv} - \frac{1}{L}V_{bat}(1 - u) \tag{3}$$

where C is the capacitance, r_C is its Equivalent Series Resistance (ESR), L is the inductance, r_L being it DC resistance, v_C is the capacitor voltage, V_{bat} is the voltage of the output battery, i_L is the inductor current and i_{pv} is the PV panel current and finally u is the switched command signal of the active switch S.

The aim of the controller is to force the input inductor current to follow a reference signal derived from the input voltage. A simple PI controller is used for this purpose whose description in the time domain is as follows

$$v_{con} = \kappa_p(i_{ref} - i_L) + \frac{\kappa_p}{\tau}\int(i_{ref} - i_L)dt \tag{4}$$

where $i_{ref} = gv_{pv}$ is the current reference as described previously, κ_p is the proportional gain of the PI controller and τ is its constant time and v_{con} is the control signal which is compared with the ramp to decide the state of the switch.

During each phase, the state space evolution is given by the following equation

$$\dot{x} = A_i x + B_i \tag{5}$$

where $i = 1$ for the on phase ($u = 1$) and $i = 2$ for the off phase ($u = 0$) and x is the state vector composed by the capacitor voltage, the inductor current and the integral of the current error, $x = (v_C, i_L, \int i_{ref} - i_L)^T$. The matrices A_i and B_i are as straightforward from (2)-(3). The converter switches from the ON configuration to the OFF one when the following switching condition is satisfied

$$v_{ramp}(t) - v_{con}(t) = v_{ramp}(t) - Kx(t) - g\kappa_p r_C i_{pv} = 0 \tag{6}$$

where $K = (g\kappa_p, -\kappa_p(1 + gr_C), \kappa_p/\tau)$ is the vector of feedback coefficients.

5 Stability Analysis

Let $t_n = DT$ in steady-state. Let $x(0)$ be the steady-state value of the periodic orbit of the system at the beginning of the period and $x(DT)$ be the steady-state value of this orbit at time instant DT. Let $\Phi_1 = e^{A_1 DT}$ and $\Phi_2 = e^{A_2(1-D)T}$, $\Psi_1 = \int_0^{DT} e^{A_1 t}dtB_1$ and $\Psi_2 = \int_0^{(1-D)T} e^{A_2 t}dtB_2$. In steady-state, the vector of state variables $x(DT)$ at the instant DT is given by

$$x(DT) = (\mathbf{I} - \mathbf{\Phi})^{-1}\mathbf{\Psi} \tag{7}$$

where $\mathbf{\Phi} = \mathbf{\Phi}_1\mathbf{\Phi}_2$ and $\mathbf{\Psi} = \mathbf{\Phi}_1\mathbf{\Psi}_2 + \mathbf{\Psi}_1$. The matrix $(\mathbf{I} - \mathbf{\Phi})$ is assumed to be nonsingular. Because an integrator exists in the system controller, this matrix is singular. To avoid this singularity, a small perturbation is introduced to the integrator dynamics. In the same way, one can obtain that $x(0)$, the value of the periodic orbit at $t = nT$, $n \in \mathbb{N}$ is given by

$$x(0) = (\mathbf{I} - \overline{\mathbf{\Phi}})^{-1}\overline{\mathbf{\Psi}} \tag{8}$$

where $\overline{\mathbf{\Phi}} = \mathbf{\Phi}_2\mathbf{\Phi}_1$ and $\overline{\mathbf{\Psi}} = \mathbf{\Phi}_2\mathbf{\Psi}_1 + \mathbf{\Psi}_2$. To study the stability of the system the monodromy matrix is used [6]. This matrix can be expressed as follows [6]

$$\mathbf{M} = \mathbf{\Phi}_2\mathbf{S}\mathbf{\Phi}_1 \tag{9}$$

where $\mathbf{S}$ is the saltation matrix defined as follows [6]

$$\mathbf{S} = \mathbf{I} + \frac{(\mathbf{f}_2(x(DT)) - \mathbf{f}_1(x(DT)))\mathbf{K}}{\mathbf{K}\mathbf{f}_1(x(DT)) - m_a} \tag{10}$$

where $\mathbf{f}_1(x) = \mathbf{A}_1x + \mathbf{B}_1$, $\mathbf{f}_2(x) = \mathbf{A}_2x + \mathbf{B}_2$ and $m_a = V_M f_s$ is the slope of the ramp modulator, $V_M = V_u - V_l$ being its amplitude. First, periodic orbits are calculated and their stability are checked by the eigenvalues of the monodromy matrix. The parameter g is varied within the range $(0.1, 0.24)$ with other parameters fixed as in Table 1. The variation of g simulated the behavior of an MPPT controller which can only provide a variable g within a certain range.

The PV panel is approximated by a current source $I_{mpp} \approx 3.1$ A which correspond to a temperature $\theta = 25$ °C and and irradiance $S = 700$ W/m^2. Under these conditions, the eigenvalues of the monodromy matrix are calculated by using the expression in (9) and (10). Figure 4 shows the loci of the eigenvalues as g is varied. It is observed that when $g \approx 0.22$, the system undergoes a period doubling since one of the eigenvalues of the monodromy matrix crosses the unite circle from -1. This is in perfect agreement with the numerical simulations obtained from the circuit-level based simulations results obtained from PSIM in Sect. 3.

According to Fig. 4, $g = 0.22$ guarantees the stability of the system. Let us limit the variation of g in such a way that this parameter cannot be larger than $g_{max} = 0.22$. Figure 5 shows the system start-up and steady state under this condition. It can be shown that the system is stable and the MPP is fully reached.

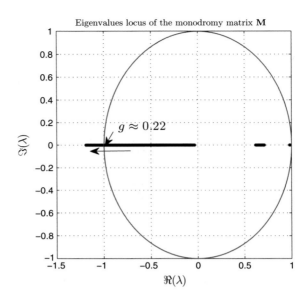

Fig. 4 Eigenvalues of the monodromy matrix when g is varied within $(0.1, 0.24)$

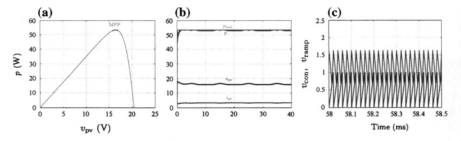

Fig. 5 Waveforms of the system power supplied by a PV panel with maximum power of 53 W and detailed view of the control signal with g limited to $g_{\max} = 0.22\ S$

6   Conclusions

The stability of a boost converter supplied by a PV panel and under an MPPT controller has been studied. Approximate analysis based on the discrete-time model has been used to uncover possible instability phenomena of the system. To make an analytical study possible the PV generator has been approximated by a constant current source and the MPPT controller has been emulated by a triangular signal generator. It has been shown that the system can exhibit period doubling phenomenon leading to intermittent subharmonic oscillation due to the variation of some parameters imposed by the MPPT algorithm. By using the monodromy martrix, it has been shown that the period doubling can be detected accurately and limits on the variation of the conductance to avoid undesired subharmonic oscillation can be obtained. Based on the approach used in this study, the critical value of the system parameters for losing stability can be located for design purpose. Therefore, the results here reported could help in selecting the parameter values of the system for avoiding this kind of behavior in a practical design. A simple way to avoid subharmonic oscillation that was used in this work was by limiting the variation imposed by the MPPT controller within a specific range. Future work will validate the results experimentally.

References

1. Haroun, R., Cid-Pastor, A., El Aroudi, A., Martinez-Salamero, L.: Synthesis of canonical elements for power processing in dc distribution systems using cascaded converters and sliding-mode control. IEEE Trans. Power Electr. **29**(3), 1366–1381 (2014)
2. Haroun, R., El Aroudi, A., Cid-Pastor, A., Garcia, G., Martinez-Salamero, L.: Large-signal modeling and stability analysis of two-cascaded boost converters connected to a PV panel under SMC with MPPT. In: IECON 2013—39th Annual Conference of the IEEE Industrial Electronics Society, pp. 949–954, 10–13 (2013)
3. El Aroudi, A., Debbat, M., Giral, R., Olivar, G., Benadero, L., Toribio, E.: Bifurcations in DC-DC switching converters: review of methods and applications. Int. J. Bif. Chaos **15**(05), 1549–1578 (2005)
4. Banerjee, S., Chakrabarty, K.: Nonlinear modeling and bifurcations in the boost converter. IEEE Trans. Power Electr. **13**(2), 252–260 (1998)
5. Al-Hindawi, M., Abusorrah, A., Al-Turki, Y., Giaouris, D., Mandal, K., Banerjee, S.: Nonlinear dynamics and bifurcation analysis of a boost converter for battery charging in photovoltaic applications. Int. J. Bif. Chaos **24**(11), 1450142, 12 p (2014)
6. Giaouris, D., Banerjee, S., Zahawi, B., Pickert, V.: Stability analysis of the continuous-conduction-mode buck converter via Filippov's method. IEEE Trans. CAS I: Reg. Pap. **55**(4), 1084–1096 (2008)
7. bp solar BP 585, available at: http://www.oksolar.com/pdfiles/Solar

A PV System Equipped with a Commercial and Designed MPPT Regulators

Kamal Hirech, Mustapha Melhaoui, Rachid Malek and Khalil Kassmi

Abstract In this work, we present the results concerning the design, implementation and test of a photovoltaic installation with two regulation systems, one is a commercial regulator and the other is designed in this work. This latter optimizes the installation operation by a new approach with improved control blocks: MPPT, charging/discharging process, estimates the state of charge and manages the energy between blocks. The regulation systems experiment during 4 days with weather disturbances shows the robustness of the designed one which improves the battery life: the accuracy and speed on the maximization of the power supplied by panels, precision of the state of charge estimation, improvement of the energy management at the charge/discharge.

Keywords Photovoltaic system · Batteries · Charge/discharge · Soc · Grid · MPPT

1 Introduction

The photovoltaic installations generally exist in two structures:

- The stand-alone installations whose the whole production consumed locally require the presence of the storage elements which are generally lead acid batteries. This latter are exposed to several types of degradation related to the manner of use, aging, temperature influence [1].
- Grid connected systems [2] that transfer the energy produced to the grid. Since the injection periods on the grid are not controlled, recent research [3] have been devoted to study the integration of the batteries to overcome the problems

K. Hirech · M. Melhaoui · R. Malek · K. Kassmi (✉)
Faculty of Science, Department of Physics, Laboratory LETAS,
Mohamed First University, 60000 Oujda, Morocco
e-mail: Khkassmi@yahoo.fr

© Springer International Publishing Switzerland 2016
A. El Oualkadi et al. (eds.), *Proceedings of the Mediterranean Conference on Information & Communication Technologies 2015*, Lecture Notes in Electrical Engineering 380, DOI 10.1007/978-3-319-30301-7_52

related to connections to the grid, including [2]: the availability of solar energy, the high losses and the perturbations introduced by the non-linear behavior [2].

The optimization of these two structures operation requires the optimization of energy flows between these elements and mainly that of the batteries, which makes the presence of solar regulator primordial to these installations [4].

Commerce regulators, though they are many, they have limitations in their operation, as control parameters (voltage, current) are fixed by the manufacturers that make it impossible to optimize the installation operation following the user needs or characteristics of elements [5], nor charge the batteries completely or optimizes their lifetimes [5]. In addition, they don't include possibility of using the surplus energy.

In this work, we propose to study the stand-alone PV system structure equipped with two solar regulators: the first is a commercial one [6] and the second is designed and realized in this work. The designed one objective is to: optimize the overall system operation, manage the energy supplied by the PV generator and the batteries charge/discharge and include the option of injecting the energy surplus into the grid.

2 System Topology

The battery charging is ensured by the two regulators connected in parallel with the battery, so they function independently and each one is connected to two PV panels [7] in series. These regulators are presented as blocks in Fig. 1. The commercial regulator transfers a P_{REG} power and that designed transfers $P_{CNV-BAT}$ power, then the charging power of the battery P_{BAT} is calculated by the Eq. 1:

$$\mathbf{P_{BAT}(t) = P_{CNV-BAT}(t) + P_{REG}(t)} \tag{1}$$

The commercial regulator, which is used for the battery charging, is a closed box that the documentation or the manual does not give details on its basic circuit [6].

The designed and realized regulator in this work consists of various blocks:

- A DC/DC convertor Buck type.
- A 3 power circuits to control the charge, discharge and injection of energy surplus into the grid by a new technique, these circuits are controlled by PWM signals
- A Management and Supervision System (MSS) to realize all tasks: the optimization of the system operation by the MPPT control, the regulation of the charge/discharge process, the estimation of the state of charge, the injection of energy into the grid and the control of the data transfer on a PC or/and LCD.

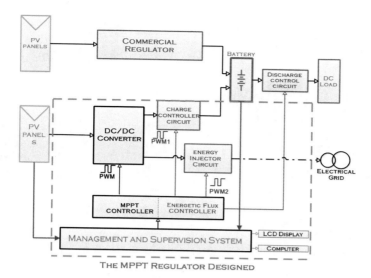

Fig. 1 Global diagram of the photovoltaic system

3 The System Operation

3.1 The Commercial Regulator Operation

The most of commercial regulators and specially the one cited in the reference [6] used in this work control the discharge by two voltage thresholds: the first when V_{BAT} = LVD = 10.7 V allows stopping the discharge and the second when V_{BAT} = LVR = 12.6 V to ensure the discharge again. The battery charging is ensured by an algorithm using two phases. In the first, the battery is charged to reach 13.7 V and in the second, the voltage V_{BAT} will be fixed around 13.7 V to continue charging.

3.2 The Designed Regulator Operation

The MPPT regulator designed in this work has as main purpose solar battery charging with the maximum of power supplied by panels. This regulator guarantees:

- **The MPPT control** is based on an improved algorithm of hill climbing shown in Fig. 2a [8]. The improvement provided is to search the PPM in an optimal voltage range (Vmin, Vmax) determined according the panels and the weather conditions.

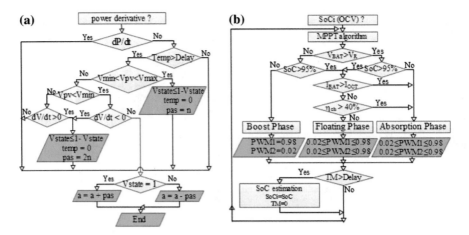

Fig. 2 Algorithm of: **a** the improved MPPT of Hill Climbing [8], **b** the battery charging

- **The control of charge process** according to the algorithm of Fig. 2b, the charging occurs in three phases [5]. In our algorithm, the charging is done by using the battery characteristics shown in Table 1.
- **The control of discharge process** is done by a new approach, which represented in the algorithm shown in Fig. 3a. The discharge is activated/deactivated on the basis of batteries SoC, that if the SoC lower than SoCmin.D, the discharge process will stop, and if the SoC is higher than SoCmin.R, the discharge process is activated, but when SoCmin.D < SoC < SoCmin.R, the situation of the installation determines the charge/discharge process following the algorithm of Fig. 3a. If the discharge is activated then d SoC/dt ≤ 0; it will be stopped when SoC reached SoCmin.D (Table 1) for protection against deep discharge. If a new charge deactivated then d SoC/dt > 0; the discharge will be blocked until the SoC reached SoCmin.R to ensure the recovery of a minimum quantity of charge.

Table 1 Electrical characteristics and values of charge/discharge parameters of battery at 25 °C

Characteristics and conditions of charge	Batteries of 6 cells
Nominal voltage V_{BAT}	12 V
Nominal capacity Qo	110 Ah
Voltage of regulation V_R	14.4 V
Voltage of floating V_{FLT}	13.3 V
Current of over charge terminate I_{OCT}	1 A
State of charge. minimum to disconnect SoCmin.D	40 %
State of charge. minimum to reconnect SoCmin.R	45 %

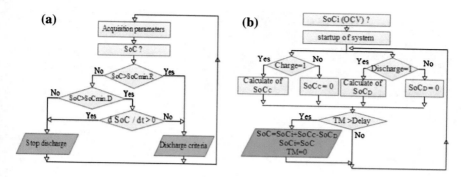

Fig. 3 The control algorithm of: **a** the battery discharging **b** the state of charge estimation

- **The estimation of state of charge** is ensured with two methods, the first is based on the calculation of the initial state of charge SoCi from the open-circuit voltage [9] (Eq. (2)). The second method is that of Coulomb counting [1] to calculate the quantity of charge injected by integration of charging current I_{BAT} (relating to SoCc Eq. (3)) or subtracted from the battery by integration of discharging current Idech (relating to SoCd Eq. (4)), taking into consideration the faradic efficiency η. The algorithm of Fig. 3b illustrates all possible calculation of SoC.

$$\text{SoCi } (25\ ^\circ\text{C}) = 100 \times \text{OCV} - 1170 \tag{2}$$

$$\text{SoCc(t)} = [100\ \%/\text{Qo}] \times \int \eta \times Ibat(t)dt \tag{3}$$

$$\text{SoCd(t)} = [100\ \%/\text{Qo}] \times \int Idech(t)dt \tag{4}$$

- **The management of the energy produced** is done by the proposed method that ensures the transfer of the energy surplus into the grid during the second and third charge phases where the battery does not absorb all the energy produced. When the battery is fully charged; the totally of energy produced is injected into the grid. The maximum of power is extracted from panel during the operation of the system. The control of energetic flow is carried by hashing current supplied to the output of the converter via circuits 1–3 controlled by PWM signals.

Fig. 4 **a** Designed PV system and digital **b** visualization of the electrical quantities

4 Results

4.1 Experimental Process

A PV System and automated electrical test bench used (Fig. 4a) are formed by:

- Four monocrystalline photovoltaic panels. Each one provided a power of 60 W at a voltage of 14.2 V and a current of 4.2 A [5–7].
- A solar battery type lead acid of nominal voltage and capacity: 12 V and 110 Ah.
- A Buck converter sized to operate at 10 kHz, 250 W and between 1 and 15 A.
- A commercial regulator for 12 V or 24 V battery, dedicated for PV installations.
- The regulator designed consists of a management and supervision system (MSS), which provides communication between different blocks, transfer and presentation the results on PC (Fig. 4b) and ensures the tasks mentioned in Sect 3.2.

5 Results and Discussion

5.1 Battery Charging

The charging of the solar battery used in this work is done by using two regulators already described during 4 days (D1–D4) when the irradiance (Fig. 5a) suffered by sudden variations particularly in the first day (D1). At first, $V_{BAT} = 12$ V, and the calculation of SoCi from Eq. (2) gives 29 %. The different results in Fig. 5 show:

- During the boost phase (0–17 h): the battery is charged (Fig. 5b) with a current I_{BAT} that comes from both regulators. This causes a rapid increase of V_{BAT} until 13.7 V and SoC until 45 %. At this threshold ($V_{BAT} = 13.7$ V), the commercial regulator [6] cuts the charging. On the contrary the regulator designed continuous the charging until $V_{BAT} = V_R = 14.4$ V and SoC reach 82 % (Fig. 5c).

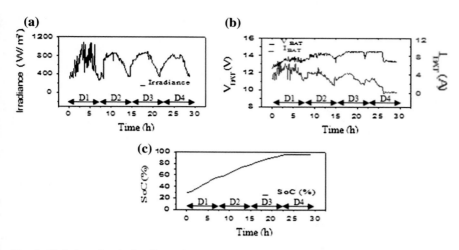

Fig. 5 Variation of **a** the irradiance, **b** V_{BAT} and I_{BAT} **c** the SoC in function of time

- During the absorption phase (17–26 h): the battery is charged (Fig. 5b) by I_{BAT} current that comes from the regulator designed around $V_{BAT} = 14.4$ V to complete the charging to the SoC = 95 %. The commercial regulator provides nothing since it cuts the charging at 13.7 V. The end of this phase is determined by $I_{BAT} = I_{OCT} = 1$ A.
- During The floating phase (26–29 h) the designed regulator indicates that the battery is fully charged and then compensates the self-discharge due to the internal resistance by a current pulses of low value (0.3 A) around $V_{FLT} = 13.3$ V(Fig. 5b).

These results show the good control and the reliability of the algorithm adopted by the regulator designed to charge the solar batteries contrary to that of the commerce.

5.1.1 Battery Discharging

To test the protection performances of the battery against deep discharge we'll experienced the two regulators, each one is connected to a resistive load and a solar battery which supports 60 % of discharge. The results in Figs. 6 and 7 show:

- The designed system of regulation has charged the battery up to a 67 % of SoC, and a resistive load was connected, a decrease of V_{BAT} and SoC (Fig. 6a) is observed with Idech = 6.5A (Fig. 6b), when SoC = SoCmin.D = 40 %, the discharge is stopped to protect the battery. As long as the SoC is lower than SoCmin.R = 45 % (Fig. 6a) the discharge is blocked In order to recover a minimum of capacity.

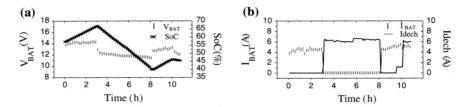

Fig. 6 The evolution of.**a** V_{BAT} and SoC. **b** I_{BAT} and Idech, of regulator designed

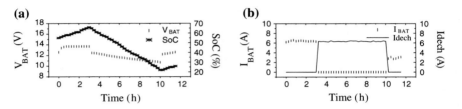

Fig. 7 The evolution of.**a** V_{BAT} and SoC. **b** I_{BAT} and Idech, of commercial regulator

• The commercial regulator has charged the battery until SoC = 66 %. In the same conditions the Idech = 6.5 A (Fig. 7b) fixed by resistive load which leads to a decrease of VBAT and SoC. After 7 h of discharge, V_{BAT} reaches LVD = 10.7 V and the discharge is stopped, the calculation generates SoC = 21 % (Fig. 7a). The reconnection can only take place when V_{BAT} reaches LVR = 12.6 V (Fig. 7a).

These results show that the commercial regulator doesn't protect the battery against deep discharge, as it discharged almost 80 % of the battery capacity on the contrary of the designed regulator that stopped the discharge to 40 %, which protects the battery and optimize its use in order to increase its lifetime.

5.1.2 Functioning of the Complete PV System

The system operation during the 4 days allowed the recovering the typical results of the duty cycle and the different electrical quantities. From these results, the efficiencies are inferred. A particular attention was given to compare the performance of the regulators used. The analysis of the results presented in Figs. 8, 9 and 10 illustrates that the optimal operating of the regulator designed is deducted from the comparison between the experimental and simulation (Optimum) results evaluated using Pspice simulator [10] under the same conditions. The results obtained show:

• The duty cycle depends on the variation of the irradiance (Fig. 8).
• The electrical quantities of the PV panels (Fig. 8) are close to optimal values.

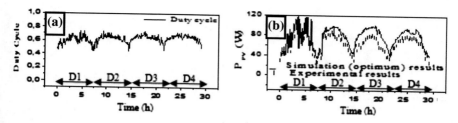

Fig. 8 Variation of **a** the duty cycle, **b** electrical quantities of PV panels

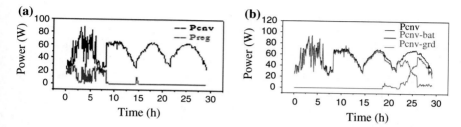

Fig. 9 Variation of powers **a** Pcnv and Preg, **b** P_{CNV}, $P_{CNV-BAT}$ and $P_{CNV-GRD}$

- During the first day, the irradiance reach the value 1000 W/m². The commercial regulator transfer a low power P_{REG} = 20 W (Fig. 9a) to the battery. however, the designed regulator transfer a power P_{CNV} = $P_{CNV-BAT}$ reach 80 W (Fig. 9b) in the middle of the day to charge the battery.
- During the rest of the functioning period, and when V_{BAT} = 13.7 V, the commercial regulator stops charging (Fig. 9a) and decides to start the 2nd phase of charge. To this voltage, the measuring of SoC has shown that not exceed 45 % (Fig. 5c). By contrary, the designed one continues to operate in optimum conditions until the 4th day when V_{BAT} = V_R = 14.4 V (Fig. 9b) and the absorption phase begins. At this voltage, the SoC shown that 82 % of capacity (Fig. 5c) is recovered. The energy transferred to the battery by the system designed is 1.25 Kwh (Fig. 9a); it's 10 times greater to this transferred by the commercial one (0.13 Kwh) (Fig. 9b).

The study and comparison of different efficiencies calculated and deducted from various results are presented in Fig. 10. These results show:

- The Buck converter efficiency was around 82 % (Fig. 10a) which witness the good operation, since it is close to that simulated [10] from PSPICE (87 %) (Fig. 10a).
- During the Boost phase (0–17 h), as the V_{BAT} is lower than 13.7 V, the commercial regulator efficiency doesn't exceed 40 % (Fig. 10a) and when V_{BAT} exceeds 13.7 V the efficiency is 0. We conclude that the system not operate in

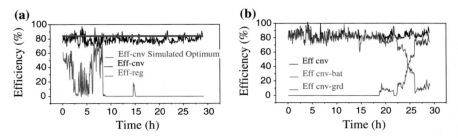

Fig. 10 The variation of **a** η_{CNV} and η_{REG} **b** η_{CNV}, $\eta_{CNV\text{-}BAT}$ and $\eta_{CNV\text{-}GRD}$ during functioning

the optimal conditions. However, the designed one efficiency was about 82 % (Fig. 10a), it is satisfactory and close to optimum. No power is transferred (the efficiency is 0).

- During the second and third phase of charging (17–29 h) the efficiency of the injecting into the grid increases and that of the batteries charge decreases (Fig. 10b). This is so due to the decrease (increase) of the power absorbed (injected) by the battery (grid). The overall efficiency of the designed regulator (around 82 %) is always equal to the sum of the efficiencies of charging and injection into the grid (Fig. 10b). In this phase the efficiency of the commercial regulator is 0 (Fig. 10a).

These results show, that the regulator designed allows optimizing and speeding up the charging and injecting the energy surplus not absorbed by the battery into the grid. All that is under the optimal conditions. By contrary the commercial regulator guarantees only a direct connection between panels and battery.

6 Conclusion

In this work, we studied the charging of a solar battery using two regulators, the first is commercial and the second is a designed MPPT regulator that is realized in this work. Experimentation of the two regulators for whole days of sunshine shows the limitation of the commerce regulator performance. It stops the battery charging to 45 % of their charge to skip to the 2nd charge phase. Contrary; the MPPT controller designed ensures full battery charge (over 95 %) with remarkable precision. In addition, it optimizes the operation of the PV system and provides a very good management of the energy produced. Through the power switches designed in this work, and DC/DC converters, it exploits all the energy produced by the panels and the surplus is injected into the grid, for example.

Acknowledgment *This work is supported by:*

- Belgian Development Agency CTB (Project MIP/012/010), Morocco.
- United Nations Development Programme UNDP Art Gold Morocco,ENV2008 2 oo.
- Cooperation Moroccan-Belgian "Institutional University Commission", IUC, Oujda, 2008-2012 (Water and Environment Activity/ Sub-Activity Renewable Energy).
- Moroccan-Tunisian Cooperation, SCIENTIFIC RESEARCH AND TECHNOLOGY Project (11/MT/38).

References

1. Zhou, W., Yang, H., F, Zhaohong: Battery behavior prediction and battery working states analysis of a hybrid solar–wind power generation system. Renew. Energy **33**, 1413–1423 (2008)
2. Riffonneau, Y., Bacha, S.: Optimal power flow management for grid connected PV systems with batteries. IEEE Trans. Sustain. Energy **2**(3), 309–320 (2011)
3. Daud, M.Z., Mohamed, A., Hannan, M.A.: An optimal control strategy for DC bus voltage regulation in photovoltaic system with battery energy storage. Hindawi Pub. Corp. Sci. World J. **2014**(271087), 16 (2014)
4. Tesfahunegn, S.G., Vie, P.J.S., Ulleberg, O., Undeland, T.M.: A simplified battery charge controller for safety and increase dutilization in standalone PV applications (978-1-4244-9965-6/11/$26.00 2011 ieee)
5. Hirech, K., Melhaoui, M., Yaden, F., Baghaz, E., Kassmi, K.: Design and realization of an autonomous system equipped with a regulator of charge/discharge and digital MPPT command. The Mediterranean Green Energy Forum 2013 (MGEF-13)
6. http://www.e-energieverde.ro/CM5024Z,%20CM5048.pdf (2014)
7. Web-page: www.solarcellsales.com/techinfo/docs/Shell_SP75.pdf (2014)
8. Melhaoui, M., Baghaz, E., Hirech, K., Yaden, F., KASSMI, K.: Contribution to the improvement of the MPPT control functioning of photovoltaic systems. Int. Rev. Electr. Eng. (IREE) **9**(2), 393–400 (2014). ISSN 1827-6679
9. Cheng, K.W.E., Senior Member, IEEE, Divakar, B.P., Wu, H., Ding, K., Ho, H.F.: Battery-management system (BMS) and SOC development for electrical vehicles. IEEE Trans. Veh. Technol. **60**(1), 76–88 (2011)
10. Web-page: http://www.cadence.com (2014)

A Comparative Study Between Two MPPT Controllers Based on the Principe of Sliding-Mode Control Theory and Intelligent Control Technique in Photovoltaic Systems

Radhia Garraoui, Abdelali El Aroudi, Mouna Ben Hamed,
Lassaad Sbita and Khalifa Al-Hosani

Abstract This chapter presents a PV system composed by a photovoltaic array connected to a DC-DC boost converter with resistive load in order to extract the maximum power generated by the PV panel. This paper presents a model of photovoltaic (PV) system to investigate the P-V characteristics under varying irradiation and temperature conditions. The proposed photovoltaic model is used for evaluating the techniques of maximum power point tracking (MPPT). For this reason, a comparative study between two control methods for maximum power point tracking (MPPT) algorithms in photovoltaic (PV) systems is investigated and well presented. The two MPPT controllers presented in this work are: The Fuzzy Logic Controller (FLC) and the Sliding Mode Controller (SMC). The MPPT controller based on the fuzzy-logic-algorithm is considered as an intelligent technique and it uses directly the DC-DC converter duty cycle as a control variable and it provides a fast response and good performances against the climatic and load changes. The SMC presents, also a very good response for tracking the maximum power point (MPP) for photovoltaic systems. The input parameters that were

R. Garraoui (✉) · M. Ben Hamed · L. Sbita
Photovoltaic, Wind and Geothermal Systems Research Unit,
National Engineering School of Gabes, Gabes, Tunisia
e-mail: radhiagarraoui@gmail.com

M. Ben Hamed
e-mail: mouna.benhamed@enig.rnu.tn

L. Sbita
e-mail: lassaad.sbita@enig.rnu.tn

A. El Aroudi
Department of Electronics, Electrical Engineering and Automatic Control,
Universitat Rovira I Virgili, Tarragona, Spain
e-mail: abdelali.elaroudi@urv.cat

K. Al-Hosani
Petroleum Institute, Abu Dhabi, UAE
e-mail: khalhosani@pi.ac.ae

© Springer International Publishing Switzerland 2016
A. El Oualkadi et al. (eds.), *Proceedings of the Mediterranean Conference
on Information & Communication Technologies 2015*, Lecture Notes
in Electrical Engineering 380, DOI 10.1007/978-3-319-30301-7_53

considered are the voltage and the current. The duty cycle determined by the MPPT controller is used to generate the optimal MPP under different operating conditions. Simulation results show that both algorithms can effectively perform the MPPT hence improving the efficiency of PV systems and via the comparison it is possible to understand the advantages of every controller.

Keywords Photovoltaic systems · MPPT · Sliding mode controller · Fuzzy logic controller

Nomenclature

G:	*Global insulation* (W/m^2)
G_n:	*Reference insulation* (W/m^2)
T:	*Cell Junction temperature* (°C)
T_r:	*Reference cell temperature* (°C)
i_{pv}:	*Output PV current* (A)
i_L:	*Inductance current* (A)
i_{ph}:	*Light-generated current* (A)
i_{rr}:	*Saturation current* (A) at Tref
V_{pv}:	*PV output voltage* (V)
A:	*Ideality factor*
α:	*Duty cycle*
K_b:	*Boltzmann constant* (1.3806×10^{-23})
E_g:	*Band gap energy* (eV)
q:	*Charge of an electron* (C)
C:	*Capacitors* (F)
L:	*Inductance* (H)
K_i:	*A i_{sc}/T Coefficient* (A/°K)
r_{ci} and $r_{l,\ pv}$:	*Parasitic resistance* (Ω)

1   Introduction

The conventional energy resources in the world are threatened today. Moreover, they emit large amounts of greenhouse gas emissions responsible for climate change and health problem. Many pollution problems can be avoided if electrical power is generated from renewable energy sources rather than traditional fossil fuels [1]. The use of some types of renewable energy increased notably in recent years, with its advantages of being pollution-free. In particular, wind, solar, and hydroelectric systems can generate electricity without air pollution emissions and can help to reduce energy prices in the future. For these reasons, renewable energy resources are receiving increasing interest in recent years being photovoltaic (PV) energy one of the most attractive sources due to its abundance. PV generators have a nonlinear current-voltage or power-voltage characteristic with a maximum power point

(MPP) depending on the atmospheric conditions, namely the irradiance and the temperature. Therefore, a maximum power point tracking (MPPT) system is required to ensure that the PV always operates at its MPP regardless of the temperature, insulation and load changes. The output of this controller is fed-back to a DC-DC switching converter so that the array delivers a maximum of energy. Generally speaking, a PV system is composed of three key elements: The first one consists of the PV array as a source of energy, the second one is the DC-DC converter used as an interface which allows the adaptation of the energy flow between the PV array and the load. The third block is the control system carrying the regulation of some variables of interest with the aim to extract the MPP from the PV array. For an appropriate operation of PV systems at maximum power, there are a variety of methods that can be used [2].

1.1 Description of the PV System

In order to investigate the reliability of MPPT controllers, a PV power system with a boost converter is considered. The system under study is presented in Fig. 1.

It presents also a proposition of an equivalent circuit model of a PV cell which is composed of a light generator source i_{ph}, a parallel resistor R_p expressing a leakage current, a diode and a series resistor R_s describing an internal resistance to the current flow. The equivalent model of a PV cell can be described by the mathematical expression (1–3). The parallel resistor R_p is neglected because of its large resistance value and the series resistor R_s is also neglected due to its very small resistance value. Neglecting these two parameters simplifies significantly the

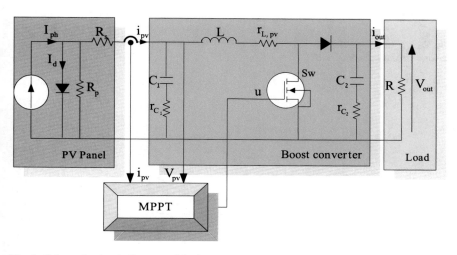

Fig. 1 Schematic circuit diagram of the boost converter supplied by a PV panel

numerical simulation without affecting the accuracy and makes it faster. The first equation describes the output current of the photovoltaic cell:

$$i_{pv} = i_{ph} - i_d \left[\exp(\frac{qV_{pv}}{k_b TA}) - 1 \right] \qquad (1)$$

The equation of the PV current as a function of temperature and irradiance can be written as follows:

$$i_{ph} = \frac{G}{G_n} [i_{scr} + k_i(T - T_r)] \qquad (2)$$

The saturation current equation is described by:

$$i_d = i_{rr} \left[\frac{T}{T_r} \right]^3 \exp(\frac{qE_g}{k_b A} \left[\frac{1}{T_r} - \frac{1}{T} \right]) \qquad (3)$$

The PV cell parameters used in this study are shown in Table 1:

Figure 2 and 3 show the effect of varying weather conditions on MPP locations at P-V curves. Figure 2 shows that The open circuit voltage and the power increase with the important solar irradiance value. However, in Fig. 3 with increasing temperature, the opposite occurs, i.e. the PV generator's power decreases when the temperature increases.

Table 1 The photovoltaic cell parameters

q	1.6022×10^{-19} C	K_b	1.3806×10^{-23}	T_r	298 K
E_g	1.1557 (eV)	G_n	1000 (W/m^2)	i_{scr}	3.45 A
K_i	0.60095 (A/°K)	i_{rr}	5.98×10^{-8} A	A	1.2

Fig. 2 P-V characteristics under different irradiance and T = 25°C

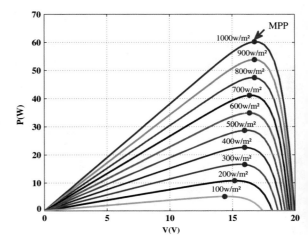

Fig. 3 P-V characteristics under different temperature and G = 1000 W/m^2

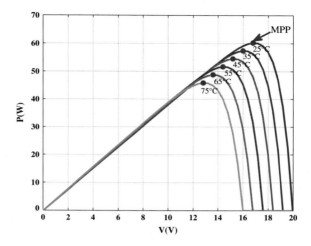

2 Modeling the Boost Converter

Boost converters are used for performing MPPT controllers in PV systems thanks to their capability of delivering an output voltage larger than the low input voltage of the PV panel. The DC-DC converters are largely used in regulated switch mode DC power supplies. From the energy point of view, the output voltage regulation in the DC-DC converter is achieved by constantly adjusting the amount of energy absorbed from the source and that injected into the load, which is in turn controlled by the relative durations of the absorption and injection intervals. These two basic processes of energy absorption and injection constitute a switching cycle. The conventional DC-DC boost converter circuit is shown in Fig. 1. Depending on the position of the switch Sw, the system has two operating topologies. By applying Kirchhoff's voltage law we get the mathematical model of the system. It is important to take the effects of parasitic resistances in the capacitors and inductance into consideration. α stands for the ratio between the ON time duration to the entire switching period. In order to simplify the expression we suppose that:

$$\rho = \frac{r_{C_2}.R}{r_{C_2}+R}, \kappa = \frac{R}{R+r_{C_2}}$$

When the switch is OFF, the dynamics of the circuit are governed by the following equation:

$$\begin{cases} \frac{di_L}{dt} = \frac{1}{L}.\left(V_{pv} - \left(r_{L,pv} + \frac{r_{C_2}.R}{r_{C_2}+R}\right).i_L - \frac{R}{R+r_{C_2}}v_{C_2}\right) \\ \frac{dv_{C_1}}{dt} = \frac{1}{C_1}\left(i_{pv} - i_L\right) \\ \frac{dv_{C_2}}{dt} = \frac{R}{C_2(r_{C_2}+R)}.i_L - \frac{v_{C_2}}{C_2(r_{C_2}+R)} \end{cases} \quad (4)$$

When the switch is ON, the circuit is described by:

$$\begin{cases} \frac{di_L}{dt} = \frac{1}{L}.(V_{pv} - r_{L,pv}.i_L) \\ \frac{dv_{C_1}}{dt} = \frac{1}{C_1}(i_{pv} - i_L) \\ \frac{dv_{C_2}}{dt} = -\frac{v_{C_2}}{C_2(r_{C_2} + R)} \end{cases} \tag{5}$$

The average model of the system is given by:

$$\begin{cases} \frac{di_L}{dt} = \frac{1}{L}.(V_{pv} - r_{L,pv}.i_L - (1 - \alpha).\kappa.(r_{C_2}.i_L + v_{C_2})) \\ \frac{dv_{C_1}}{dt} = \frac{1}{C_1}(i_{pv} - i_L) \\ \frac{dv_{C_2}}{dt} = (1 - \alpha)\frac{\kappa}{C_2}.i_L - \frac{v_{C_2}}{C_2(r_{C_2} + R)} \end{cases} \tag{6}$$

3 The Intelligent MPPT Controller: FLC

Fuzzy logic uses general information on the system to be controlled and the required performances are achieved by establishing some rules in the form of 'if-then' sentences according to the appropriate decision that the controller will take later. The Fuzzy logic control does not require a mathematical model of the system. The process of Fuzzy Logic Controller (FLC) can be subdivided into three steps: first, fuzzification, then rule evaluation, and finally defuzzification [3, 4]. FLC is applied in designing the MPPT controller in this section. The input variables of the FLC in this paper are the error and the change in error. The error is calculated as the change in the PV power to the change in the PV voltage method. The Fuzzy inference is determined by using Mandamni's method, and the defuzzification uses the center of gravity method to compute the output of this FLC which is the duty cycle α. The error E and the change in error dE are represented as follows:

$$E(k) = \frac{P(k) - P(k-1)}{V(k) - V(k-1)} \tag{7}$$

$$dE = E(k) - E(k-1) \tag{8}$$

The position of the operating point at time E is determined by the sign of dE. The duty cycles from 25 rules must be computed and combined for a specified value. The fuzzy system rules can be designed as shown in Table 2. Several linguistic variables [5] by using five fuzzy subsets are denoted by: NB: negative big, NS: negative small, ZO: zero, PS: positive small, PB: positive big.

Table 2 The fuzzy system rules

E/dE	NB	NS	ZO	PS	PB
NB	ZO	ZO	PB	PB	PB
NS	ZO	ZO	PS	PS	PS
ZO	PS	ZO	ZO	ZO	NS
PS	NS	NS	ZO	ZO	ZO
PB	NB	NB	ZO	ZO	ZO

4 MPPT Based on Sliding Mode Control Theory

Sliding Mode Control is known to be a robust control method appropriate for controlling switched systems. High robustness is maintained against various kinds of uncertainties such as external disturbances and measurement error [6]. In classic Sliding Mode Control (SMC) design, the sliding variable is selected such that it has relative degree one with respect to the control. For the case of the photovoltaic system the design procedure can be divided into two steps: First, finding the switching function, such that the internal dynamics in the sliding mode are stable. Then, designing a control that will drive the plant state to the switching surface and maintain it there. The sliding surface will be selected by imposing the maximum power point identifier factor equal to zero as shown in Eq. (9).

$$\frac{\partial P}{\partial I} = I\left(\frac{\partial V}{\partial I} + \frac{V}{I}\right) = 0 \tag{9}$$

The non trivial solution of (9) is $(\partial V/\partial I + V/I) = 0$
Thus, a proper sliding manifold in the state space can be defined as:

$$\sum = \left\{ x \backslash S(t,x) = \frac{\partial V}{\partial I} + \frac{V}{I} = 0 \right\} \tag{10}$$

The switching control u of the SMC is presented as follows:

$$u = 0.5.(1 + sign(S)) \tag{11}$$

5 Simulation Results

We suppose that irradiance trajectory used in this simulation work takes the following form: it increases from 500 W/m^2 to 1000 W/m^2 in steps of 500 units every 0.06 s. The photovoltaic system has responded correctly and rapidly, the MPPT controller based on SMC provides the fastest response against trade lighting, the two MPPT algorithms resisted with success the abrupt change of irradiance, this is clear via the answers provided by the FLC and the SMC on the power, which are shown in Fig. 4a. This is clear if we see the generated duty cycle while observing

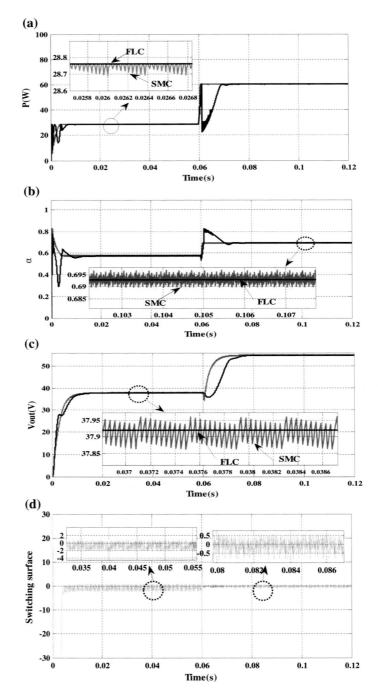

Fig. 4 The power under load change and variable irradiance **a** Irradiance change **b** Load change

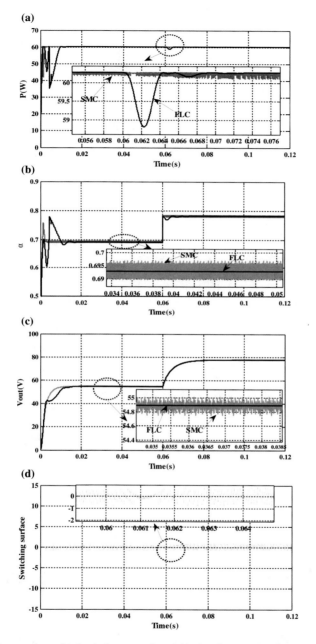

Fig. 5 The duty cycle under load change and variable irradiance **a** Irradiance change **b** Load change

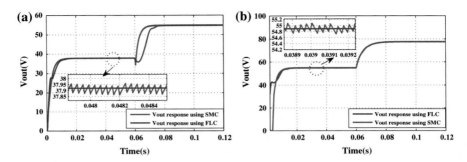

Fig. 6 The output voltage under load change and variable irradiance **a** Irradiance change **b** Load change

the simulation result in Fig. 4b. The influence of the good nature of command generated by the FLC is clear at the output voltage V_{out} in Fig. 4c. We could confirm by these plots that the MPPT based on FLC has a very small error due to its special structure that is why it has the most precise response and it doesn't cause a power loss. Moreover, the MPPT controller based on Sliding mode Principe ensure some good dynamics around his switching surface which is stable and and attractive, as it is mentioned in Fig. 4d. The other test consists on load change. We assume that the resistance is suddenly changed from 50 Ω to 100 Ω at 0.06 s, the MPPT system is simulated for such variation of R, to obtain the result on power as shown in Fig. 5a. Obviously, the MPPT system still tracking the optimal power. So, by using our two MPPT controllers, the system still holds on the MPP. Under the variation of load resistance, and according to the responses on duty cycle and output voltage presented respectively in Fig. 5b, c we prove the performances of these two controllers. There is a difference between the natures of the responses occurred by FLC and the command based on the SMC. The switching control SMC is more robust against the load change because of its very low sensitivity in comparison with FLC. This type of control low provides a problem which is characterized by

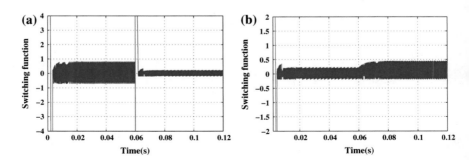

Fig. 7 The switching function under load change and variable irradiance **a** Irradiance change **b** Load change

the high oscillation's frequency. This is clear when observing the plot of the sliding function in Figs. 4d and 5d. However, the FLC seems to be the most precise.

6 Conclusions

This work presents the Principe of two MPPT controllers which are the Sliding Mode Controller (SMC) and the Fuzzy Logic Controller (FLC). Via an extended MATLAB simulation we propose some tests to validate the performances of these controllers against the climatic and load change. The FLC makes the output PV system value close to the theoretical maximum value. The simulation results show that in the case of irradiance and load abrupt change, the SMC can quickly find a new maximum power point, and has high power output efficiency. The FLC has also performing results. The experimental validation of these two MPPT controllers will be our perspectives.

References

1. Bose, B.K.: Global warming: Energy, environmental pollution, and the impact of power electronics. IEEE Ind. Electron. Mag. 4(1), 6–17 (2010)
2. Agorreta, J.L., Reinaldos, L., Gonzalez, R., Borrega, M., Balda, J., Marroyo, L.: Review of maximum-power-point tracking techniques for solar-photovoltaic systems. Energy Technol. 1 (8), 438–448 (2013)
3. Garraoui, R., Ben Hamed, M., Sbita, L.: MPPT controller for a photovoltaic power system based on fuzzy logic. In: 10th IEEE International Conference on International Multi-Conference on Systems, Signals and Devices (SSD), Hammamet, Tunisia, pp. 18–21 (2013)
4. Lalouni, S., Rekioua, D., Rekioua, T., Matagne, E.: Fuzzy logic control of stand-alone photovoltaic system with battery storage. J. Power Sour. 193(2), 899–907 (2009)
5. Patcharaprakiti, N., Premrudeepreechacharnb, S., Sriuthaisiriwong, Y.: Maximum power point tracking using adaptive fuzzy logic control for grid-connected photovoltaic System. Renew. Energy 30(11), 1771–1788 (2005)
6. Il-Song, K.: Robust maximum power point tracker using sliding mode controller for the three-phase grid-connected photovoltaic system. Sol. Energy 81(11), 405–414 (2007)

Part IX
RF Devices and Antennas for Wireless Applications

Design & Simulation of Rectangular Dielectric Resonator Antenna for UMTS Application

Kaoutar Allabouche, Mohammed Jorio, Tomader Mazri
and Najiba El Amrani El Idrissi

Abstract This paper represents a design and simulation of Rectangular Dielectric Resonator Antenna (RDRA) for base stations operating in the UMTS (Universal Mobile Telecommunications System). The Dielectric resonator is planned to be used as a radiating element, fed by a 50 Ω Microstrip transmission line technique at frequency around 2 GHz. The return loss, input impedance, VSWR and radiation patterns are studied. The simulation results were performed using two numerical methods. The conception was realized using finite integration method (CST microwave studio) and verified by finite element method (HFSS: Ansoft high frequency structure simulator).

Keywords Rectangular dielectric resonator antenna (RDRA) · UMTS band · Base stations · CST microwave studio · HFSS · MATLAB

1 Introduction

The new tendencies in wireless communications aim to deliver a global mobility with wide variety of services. Therefore, in order to satisfy these requirements, antennas operating on wireless communication, should satisfy three primordial requests:

K. Allabouche (✉) · N. El Amrani El Idrissi
Laboratory Signals, Systems and Components, FST Fez,
Fez, Morocco
e-mail: kaoutar.allabouche1@gmail.com

K. Allabouche · M. Jorio
Laboratory of Renewable Energies and Intelligent Systems, FST Fez,
Fez, Morocco

T. Mazri
Laboratory of Engineering Systems, ENSA Kenitra, Kenitra, Morocco

© Springer International Publishing Switzerland 2016
A. El Oualkadi et al. (eds.), *Proceedings of the Mediterranean Conference on Information & Communication Technologies 2015*, Lecture Notes in Electrical Engineering 380, DOI 10.1007/978-3-319-30301-7_54

519

compact size, omnidirectional radiation pattern and wide impedance bandwidth [1, 2]. Dielectric Resonator Antennas (DRAs) is one of these solutions. It has been demonstrated that dielectric masses of cylindrical [3], rectangular parallelepiped [4], hemispherical [5], half-split cylindrical [6], equilateral triangular [7] shapes can be conceived to radiate through appropriate choices of feed location and dimensions. Diverse types of feeding structures like coaxial probe [1], Microstrip line [8], microstrip fed aperture [9], and coplanar waveguide [10] have been recommended. As compared to the Microstrip Antennas, DRAs has a much wider impedance bandwidth. However, all excitation methods applicable to the micro strip antenna can be used also for the DRA [11]. Some of its interesting characteristics are enumerated below that truly speak of their capability and applicability:

Ease of integration with other antennas, Negligible dielectric losses [11], High temperature tolerance [12], Design & shape flexibility…

All these advantages of DRAs make them elevant elements for antenna applications at microwave frequencies. Development of mobile and wireless communication systems such as UMTS network working at 2 GHz needs broadband antennas. Design of Rectangular Dielectric Resonator Antenna operating in the UMTS band (1920–2170 GHZ) is explored in this paper. The antenna is fed by Microstrip feed line technique (50 Ω). An appropriate structure is obtained that illustrates an important impedance bandwidth (for S11 > 10 dB) at 1.9–2.1 GHz. Through this paper, the analysis is performed using CST microwave studio [13] and HFSS [14].

2   Antenna Theory and Parametric Study

Dielectric waveguide model [15] is used to design a Rectangular Dielectric Resonator antenna (RDRA) operating in the UMTS band at 2 GHz. The dimensions of the radiated element (DR) are resulted by using the Eqs. (1) of the dielectric waveguide model (DWM) for RDRAs in free-space. In fact, the equation indicates that the resonance can be generated inside a Rectangular Resonator of length b, width a and height d (Fig. 1):

$$f_r = \frac{c}{2\pi\sqrt{\varepsilon_r}} \times \sqrt{k_x^2 + k_y^2 + k_z^2} \tag{1}$$

$$k_x = \frac{m\pi}{a}; \quad k_y = \frac{n\pi}{b}; \quad k_z = \frac{l\pi}{2d}$$

Where: ε_r: Permittivity of the dielectric resonator.
k_x, k_y & k_z: Denote the wave-numbers along the x, y, and z directions within the RDRA structure respectively.
a, b & d: Dimensions of the RDRA along the x, y and z axis.

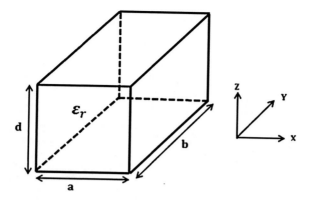

Fig. 1 Configuration of the rectangular dielectric resonator antenna

Table 1 Summary of the DRA's parametric study

Configuration	Dimensions of the rectangular DRA			Resonance frequency obtained in simulation
	a (mm)	b (mm)	d (mm)	
DR-1	37	30	1	2 GHz
DR-2	37.5	30	40	1.74 GHz
DR-3	68.8	25	20	1.67 GHz
DR-4	29	40	20	1.85 GHz
DR-5	38	30	6	1.85 GHz

   MATLAB software has been used for calculation of antenna parameters, because the dimensions of the DR are a frequency tuning parameters, determining the resonant behavior of the antenna. The optimization of the antenna design was done by developing a script. Different configurations were proposed by resolving the Eq. (1). The appropriate configuration for our application is the one providing a resonance at 2 GHz. Table 1 presents some configurations that have been generated by the script developed in MATLAB and simulated in CST microwave studio to predict their resonance frequencies.

   In Fig. 2, we present the return loss versus frequency characteristic of each configuration. Note that the first configuration presents a resonance at 2 GHz. Which confirms the choice of the first configuration that will be retained for further work.

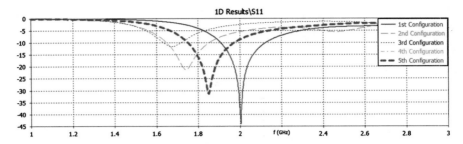

Fig. 2 Reflection coefficient depending on the antenna's size from CST microwave studio

3 The RDRA Design and Structure

The geometry of the proposed RDRA is shown in Fig. 3. It consists of a dielectric resonator antenna, which is directly excited by 50 Ω Microstrip line. The feed line is printed on a substrate of dielectric constant $\varepsilon_r = 4.3$ and thickness of 1 mm. The resonator's dimensions and characteristics are detailed in Table 2.

Fig. 3 Geometry of the proposed RDRA fed by Microstrip feed line

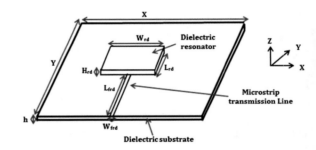

Table 2 Dimension of the proposed rectangular DRA

Antenna characteristics	Value (mm)
Length of the dielectric resonator, L_{rd}	30
Width of the dielectric resonator, W_{rd}	37
Height of the dielectric resonator, H_{rd}	1
Length of the feed line, L_{frd}	34.5
Width of the feed line, W_{frd}	2
Substrate's height, H_{srd}	1
Length of the substrate, y	60
Width of the substrate, x	70

4   Results and Discussions

4.1   Return Loss versus Frequency Characteristics and VSWR

Figure 4 shows the simulation results for return loss characteristics vs. Frequency characteristics using CST and HFSS. Lowest value of return loss is obtained at 2 GHz with a value of −44 dB. Simulation results are in good. Moreover the proposed RDRA provides an interesting bandwidth (0.2244 GHz), which corresponds to the range of frequency of interest.

From Fig. 5 the curve of VSWR versus frequency that was simulated using CST and HFSS is shown in Fig. 6. Its value is less than 2 in the whole antenna bandwidth and equal to 1.01 at 2 GHz.

4.2   Far Field Simulation

The measurement of far field radiation pattern of the antenna at 2 GHz using finite integration method (CST Microwave Studio) is presented in Fig. 6. We can observe

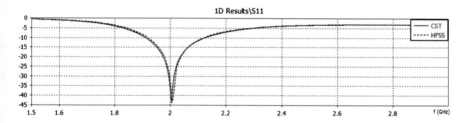

Fig. 4 Simulated return loss versus frequency graph CST and HFSS

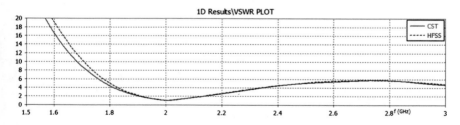

Fig. 5 VSWR versus frequency graph

Fig. 6 3D far field pattern of
microtsrip line fed RDRA
from CST studio software

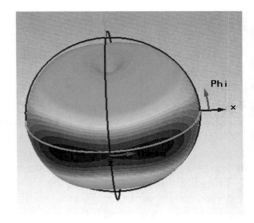

that the RDRA provides an omnidirectional behavior following the axis XY and
bidirectional along the YZ and XZ axis. Also the efficiency of the proposed antenna
achieved a value of 90.42 % at 2 GHz.

5 Conclusion

The paper has focused on the development of Rectangular Dielectric Resonator
Antenna dedicated to mobile communications, in particularly third generation
(UMTS). The simulation results prove that the structure designed has an efficient
reflection coefficient which is equal to −44 dB at 2 GHz, an omnidirectional
radiation pattern in the XY Plane and a very good adaptation.

References

1. Fujimoto, K., James, J.R.: Mobile Antenna Systems Handbook. Artech House, Boston, UK
   (1994)
2. Wong, K.L.: Planar Antennas for Wireless Communications. John Wiley & Sons, USA (2003)
3. Long, S.A., McAllister, M.W., Shen, L.C.: The resonant cylindrical dielectric cavity antenna.
   IEEE Trans. Antennas Propag. **31**(3), 406–412 (1983)
4. McAllister, M.W., Long, S.A.: Rectangular dielectricresonator antenna. IEE Electron. Lett. **19**,
   218–219 (1983)
5. McAllister, M., Long, S.A.: Resonant hemispherical dielectric antenna. IEE Electron. Lett. **20**,
   657–659 (1984)
6. Mongia, R.K.: Half-split dielectric resonator placed on metallic plane for antenna applications.
   IEE Electron. Lett. **25**, 462–464 (1989)
7. Lo, H.Y., Leung, K.W., Luk, K.M., Yung, E.K.N.: Low profile equilateral-triangular dielectric
   resonator antenna of very high permittivity. Electron. Lett. **35**(25), 2164–2166 (1999)

8. Kranenburg, R.A., Long, S.A.: Microstrip transmission line excitation of dielectric resonator antennas. IEE Electron. Lett. **24**(18), 1156–1157 (1988)
9. Robert, E., Collin: Foundations for Microwave Engineering. IEEE Press (2001)
10. Kranenburg, R.A., Long, S.A., Williams, J.T.: Coplanar waveguide excitation of dielectric resonator antennas. IEEE Trans. Antennas Propag. **39**(1), pp. 119–122 (1991)
11. Luk, K.M., Leung, K.W.: Dielectric resonator antennas (2002)
12. Holzwarth, S., et al.: Active antenna arrays at Ka-band: Status and outlook of the SANTANA project. In: Proceedings of the Fourth European Conference on Antennas and Propagation (EuCAP), pp. 1–5 Barcelona, Spain (2010)
13. Ansoft HFSS Technical Notes, http://wenku.baidu.com/view/ae71f9eef8c75fbfc77db26b.html (2010)
14. Ni, M.Y.: Learning how to use Ansoft and CST. Worcester Polytechnic Institute, (2008)
15. Ittipiboon, A., Mongia, R.K.: IEEE Trans. Antennas Propag. **45**(9), 1348–1356 (1997)

Experimental Analysis of the Effect of Velocity on Vertical Handoff Latency Using IEEE 802.21

Asmae Ait Mansour, Nourddine Enneya, Mohamed Ouadou and Driss Aboutajdine

Abstract The concept of heterogeneous networks has already begun with the 4G that will allow communications using all types of services based on IP, and perform a soft handoff between heterogeneous networks. Decisions for vertical handoff in heterogeneous, affect the QoS of the mobile users, which are also dependent upon the mobility. Consequently the choice of the right radio interfaces, at a given moment when mobile terminal has more than one available wireless or mobile network can be based on various factors beside signal strength. In this paper using the IEEE 802.21 standard, we propose to study the impact of handoff latency in lost packet with various speed of the mobile.

Keywords Vertical handoff · IEEE 802.21 · Mobility · Handoff latency

1 Introduction

As mobile wireless networks increases in popularity and pervasiveness, we are facing the challenge of integration of diverse wireless networks such as WLANs and WWANs. Therefore interesting issues such as vertical handoff takes a certain extent in the coexistence of heterogeneous networks and mobility. However, there are several challenging issues on vertical handoff support. Such as the decision in heterogeneous networks, that depends on different system characteristics [1].

Therefore lot of research activities are carried on in heterogeneous handoff. One of this in Ref. [2] proposes a method that triggers and combines data rate and channel occupancy in order to fairly balance users among the two networks. Authors of [3]

A. Ait Mansour (✉) · M. Ouadou · D. Aboutajdine
LRIT-CNRST (URAC No29), Faculty of Sciences, Mohammed V University, Rabat, Morocco
e-mail: asmae.ait@gmail.com

N. Enneya
Faculty of Sciences, Ibn Tofail University, Kenitra, Morocco

© Springer International Publishing Switzerland 2016
A. El Oualkadi et al. (eds.), *Proceedings of the Mediterranean Conference on Information & Communication Technologies 2015*, Lecture Notes in Electrical Engineering 380, DOI 10.1007/978-3-319-30301-7_55

aim to overcome Handoff latency by proposing a method based on dynamic region to reduce the movement detection delay. Another handoff procedure was proposed in paper [4]. But it did not consider the handoff trigger, which is an important component to reduce handoff latency.

The rest of the paper is organized as follows, the handoff process with those two types are described in Sect. 1. In Sect. 2 we present the standard IEEE 802.21. Section 3 presents a sequence of events that a MN (mobile node) and network perform in order to make a successful HO in different scenario of a mobile user. Finally we outline some conclusions and future work in Sect. 4.

2 Handoff Process

2.1 Horizontal Handoff

Horizontal handoff [5] is a transfer between two base-stations (BSs) of the same system [6] and they are typically required when the serving access router becomes unavailable due to MN s movement. It can be further classified into Link-layer handoff and Intra-system handoff. Horizontal handoff between two BS, under same foreign agent (FA) is known as Link-layer handoff. In Intra-system handoff, the horizontal handoff occurs between two BSs that belong to two different FAs and both FAs belongs to the same system and hence to same the gateway foreign agent (GFA).

2.2 Vertical Handoff

Vertical handoff refers to transfer that has occurred between base stations that are using different wireless network interfaces [6, 7], it s usually used to support node mobility and for convenience rather than connectivity reasons (e.g., according to user choice for a particular service).

The two of the major challenges in vertical handoff management are seamlessness and automation aspects in network switching. In the vertical handoff, the most important decision factors may need to be taken into consideration to maximize satisfaction of the identified user, such as cost of services, power consumption and velocity of the mobile terminal [8]. We can regroup different criteria as follows: Cost of Service, Security, Power Requirement, Proactive Handoff, Received signal strength (RSS), Velocity.

3 IEEE 802.21

To enable handoff between heterogeneous networks including both 802 and non-802 networks we can use the IEEE 802.21 [9]. MIH (Media Independent Handoff) is a middleware located between the Link layer and the IP layer; it provides information to allow handoff. The function that allows the transfer in protocol is implemented by MIHF (MIH Function). MIH defines three main mobility services the Event Service, the Command Service and The Information Service.

The standard defined in the IEEE 802.21 specification has triggers that allow higher layers to take action. The triggers are classified into two types, predictive and event triggers. Predictive triggers are a likelihood of a change in system properties in the future. Event triggers describe a definite event that has occurred. The MIH triggers [10] are implemented in the mobile station (MS) which is helpful in realizing the seamless handoff.

4 Simulation Scenario (WiMAX-WiFi)

4.1 Topology and Scenario to Simulate the System Consider

The simulation scenario as shown in Fig. 1 will use two technologies the wifi and wimax. At the first 5 s the MN try to detect available network by sending "Get Status Request" message from the MIH user to the MIHF, then "get status response" occur and inform of the existence of two interface type. The MN receive the link Detected by MN's Wimax interface in this way the MN's will use this interface to connect to the BS since there is no better network. Thereby the MN is actually connected to Wimax BS and it will trigger off a link up event toward MIH user. To begin the Wimax interface configuration MN's MIP Agent sends request RS (router solicitation) to the ND agent wimax then the BS sends RA (router advertisement). The

Fig. 1 Topology scenario (WIMAX-WIFI)

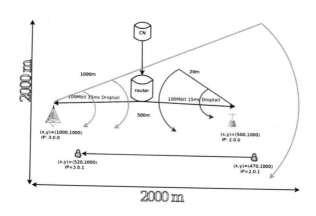

MN's user commands to the wimax interface to send MIH capability Request to the BS and after receiving the MIH Capability Response sent by the BS, the MN's MIH agent knows the identification of the remote MIHF.

At t = 6 s the CN starts to send CBR traffic to the MN and the moving towards wifi cell that starts at t = 7 s. When the MN's wifi interface starts detecting 802.11 beacon with power above the threshold value. The event link detected is triggered and sent to the MIH agent of the MN. The MIH user commands to MN's wifi interface to connect to the AP since he is a better interface. The MN's wifi interface triggers off the event "link up" after receiving the "Association Response" and send it to MIH agent that will command in its turn the MIP agent to request to the ND Agent to send RS. Once the RS sent by a MN's Wifi interface the RA is then received and the configuration of the interface could begin. MN's MIH Agents are notified. In order to inform the CN of the new MN location the wifi interface sends a redirect message to the CN. In this case the mobile node uses the both interfaces at the same time in order to perform seamless handoff and it's named make before break. The MN's MIH Agent receives the confirmation that the CN uses the link between wifi interface and AP, then he commandes the wifi interface to send a MIH capability Response to the AP with the aim to know the identification of the new remote MIHF identification.

The period for which the mobile stays in the wifi network depends on the speed. The link down event is targeted when the MN reach the Boundary of wifi. Since the MN's WIMAX interface is still active, the MN MIP agent command the wimax interface to send redirect message to CN in order to inform the CN of the MN location. The MN will make sure that there's no other wifi network available by sending a "Probe Request". When the MIH agent will receive the confirmation that the CN has been notified of the MN new address and redirects the reception of the CBR traffic, the traffic will be using the link between the wimax interface and BS.

4.2   Analysing Performance of HO WiFi-Wimax

We evaluate as showing in Fig. 2 the rate of packet loss depending on the speed of the mobile node. The graph shows three different plots for different speed, so we can conclude that the speed affects the rate of lost packet. As the speed increases the graph shows that the number of packets received goes on increasing. In this simulation scenario we can see that at the first handoff from wimax to wifi (HDA) there is no packet loss detected because we follow the principle make before. But at handoff from wifi to wimax (HDB), we find that the destruction of the packets is due to the settling time for a new location where the mobile no longer receives packets from the old base station (Fig. 3). The second handoff begins after receiving link down that explains packet loss. In addition Fig. 3 shows that the time of L2 handoff latency

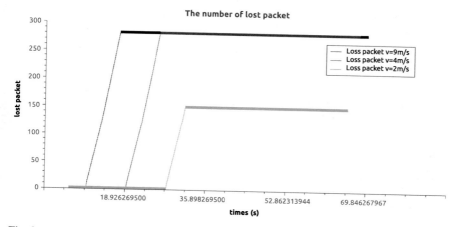

Fig. 2 Packet loss and mobility (WIFI-WIMAX)

Fig. 3 Packet loss and
execution of handoff
(WIFI-WIMAX)

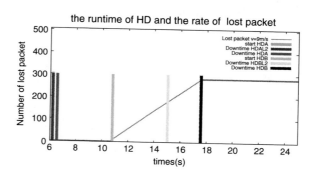

(Downtime HDBL2) is very considerable then the L3 handoff latency (Downtime HDB), which is the time used for renewing network layer settings. Therefore improving L2 performance will be very important for seamless handoff.

5   Conclusion

The simulation scenario of vertical handoff is carried out in ns2. The relation between node mobility and handoff with different speed is analysed. Obtained result has proven an acceptable simulation to what could be expected in real case scenarios. After evaluating the performances of the vertical handoff between the two wireless networks WiFi and WiMAX, we can notice that the high velocity affects the time of execution of handoff and as result the loss packet will be affected too.

References

1. McNair, J., Zhu, F.: Vertical handoffs in fourth-generation multinetwork environments. Wirel. Commun. **11**(3), 8–15 (2004)
2. Dai, Z., Fracchia, R., Gosteau, J., Pellati, P., Vivier, G.: Vertical handover criteria and algorithm in IEEE802. 11 and 802.16 hybrid networks. In: IEEE International Conference on Communications, ICC'08, pp. 2480–2484 (2008)
3. Zhao, Q.L.: A method for reducing movement detection delay based on dynamic region. J. Softw. **16**(6), 1168 (2005)
4. Zhang, Y., Zhuang, W., Saleh, A.: Vertical handoff between 802.11 and 802.16 wireless access networks. In: Global Telecommunications Conference (2008)
5. IEEE Standard for Local and Metropolitan Area Networks Part 16 Air Interface for Fixed and Mobile Broadband Wireless Access Systems Amendment 2: Physical and Medium Access Control Layers for Combined Fixed and Mobile Operation in Licensed Bands and Corri. [s.n.]
6. Stemm, M., Katz, R.H.: Vertical handoffs in wireless overlay networks. Mobile Netw. Appl. **3**(4), 335350 (1998)
7. Hong, J.W., Leon-Garcia, A.: Requirements for the operations and management for 4G networks. In: Proceedings of the 19th International Telegraffic Congress (ITC 19), p. 981990. Beijing, China (2005)
8. Lee, D., Han, Y., Hwang, J.: QoS-based vertical handoff decision algorithm in heterogeneous systems. In: 2006 IEEE 17th International Symposium on Personal, Indoor and Mobile Radio Communications, p. 15 (2006)
9. IEEE standard for local and metropolitan area networks—part 21: media independent handover. In: Techical Report (2009)
10. Gupta, V., Johnston, D.: A generalized model for link layer triggers. Submission IEEE 802 (2004)

New Design of RFID TAG Using the Non Uniform Transmission Lines

Mohamed Boussalem, Mohamed Hayouni, Tan-Hoa Vuong,
Fethi Choubani and Jacques David

Abstract The issue of our paper is to expose an original technique to improve the performance of RFID TAG in terms of its geometric dimensions and radiation pattern. This technique exploits the non uniform transmission lines (NUTLs) features to design the passive antenna element on the RFID TAG. We have yet demonstrated that geometric dimensions are reduced and the radiation pattern of the tag was improved by such use of NUTLs. Optimization of the profiles of the different elements of non uniform Lines achieves the appropriate radiation diagram and improves the radiation intensity, consequently.

Keywords Nutls · Hill equation · RFID antenna · Radiation intensity · Radiation pattern · Effective angle

1 Introduction

To improve the performance of microwave circuits and particularly the passive RFID tags suggested the use of non uniform transmission lines due to their appropriate frequency behaviour [1].

Their fundamental property allows the reduction of the geometric dimensions and improvement of the tag's radiation pattern [2, 3] .The analysis of such structures is achieved by a numerical calculation program based on the work of Hill, which consists of determining the general solution of the propagation distribution equation of the electric and magnetic fields and deducing the accurate model of the transmission line. Therefore, several non uniform transmission lines with various profiles (hyperbolic, linear and exponential) have been analyzed. Their contribution

M. Boussalem (✉) · T.-H. Vuong · J. David
Laboratoire Laplace—INP Toulouse, Rue Charles Camichel, 31071 Toulouse, France
e-mail: m.bousalem@gmail.com

M. Boussalem · M. Hayouni · F. Choubani
Innov'Com Lab., SUPCOM-Université de Carthage, Tunis, Tunisia

© Springer International Publishing Switzerland 2016
A. El Oualkadi et al. (eds.), *Proceedings of the Mediterranean Conference on Information & Communication Technologies 2015*, Lecture Notes in Electrical Engineering 380, DOI 10.1007/978-3-319-30301-7_56

to improve the pattern and reduce the dimension of passive antenna RFID TAG has been experimentally validated.

2 Analysis of NUTLs

The modelling of a non uniform structure passes by determining the explicit expression describing accurately its behaviour and deducing its scattering matrix S.

The first step is to derive the propagation equations adequately for the non uniform profile. Secondly, we perform mathematical manipulations to transform these equations in the form of a traditional Hill's equation without a first derivative term. Next, the obtained equation is solved using the Floquet theorem, to explicit the general and particular solutions, respectively. Finally, we use these solutions to calculate the different elements of the scattering matrix S of the lines. More investigations will be carried out to validate this analytic approach and by further developments and experimental prototypes. Results obtained of a simple, exponential and hyperbolic lines optimized to resonate at a fundamental frequency equal to 1 GHz (Figs. 1, 2 and 3) are showed in Table 1.

Actually, the non uniform transmission lines have a frequency behaviour which strictly depends upon their forms and their profiles of non homogeneity. While the transmission structures resonate in a regular multiple of fundamental frequencies, the non uniform lines resonate on frequencies which are different from integer multiples of fundamental.

3 Experimental Application of LNUT

Our methodology consists of replacing uniform structures present on the RFDI TAG and its passive antenna, by equivalent non uniform parts. Indeed, classical passive RFID TAGs in microwave are based on elements used as antennas [4]. Usually, these elements are designed like uniform transmission lines.

(a)   **(b)**   **(c)**

Fig. 1 a Simple $\lambda/4$ short-circuits line W1 = 2.988 mm, W2 = 2.88 mm and L = 3.14 cm substrate: epoxy 1.52 mm. **b** Non uniform $\lambda/4$ short-circuits line with linear and hyperbolic profile W1''' = 2.988 mm, W2''' = 2.88 mm and L''' = 4.73 cm Substrate: Epoxy 1.52 mm. **c** Non uniform $\lambda/4$ short-circuits line with exponential profile, W1' = 2.988 mm, W2' = 2.88 mm and L' = 4.73 cm substrate: epoxy 1.52 mm

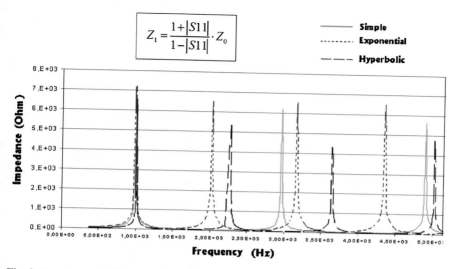

$$Z_1 = \frac{1+|S11|}{1-|S11|} \cdot Z_0$$

———— Simple
········ Exponential
— — — Hyperbolic

Fig. 2 Impedance of different NUTLs depending on the frequency

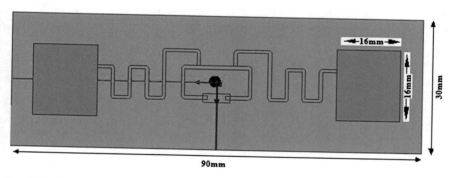

Fig. 3 Passive UHF RFID tag designed with simple lines

Table 1 Resonant frequency of simple and non uniform lines

	Fundamental frequency (GHz)	1st harmonic frequency (GHz)	2nd harmonic frequency (GHz)
Simple line	1	3	5
Exponential profile	0.95	2.14	3.43
Hyperbolic profile	1	2.3	3.6

The purpose of this work is to develop the same printed circuit structure for both types of RFID chips that operate on two separate frequencies, one at 2.45 GHz and the other at 5.8 GHz. With a Substrate: FR4 (Epoxy glass) and Tag Size: 30 × 90 mm.

Figures 4, 6 and 8 perfectly illustrates the value of using "NULTs" in this type of device. We expect RFID chips in our industrial partner to validate our concept and contribution of adjustments according to the results of measurement campaigns (Figs. 5 and 7).

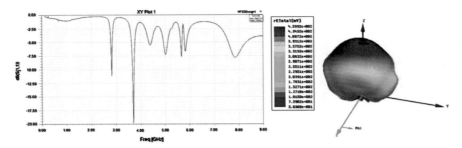

Fig. 4 Operating frequency and radiated pattern of RFID TAG designed with simple lines

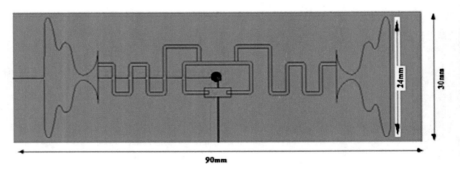

Fig. 5 Passive UHF RFID tag designed with non uniform line LNUT1

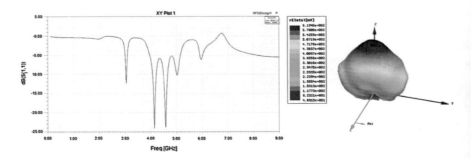

Fig. 6 Operating frequency and radiated pattern of RFID TAG designed with LNUT1

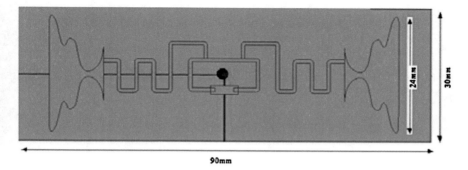

Fig. 7 Passive UHF RFID tag designed with non uniform line NUTL2

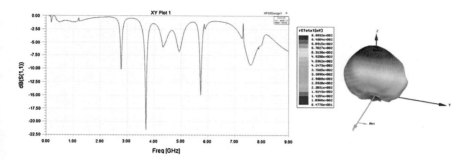

Fig. 8 Operating frequency and radiated pattern of RFID TAG designed with NUTL2

4 Conclusion

Analysis of NUTLs using Hill's equation is achieved using an efficient iterative method based on Fouquet' exponents determination. Once voltages and currents are defined over each point along the transmission structure, S parameters and other pertinent parameters s can be easily derived.

Te NUTLs have a frequency behaviour depending tightly on their geometric profiles. This fundamental property was used for reducing RFID tag dimension and improving its radiation pattern. Non uniform structures have exhibited attractive results and obtained results showed good agreement with experience.

References

1. Doan, T.N.H., Ghiotto, A., Guilloton, L., Vuong, T.P., Castelli, P.T.: Conception et realisation d'une Antenne pour un Lecteur RFID-PDA. OHD 2007, Valence, France, Sept 2007

2. Cho, C., Choo, H., Park, I.: Design of novel RFID tag antennas for metallic objects. In: IEEE International Symposium on Antennas and Propagation Digest, pp. 3245–3248 (2006)
3. Diugwu, H., Batchelor, J.C.: Analysis of the surface distribution in a dual band planar antenna for passive RFID tag. In: IEEE International Symposium on Antennas and Propagation Digest, pp. 459–462 (2005)
4. Nikitin, P.V., Rao, K.V., Lazar, S.: An overview of near field UHF RFID. In: IEEE International Conference on RFID, pp. 167–173 (2007)
5. Hassan, S.M.S., Sundaram, M., Kang, Y., Howlader, M.K.: Measurement of dielectric properties of materials using transmission/reflection method with material filled transmission lines. In: Proceeding of IEEE IMTC, vol. 1, pp. 72–77, 16, 19 May 2005

Plasmonic Analogue of Electromagnetically Induced Transparency in Detuned Nano-Cavities Coupled to a Waveguide

Adnane Noual, Ossama El Abouti, El Houssaine El Boudouti, Abdellatif Akjouj, Bahram Djafari-Rouhani and Yan Pennec

Abstract We theoretically investigate the classical analogue of electromagnetically induced transparency (EIT) in a plasmonic structure constituted by double side cavities connected symmetrically to a waveguide. The EIT is demonstrated by simply detuning the sizes of the two cavities (i.e., the length difference ΔL, keeping their width w similar). The physical mechanism behind the EIT resonance is unveiled as being caused by the destructive and constructive interference between the confined modes in the two cavities. The former play the role of two coupled radiative oscillators. The proposed structure may have important applications for designing integrated devices such as: narrow-frequency optical filters, novel sensors and high-speed switches.

Keywords Electromagnetic induced transparency · Surface plasmon-pariton · Cavity resonator

A. Noual (✉) · O.E. Abouti · E.H.E. Boudouti
Laboratoire de Physique, de la Matière et de Rayonnements, Département de Physique, Faculté des Sciences, Université Mohamed Premier, 60000 Oujda, Morocco
e-mail: noualad@yahoo.fr

O.E. Abouti
e-mail: ossama.elabouti@yahoo.com

E.H.E. Boudouti
e-mail: elboudouti@yahoo.fr

E.H.E. Boudouti · A. Akjouj · B. Djafari-Rouhani · Y. Pennec
Institut d'Electronique, de Microélectronique et de Nanotechnologie, UMR CNRS 8520, UFR de Physique, Université de Lille 1, 59655 Villeneuve d'Ascq, France
e-mail: abdellatif.akjouj@univ-lille1.fr

B. Djafari-Rouhani
e-mail: bahram.djafari-rouhani@univ-lille1.fr

Y. Pennec
e-mail: yan.pennec@univ-lille1.fr

© Springer International Publishing Switzerland 2016
A. El Oualkadi et al. (eds.), *Proceedings of the Mediterranean Conference on Information & Communication Technologies 2015*, Lecture Notes in Electrical Engineering 380, DOI 10.1007/978-3-319-30301-7_57

1 Introduction

Electromagnetically induced transparency (EIT) is a quantum interference phenomenon that renders an opaque medium transparent in a narrow spectral region with low absorption and steep dispersion [1, 2]. These properties have been exploited to show different applications in these systems such as: slow light, sensing and data storage [2, 3]. However, it was realized that EIT-like behaviors are not uniquely associated to quantum systems and can be extended to classical systems [2, 3]. In this context, several classical systems have been designed to demonstrate the classical analogue of EIT. Among these systems one can cite: plasmonic nanostructures [4–7], planar metamaterials [8, 9], photonic crystal waveguides coupled to cavities [10–12], coupled microresonators [13–16], micro and radiowave circuits [17–19] and acoustic slender tube waveguides [20, 21].

As concerns plasmonic systems [4–9], metamaterials made of split-rings and cut-wires as well as plasmonic waveguides with coupled cavities have shown EIT and Fano [22] resonances with high quality factors. Due to their deep subwavelength confinement of light at the metal-dielectric interface, plasmonic materials have been suggested as an alternative to overcome the classical diffraction limit and manipulation of light in nanoscale domain [23]. The EIT resonance can be obtained using two coupled resonators (or oscillators) with closely spaced frequencies. The resonators with high (dark resonator) and low (radiative resonator) Q factors give rise to the so-called Λ-type configuration [1] in three atomic levels, whereas the resonators with low Q factors (radiative resonators) give rise to the so-called V-type configuration [4, 6, 18–21]. The latter mechanism has been demonstrated in little works in comparison with the first one.

In few recent works, the V-type resonances in double stub resonators connected at the same site along a waveguide have been shown in plasmonic [4, 6], photonic [18, 19] and acoustic [20, 21] materials. The case of two plasmonic cavities interacting with a waveguide by means of evanescent waves (near field mechanism) through a metal gap, has been shown by Zhang et al. [24]. They have studied numerically light propagation in a metal-air-metal waveguide and two side-coupled cavities located at a symmetric position around the waveguide. The two cavities are characterized by the same size but with different dielectric permittivities. They have shown the possibility of the existence of plasmonic EIT resonance that can be detuned by varying the dielectric permittivities in the two cavities. This study has been implemented by calculating the transmission amplitude through the system using the finite difference time domain (FDTD) method. In this work, we studied numerically the plasmonic analogue of EIT in a two side-coupled cavities to a waveguide. The optical properties of the modeled structure such as transmission, reflection and absorption spectra are obtained using finite-element method (Comsol Multiphysics Package) [25]. The waveguide and the cavities are embedded in a metal and filled with air. Also, we consider that both cavities have the same width (w = 100 nm), whereas their lengths L_1 and L_2 can be detuned. We show the

Fig. 1 Structure of the plasmonic nanoscale resonator system

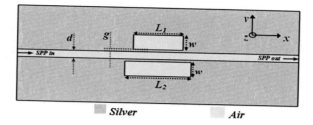

possibility of existence of EIT resonance by detuning the size of the two cavities (Fig. 1). Their separation from the waveguide is referred to as g (metallic gap that enables evanescent coupling).

2 Numerical Results

In our model the dielectric function of the metal (silver) is described by the lossy Drude model whose parameters are: $\varepsilon_\infty = 12.5$, $\omega_p = 2.05 \times 10^{16}$ rad/s and $\Gamma = 10^{14}$ rad/s [26]. The waveguide width, d, is equal to 50 nm, and the incident plane wave is a TM polarized one. In order to show the possibility of existence of EIT-resonance, i.e. a resonance squeezed between two transmission zeros, we have to take L_1 and L_2 slightly different. An example corresponding to this situation is given in Fig. 2a (pink curves) for $L_1 = 240$ nm and $L_2 = 258$ nm (i.e., $\Delta L = L_2 - L_1 = 18$ nm). One can notice the existence of deep dips in the transmission spectra around $\lambda_1 = 654$ nm and $\lambda_2 = 693$ nm and a resonance peak (transmission window) is induced at $\lambda_r = 672$ nm between the dips. The resonance does not reach unity because of the absorption in the system (see Fig. 2c), the reflection being very weak (Fig. 2b). As a matter of comparison, we have also plotted the transmission, reflection and absorption for each resonator alone (black and red curves).

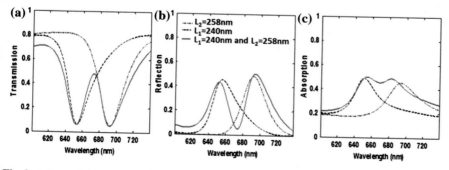

Fig. 2 **a** Transmission spectra when a single cavity is coupled to the waveguide with $L_1 = 240$ nm (*black curve*) and $L_2 = 258$ nm (*red curve*). The *pink curve* corresponds to the case when both cavities are present in the system ($L_1 = 240$ nm and $L_2 = 258$ nm). **b** The same as in (**a**) but for the reflection. **c** The same as in (**a**) but for the absorption

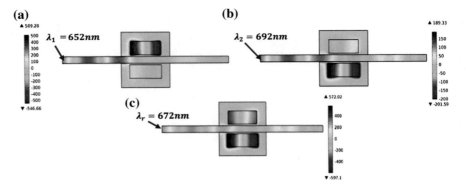

Fig. 3 **a**, **b** Magnetic field map at transmission dips around $\lambda_1 = 652$ nm and $\lambda_2 = 692$ nm respectively. **c** The same as in (**a**) and (**b**) but for the resonance peak around $\lambda_r = 672$ nm

One can notice that the transmission zeros are induced by each cavity [27], whereas the resonance is a consequence of the constructive interference between cavities. In order to show the spatial localization of the different modes in Fig. 2, we have plotted in Fig. 3 the amplitude of the magnetic field map at the two dips around $\lambda_1 = 654$ nm (Fig. 3a) and $\lambda_2 = 693$ nm (Fig. 3b). As predicted, the magnetic fields are mainly confined in each cavity and do not propagate in the system. These results are in accordance with those in Fig. 2a where the trapping and the rejection of the incident light wave is induced by the two cavities [27]. Around the resonance at $\lambda_r = 672$ nm (Fig. 3c), one can notice that both cavities are excited and the wave is transmitted along the waveguide as mentioned above. Therefore, the EIT resonance is a consequence of the constructive interference between the waves in the two cavities. However, the field is more localized in the lower cavity compared with upper one, which indicates that the latter is a bit less excited than the former. This is actually related to the asymmetry of the transmission spectra in Fig. 2a (pink curve) where the resonance peak wavelength is closer to the lower cavity eigenmode resonance wavelength than to the upper cavity. On the other hand, the field map shows that the excited cavities modes oscillate out of phase; this effect has been also observed in microwave photonic circuits [17–19].

3 Conclusion

In this paper, we have demonstrated numerically the possibility of existence of plasmonic analogue of EIT in metal-air-metal waveguide coupled to two nano-cavities filled with air but with different sizes. The waveguide and the cavities are embedded in a metal and the interaction between the incident light wave and the cavities occurs by means of evanescent waves through a small metal gap. We have shown (not presented here) that the behavior of the EIT resonance in the transmission, reflection and absorption can be detuned by means of the difference in the size of the two cavities (depending of the value of the metallic gap g). These results may have important applications for designing integrated devices such as: narrow-frequency optical filters, novel sensors and high-speed switches.

Acknowledgments One of the authors (A.N.) acknowledges the use of the IRIDIS High Performance Computing Facility, and associated support services at the University of Southampton (UK), in the completion of this work.

References

1. Fleischhauer, M., Imamoglu, A., Marangos, J.P.: Electromagnetically induced transparency: optics in coherent media. Rev. Mod. Phys. **77**, 633 (2005)
2. Harris, S.E.: Electromagnetically induced transparency. Phys. Today **50**, 36 (1997)
3. Liu, C., Dutton, Z., Behroozi, C.H., Hau, L.: Observation of coherent optical information storage in an atomic medium using halted light pulses. Nature **409**, 490 (2001)
4. Piao, X., Yu, S., Park, N.: Control of Fano asymmetry in plasmon induced transparency and its application to plasmonic waveguide modulator. Opt. Express **20**, 18994 (2012)
5. Liu, N., Langguth, L., Weiss, T., Kastel, J., Fleischhauer, M., Pfau, T., Giessen, H.: Plasmonic analogue of electromagnetically induced transparency at the Drude damping limit. Nat. Mater. **8**, 758 (2009)
6. Han, Z., Bozhevolnyi, S.I.: Plasmon-induced transparency with detuned ultracompact Fabry-Perot resonators in integrated plasmonic devices. Opt. Express **19**, 3251 (2011)
7. Zhang, S., Genov, D, A., Wang, Y., Liu, M., Zhang, X.: Plasmon-induced transparency in metamaterials. Phys. Rev. Lett. **101**, 047401 (2008)
8. Wu, J., Jin, B., Wan, J., Liang, L., Zhang, Y., Jia, T., Cao, C., Kang, L., Xu, W., Chen, J., Wu, P.: Superconducting terahertz metamaterials mimicking electromagnetically induced transparency. Appl. Phys. Lett. **99**, 161113 (2011)
9. Singh, R., Al-Naib, I.A., Yang, Y., Chowdhury, D.R., Cao, W., Rockstuhl, C., Ozaki, T., Morandotti, R., Zhang, W.: Observing metamaterial induced transparency in individual Fano resonators with broken symmetry. Appl. Phys. Lett. **99**, 201107 (2011)
10. Fan, S., Joannopoulos, J.D.: Analysis of guided resonances in photonic crystal slabs. Phys. Rev. B **65**, 235112 (2002)
11. Yang, X., Yu, M., Kwong, D.-L., Wong, C.W.: All-optical analog to electromagnetically induced transparency in multiple coupled photonic crystal cavities. Phys. Rev. Lett. **102**, 173902 (2009)
12. Sato, Y., Tanaka, Y., Upham, J., Takahashi, Y., Asano, T., Noda, S.: Strong coupling between distant photonic nanocavities and its dynamic control. Nat. Photon. **6**, 56 (2012)
13. Maleki, L., Matsko, A.B., Savchenkov, A.A., Ilchenko, V.S.: Tunable delay line with interacting whispering-gallery-mode resonators. Opt. Lett. **29**, 626 (2004)
14. Totsuka, K., Kobayashi, N., Tomita, M.: Slow light in coupled-resonator-induced transparency. Phys. Rev. Lett. **98**, 213904 (2007)
15. Raymond Ooi, C.H., Kam, C.H.: Controlling quantum resonances in photonic crystals and thin films with electromagnetically induced transparency. Phys. Rev. B **81**, 195119 (2010)
16. Ding, W., Lu´kyanchuk, B., Qiu, C.-W.: Ultrahigh-contrast-ratio silicon Fano diode. Phys. Rev. A **85**, 025806 (2012)
17. Tassin, P., Zhang, L., Zhao, R., Jain, A., Koschny, T., Soukoulis, C.M.: Electromagnetically induced transparency and absorption in metamaterials: the radiating two-oscillator model and its experimental confirmation. Phys. Rev. Lett. **109**, 187401 (2012)

18. Mouadili, A., El Boudouti, E.H., Soltani, A., Talbi, A., Akjouj, A., Djafari-Rouhani, B.: Theoretical and experimental evidence of Fano-like resonances in simple monomode photonic circuits. J. Appl. Phys. **113**, 164101 (2013)
19. Mouadili, A., El Boudouti, E.H., Soltani, A., Talbi, A., Djafari-Rouhani, B., Akjouj, A., Haddadi, K.: Electromagnetically induced absorption in detuned stub waveguides: a simple analytical and experimental model. J. Phys. Condens. Matter. **26**, 505901 (2014)
20. El Boudouti, E.H., Mrabti, T., Al-Wahsh, H., Djafari-Rouhani, B., Akjouj, A., Dobrzynski, L.: Transmission gaps and Fano resonances in an acoustic waveguide: analytical model. J. Phys. Condens. Matter. **20**, 255212 (2008)
21. Tan, W., Yang, C.Z., Liu, H.S., Wang, Z.G., Lin, H.Q., Chen, H.: Manipulating classical waves with an analogue of quantum interference in a V-type atom. Europhys. Lett. **97**, 24003 (2012)
22. Fano, U.: Effects of configuration interaction on intensities and phase shifts. Phys. Rev. **124**, 1866 (1961)
23. Akjouj, A., Lévêque, G., Szunerits, S., Pennec, Y., Djafari-Rouhani, B., Boukherroub, R., Dobrzyński, L.: Nanometal plasmonpolaritons. Surf. Sci. Rep. **68**, 1–67 (2013)
24. Zhang, Z., Zhang, L., Yin, P., Han, X.: Coupled resonator induced transparency in surface plasmon polariton gap waveguide with two side-coupled cavities. Phys. B **446**, 55 (2014)
25. Dong, M., Tomes, M., Eichenfield, M., Jarrahi, M., Carmon, T.: Characterization of a 3D photonic crystal structure using port and S-parameter analysis. In: Proceeding of Comsol Conference in Boston (2013)
26. Palik, E.D.: Handbook of Optical Constants of Solids. Academic Press, New York (1985)
27. Noual, A., Pennec, Y., Akjouj, A., Djafari-Rouhani, B., Dobrzynski, L.: Nanoscale plasmon waveguide including cavity resonator. J. Phys.: Condens. Matter **21**, 375301 (2009)

Baseband/RF Co-simulation of IEEE 802.15.3C in Outdoor Environment at 60 GHz

Tarik Zarrouk, Moussa El Yahyaoui and Ali El Moussati

Abstract This paper presents propagation of millimeter wave in order to determine the evolution of signal in outdoor environment, we design a wireless transceiver and we evaluate the performance of the High Speed Interface Physical Layer (HSI PHY) of IEEE 802.15.3c standard for investigate the coverage performance of the 60 GHz WLAN. Co-simulation techniques between heterogeneous environment have been used, Advanced Design System (ADS2011) for radio frequency and Matlab/simulink for baseband signal; we have compared the performance of three modulations schemes (64QAM, 16QAM and QPSK) by measuring the Error Vector Magnitude (EVM).

Keywords Co-simulation · HSI PHY mode · IEEE 802.15.3c · 60 GHz

1 Introduction

The need for high data rate in radio communication application leads to think about a new technology. The 60 GHz frequency band has been identified to give answer to this need. The interest major of this band is the huge unlicensed bandwidth, its available in many countries which represent a great potential in term of capacity, efficiency. Whose these performances attract a several standardization [1, 2]. Among these standards, we find the IEEE 802.15.3C, ECMA387 and alliance WiGig.

The main objective of this paper is to simulate the HSI PHY mode of IEEE 802.15.3c standard in LOS environment, by using different modulation schemes (QPSK, 16-QAM and 64-QAM). At first we have developed a co-simulation techniques for simulate and evaluate the performance of this system whatever the frequency range, we have used Matlab/Simulink to simulate the digital signal

T. Zarrouk · M.E. Yahyaoui · A.E. Moussati (✉)
Signals, Systems and Information Processing, National School of Applied Sciences, Oujda, Morocco
e-mail: aelmoussati@ensa.ump.ma

© Springer International Publishing Switzerland 2016
A. El Oualkadi et al. (eds.), *Proceedings of the Mediterranean Conference on Information & Communication Technologies 2015*, Lecture Notes in Electrical Engineering 380, DOI 10.1007/978-3-319-30301-7_58

545

processing part and ADS for the Radio Frequency (RF) one, combining these two environments was very important to properly simulate this mode.

Finally we have calculated the EVM for different modulation schemes according to distance and evaluate the feasibility of HSI PHY mode in outdoor environment of 60 GHz WLAN system and finding the impact of propagation environment using different modulation schemes (QPSK, 16-QAM and 64QAM).

In this paper, we have considered architecture Radio over Fiber mainly the electrical communication part, we have described the complete co-simulation architecture. Before concluding we have discussed all simulations and results.

2 Co-simulation Architecture

For investigating the performance of HSI PHY mode, a Matlab/Simulink-ADS co-simulation system is developed. To benefit from the easy generation and processing baseband digital I and Q signals of any type of modulation using Matlab/Simulink, and to exploit the wide range of simulation types using ADS. The general functional diagrams for the described layouts are presented in Fig. 1.

As mentioned above, the recent work is based on the IEEE 802.15.3c standard, which is the first IEEE wireless standard for data rates over 1 Gb/s, according to the report of TG3c [3], three PHYs for the mm-Wave PHY are defined; namely single carrier (SC), high speed interface (HSI) orthogonal frequency division multiplexing (OFDM) and audio video (AV). We focus our work on second mode, because it's designed for high speed bidirectional data transmission and uses OFDM techniques. The parameters of different modulation and coding schemes are given in Table 1.

We use the Matlab/Simulink environment to modeling the HSI PHY mode. Figure 2 shows the different blocks used in this mode. Random data bits are generated and coded by two LDPC encoders, the bit sequences are multiplexed, then interleaved by a block interleaver and inserted into the constellation mapper,

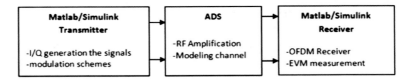

Fig. 1 General functional diagrams for the Matlab-ADS co-simulation system

Table 1 Modulation and coding schemes parameters

MCS index	Data rate (Mb/s)	Modulation scheme	FEC rate
1	1540	QPSK	1/2
5	4620	16-QAM	3/4
7	5775	64-QAM	5/8

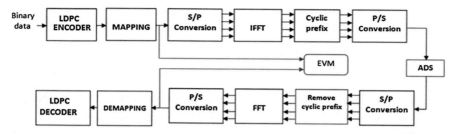

Fig. 2 HSI PHY mode of the IEEE 802.15.3c standard

the symbols generated are taken as input for tone interleaver, the IFFT block transforms these symbols in time domain (block OFDM transmitter), finally the I and Q OFDM modulation baseband generated from Simulink are simulated by ADS Agilent. The RF section up-converts the baseband signal to 60.48 GHz and amplifies it. Once the signal RF is received by the antenna, we apply a proper filtering and down-convert it.

The inverse operation is done to obtain the OFDM baseband signal, the receiver starts the processing of accumulated waveform data: the FFT Block transforms the symbols in frequency domain (OFDM receiver), then the bits are inserted into the constellation demapper for extracting the bitstream from the received complex stream, and passed into deinterleaver for synchronization with the bit interleaver of transmitter. Finally, we decoded by a LDPC decoder.

Concerning transmission in radio frequency, the microwave subjected to a lot of attenuations in free space by rain or atmospheric absorption, it can reach 16 dB/km at 60 GHz for atmospheric absorption and reach 16 dB for 25 mm/h for rain attenuation, these attenuations decrease the quality of service (QoS), why it was necessary to take into consideration these influences.

Figure 3 presents the different components used for transmit the signal in RF. The model RF realized in ADS is shown in Fig. 3, the I and Q OFDM components are generated from Simulink model, amplified and shifted to 6 GHz by intermediate frequency, the transmission power delivered is 27 dBm [2], after we up convert the signal to 60 GHz using a quadrature injection locked oscillator, then the signal

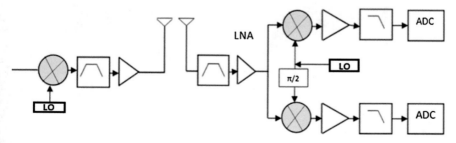

Fig. 3 RF architecture at 60 GHz for LoS environment

passes through a channel modeled for LoS environment, we only consider the atmospheric attenuation, Eq. (1) present the path loss is:

$$PL(d)[dB] = PL(d0)[dB] + 10 * n * \log10(d/d0) + \Sigma Xq \qquad (1)$$

PL (d0): is the path loss at reference distance, 10n * Log10 (d/d0) is the path loss at relative distance, d0 is the reference distance, n refers the path loss exponent and Xq is the additional attenuation due to specific obstruction by object.

At reception, the signal captured by antenna at 60 GHz, a direct conversion receiver made for down convert the signal, it is filtered by band selection filter and the low noise amplifier (LNA) amplified the signal, then a quadrature mixer to a local oscillator having a frequency 60 GHz extracted the bit stream I and Q; finally the I and Q are filtered by pass band filter.

3 Simulation and Results

We have using a co-simulation techniques composed of Matlab/Simulink for baseband and ADS for radio link. Firstly, we validate the transmit power (Fig. 4a), Fig. 4b shows the received spectrum in outdoor environment. We analyses the quality of received signal, by simulation the EVM in different modulation schemes, at first we consider the modulation QPSK, 16-QAM and 64-QAM, Table 1 presents the modulation and coding scheme (MCS) used and their parameters:

Figure 5 presents the variation of EVM measurement varying the wireless link for different modulation schemes.

Analyzing the graph, an EVM = 20 % and a distance lower than 100 m we notice that is preferable to use the modulation schemes 64QAM, cause of high throughput compared to other modulations and the modulation QPSK supports attenuations more than other modulation, the EVM reach 20 % for a distance 200 m.

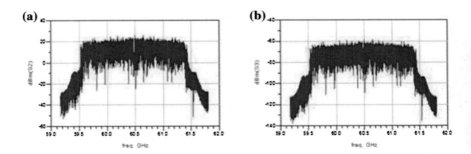

Fig. 4 Transmit and received spectrum in outdoor environment (LoS)

Fig. 5 EVM measurement
according to distance link

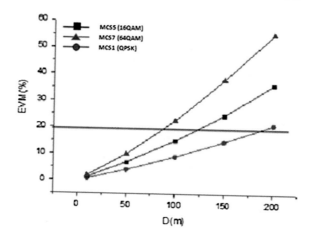

As a perspective of this work, we can be implemented an algorithm which allows adaptation in terms of modulation and coding of the transmitter based on the quality of the transmission link [4].

4 Conclusion

In this paper we have evaluate the performance of PHY HSI mode of IEEE 802.15.3C at 60 GHz, using two heterogeneous environment using two software, design the complete architecture and simulation the EVM of different modulation schemes to predict the modulation schemes appropriate for a defined distance.

References

1. Smulders, P., Yang, H., Akkermans, I.: On the design of low-cost 60 GHz radios for multigigabit-per-second transmission over short distances. IEEE Commun. Magazine (2007)
2. Collonge, S., Zaharia, G., El Zein, G.: Wideband and dynamic characterization of the 60 GHz indoor radio propagation-future home WLAN architectures. Ann. Telecommun. **58**(3, 4), 417–447 (2003)
3. IEEE 802.15.3C Wireless medium access control (MAC) and physical layer (PHY) specifications for high rate wireless personal area networks (WPANs), amendment 2: millimeter-wave-based alternative physical layer extension. IEEE STD 802.15.3c-2009
4. Azza, M.A., El Moussati, A., Mekaoui, S., Ghoumid, K.: Spectral management for a cognitive radio application with adaptive modulation and coding. Int. J. Microwave Opt. Technol. **9**(6) (2014)

Extensive Simulation Performance Analysis of Variable Transmission Power on Routing Protocols in Mobile Sensor Networks

Mehdi Bouallegue, Ridha Bouallegue and Kosai Raoof

Abstract A wireless mobile sensor network is a group of independent wireless mobile sensor nodes which forms a temporary network without the use of any centralized management or fixed infrastructure. Communication protocols are responsible for maintaining the routes in the network and guarantee reliable communication. On the other hand, appropriately adjusting the sensors transmission power is crucial for reducing network energy consumption. This paper proposes a comparison of routing strategies and the impact of variable transmission power for each mobile sensor node on the performance of these communication techniques for mobile wireless sensor networks with the aim of outlining design considerations of protocols for mobile environments. We analyze the performance of both reactive routing protocols Ad hoc On Demand Distance Vector protocol (AODV), Dynamic Source Routing (DSR) protocol and proactive protocol Destination-Sequenced Distance Vector routing protocol (DSDV) in different scenarios. The selected protocols are compared on the basis of various parameters, which include packet delivery ratio, total packet loss, network lifetime, and control overhead using variable number of nodes and speeds.

Keywords Mobile node · Wireless sensor networks · AODV · DSDV · DSR · Transmission power · NS2

M. Bouallegue (✉) · R. Bouallegue
System of Communication Sys'Com, ENIT, Tunis, Tunisia
e-mail: mehdi.bouallegue.etu@univ-lemans.fr

R. Bouallegue
e-mail: ridha.bouallegue@ieee.org

M. Bouallegue · K. Raoof
Laboratory of Acoustics at University of Maine, LAUM UMR CNRS n°, 6613 Le Mans, France
e-mail: kosai.raoof@univ-lemans.fr

© Springer International Publishing Switzerland 2016
A. El Oualkadi et al. (eds.), *Proceedings of the Mediterranean Conference on Information & Communication Technologies 2015*, Lecture Notes in Electrical Engineering 380, DOI 10.1007/978-3-319-30301-7_59

1 Introduction

A wireless sensor network is composed of a large number of node sensors. Nodes mobility is the main added value of WSNs due to their locomotion capability in addition to their ability to collecting data, computation and communication. These networks are characterized by dynamic topology. Indeed, mobile node can joint or leave the network as well as being able to change of their transmission. The information is transmitted hop by hop through the network to a collector node. On the other hand, sensor nodes are likely to operate on limited battery life, so power conservation is a crucial issue. Due to their small size, these sensors can be widely used in structural health monitoring, environmental protection or military support, as well as in many other applications. Sensor nodes consist of processing capability, containing a transceiver block, memory and a power source. Due to the finite power available to each wireless node, increasing the network lifetime has been of a great interest to developers. On the other hand, routing protocols in wireless sensor networks has also attracted a lot of attention in the recent years. Therefore, different routing techniques have been developed for wireless sensor networks and each one has its own unique characteristics. The authors of [1] compared the performance of three protocols AODV, DSR and DSDV based on PDR, end-to-end delay and throughput metrics. The simulations are performed under various situations (when packet size changes and when time interval between packet sending changes). This work concludes that AODV and DSR protocols perform better at less packet size. Performance analyses of three communication protocols are analyzed and compared in [2], under high mobility case and in high density scenario. The study concludes that AODV protocol is a viable choice for MANETs.

In this paper, we continue in same trend and we concentrate on evaluating the performance of AODV, DSR and DSDV in MANET environment with varying the transmission range and density. We investigate the impact of variable transmission power for mobile nodes on different communication protocols considered. Then, we analyze the effect on the total energy consumption of the network. Protocols were examined based on throughput, energy consumption, packet delivery fraction, end-to-end delay and packet lost.

2 Overview of Routing Protocol

A routing technique [3] in WSN presents many challenges compared to data routing in wired network. These protocols are classified according to many parameters and to the strategies of discovering and maintaining routes. Protocols can be classified as reactive, proactive and hybrid, depending on their operation and type of requests. Proactive protocols control peer connectivity to ensure the availability of any path between the active nodes. On the other hand, reactive protocols establish paths only on request. Meanwhile, the sensors are inactive in terms of routing behavior [4].

2.1 DSDV Routing Protocol

Destination Sequenced Distance Vector (DSDV) [5] is a hop-to-hop distance vector routing protocol. It is characterized by each host maintaining a table consisting of the next-hop neighbor and the distance to the destination in terms of number of hops. In order to obtain the optimal path, the protocol DSDV guarantees loop free routes to each destination node, this is based on an average settling delay, which is a delay before advertising a route. All the hosts periodically broadcast their tables to their neighboring nodes in order to maintain an updated view of the network.

2.2 DSR Routing Protocol

The DSR protocol [6] is a reactive protocol that aims to limit the bandwidth consumed by packet routing in wireless ad hoc wireless networks. Dynamic source routing protocol is based on the concept of a routing algorithm from the source node to discover routes. This means that every node needs only forward the packet to its next hop specified in the header and need not check its routing table as in a table-driven algorithm. Determining source routes requires accumulating the address of each device between the source and destination during the route discovery.

2.3 AODV Routing Protocol

The ad hoc on demand distance vector is an on demand algorithm [7, 8], meaning that it builds routes between nodes only as desired by source nodes. It maintains these routes as long as they are needed by the sources. AODV uses sequence numbers to ensure the freshness of routes. This routing protocol builds routes using a route request on a route reply query cycle. AODV uses a reactive approach for finding routes and a proactive approach for identifying the most recent path. This protocol uses the same route discovery process to DSR protocol for finding fresh routes [1, 4].

3 Performance Metrics

3.1 Energy Consumption

The energetic consumption is the average of the total energy consumption of the entire network to transmit data packets. We obtain the energy consumption by calculating the ratio of the sum of the total energy consumed by each node to the total number of nodes. So a protocol that uses less energy during the simulation is considered more effective [9].

3.2 Packet Delivery Fraction (PDF)

The packet delivery fraction represents the number of arriving data packets successfully delivered over the total number of packets from all sources on the network. Using this value as analysis of ad hoc network on the different parameters involves the accuracy and completeness of the routing technique.

$$PDF = \frac{\sum Number\_of\_Packets\_received}{\sum Number\_of\_Packets\_sent} \tag{1}$$

3.3 End-to-End Delay (EED)

The parameter end to end delay is the average time taken by a data packet from a source node to arrive at a destination node. It also contains the delay caused by the route discovery process, the queue in data packet transmission and retransmissions times at the MAC layer.

$$EED = \frac{\sum (receiveTime - SendTime)}{\sum Number\_of\_Packets\_receive} \tag{2}$$

3.4 Throughput

This value represents the ratio of the total number of data packets provided to the total duration of simulation time. This metric measures how the network can continuously provide data to the sink.

$$Thrgh = \frac{\sum Number\_of\_Bit\_received}{Simulation\_Time} \tag{3}$$

3.5 Packet Lost

It represents the total number of data packets dropped during the simulation. The loss of a packet may be due to a collision during transmission process.

$$Pkt\_Lost = Nb\_Packet\_send - Nb\_Packet\_received \tag{4}$$

4 Mobility Model

Several mobility models (MM) can be considered to simulate the movement of mobile sensors in WSN e.g. Manhattan model, Random Way Point model, Gauss Markov mobility model. Broadly, the Random Way Point mobility approach is used to model the node movement in the NS-2. This model is a variation of Random walk model with spatial dependence. It includes pause times between changes in direction and/or speed. When the pause time expires, the node chooses a random destination in the simulation field with some metric such as pause time between T_{min} and T_{max}, speed value between 0 and Sp_{max}. The values of these parameters are uniformly distributed [10].

5 Simulation and Comparative Results

The simulations were done using network simulator NS-2 version 2.34 under Linux environment [11]. In realistic scenarios, all mobile sensor nodes can send information to a chosen destination for this reason; we assume in our simulation that we have a single sink. All other nodes are considered as sources (Table 1).

In Table 2, we showed the correspondence between the transmission power and the transmission range.

In this study, we consider a sensor network based on 50 mobiles nodes randomly placed in a 1000 m × 1000 m field as represented in Fig. 1, using the DSR routing protocol. The packet size is set to 512 bytes. The initial battery energy level of each sensor is 30 J.

5.1 Energy Consumption

For the energy consumption, we note that the more transmission range increases the more the energy consumption is important. We also see that for the envisaged routing techniques, total energy consumption increases with the number of sensor nodes. On the other hand, simulation results show that AODV and DSDV protocols permits better energy consumption compared to DSR protocol (Fig. 2).

Table 1 Simulation parameters

Parameters	Values
Routing protocols	AODV, DSR, DSDV
Number of nodes deployed	50, 70 and 90
Environment size	1000 m x 1000 m
Nodes placement strategy	Random
Transmission range	100, 200, 300 m and variable
Initial node energy	30 J
Rx power	0.1 mw
Idle power	0.05 mw
Sleep power	0.03 mw
Simulation time	150 s
Node speed (m/s)	[1.0, 3.0]
Antenna model	Omni antenna
Propagation model	Two ray ground
Transport protocol	TCP

Table 2 Transmission power needed for each distance

Distance (m)	Transmission power required
100	0.0072
200	0.115
300	0.584
Variable	In [0.0072, 0.584]

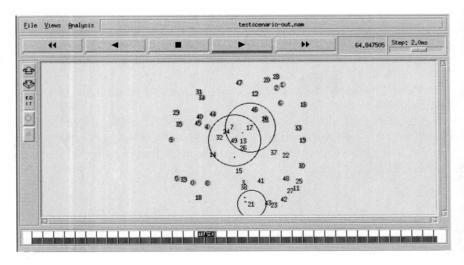

Fig. 1 Example of wireless sensor network with NS2

Fig. 2 Total energy consumption as a function of transmission range

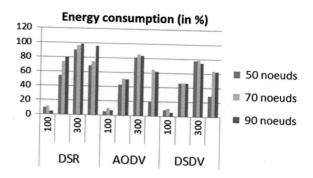

5.2 Packet Delivery Ratio

It is clear that the greater value of PDF means the better performance of the routing technique. Therefore, we deduce from Fig. 3 that for short transmission range (100 m) the packet delivery fraction is better than 200 and 300 m transmission range. Compared to the 200 and 300 m communication range, the variable transmission range has good results.

5.3 End-to-End Delay

The following graph shows the impact of different transmission range on the end-to-end delay metric with DSR, AODV and DSDV routing strategies. The variable transmission range remains low at almost all communication ranges in different nodes scenarios (Fig. 4).

5.4 Throughput

The throughput by the three routing techniques with 100, 200, 300 m and variable communication ranges in 50 nodes, 70 nodes and 90 nodes scenario.

Fig. 3 Packet delivery fraction as a function of transmission range

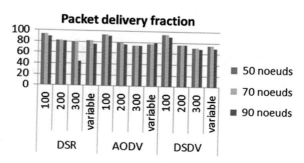

Fig. 4 End-to-end delay
versus transmission range

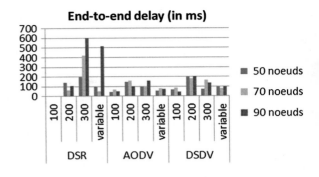

Fig. 5 Average throughput
as a function of sensor nodes
number and transmission
range

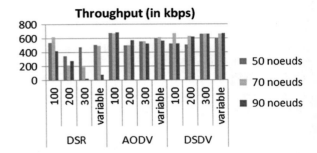

All routing protocols have good results for short range (100 m) and also for a
variable communication range. In addition, we note that for the DSDV technique, it
is even preferable to use a variable transmission range that provides better
throughput compared to the other transmission ranges used (Fig. 5).

5.5 Packet Lost

The performance of the three routing strategy in terms of total packet lost during the
simulation time. It is clear that the DSDV is better in terms of total packet lost for

Fig. 6 Total number of
packets lost during simulation

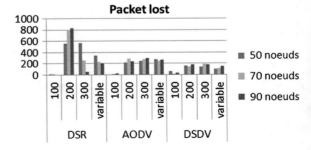

all different transmission range scenarios. In the case of variable transmission range, the number of packets lost is lower than the 200 m and 300 m communication range for all routing protocols used (Fig. 6).

6 Conclusion

The communication range of the sensor node is a very important metric which can influence network connectivity. This paper has offered a comprehensive analysis of extensive simulation analyses of DSDV, DSR and AODV MANET routing protocols under various traffic scenarios when the size of the network, the transmission range and node mobility are varied. By comparing these routing mechanisms on the basis of various performance metrics, we conclude that the variable transmission range offers a good performance for all metrics and is more preferable especially for high communication range. The results also disclose that AODV routing technique becomes more effective in providing better performance when the studied metrics are simulated.

References

1. Tuteja, A., Gujral, R.: Comparative performance analysis of DSDV, AODV and DSR routing protocols in MANET using NS2. In: International Conference on Advances in Computer Engineering, June 2010
2. Rajeshkumar, V., Sivakumar, P.: Comparative study of AODV, DSDV and DSR routing protocols in MANET using network simulator-2. Int. J. Adv. Res. Comput. Commun. Eng. 2(12) (2013)
3. Al-karaki, J.N., Kamal, A.E.: Routing techniques in wireless sensor networks: a survey. IEEE Wirel. Commun. 11, 6–28 (2004)
4. Guo, F., Wu, M., Liao, W., Wang, D.: Channel quality based routing protocol (CQBR) and realization on MANET platform. In: International Conference on Telecommunications (ICT 2014), Portugal, May 2014
5. Perkins, C., Bhagwat, P.: Highly dynamic destination-sequenced distance-vector (DSDV) for mobile computers. In: Proceedings of the ACM SIGCOMM, Oct 1994
6. Almutairi, A., Hendawy, T.: Performance comparison of dynamic source routing in ad-hoc networks. In: IEEE-GCC Conference, Dubai, 19–22 Feb 2011
7. El-azhari, M.S., Al-amoudi, O.A., Woodward, M., Awan, I.: Performance analysis in aodv based protocols for manets. In: Proceedings of the 2009 International Conference on Advanced Information Net-working and Applications Workshops, WAINA '09, pp. 187–192. IEEE Computer Society, Washington, DC, USA (2009)
8. Chen, C., Ma, J.: Simulation study of aodv performance over ieee 802.15.4 mac in wsn with mobile sinks. In: AINA Workshops(2), pp. 159–164 (2007)
9. Arefin, M., Tawhiddul, M., Toyoda, I.: Performance analysis of mobile ad-hoc networks routing protcols. In: International Conference on Informatics and Vision IEEE, pp. 535–539 (2012)

10. Martyna, J.: Simulation study of the mobility models for the wireless mobile ad hoc and sensor networks. Commun. Comput. Inf. Sci. **291**, 324–333 (2012)
11. Bouallegue, M., Raoof, K., Ben Zid, M., Bouallegue, R.: Impact of variable transmission power on routing protocols in wireless sensor networks. In: 10th International Conference on Wireless Communications, Networking and Mobile Computing (Wicom'14), Beijing, China, Sept 2014

Medium Access Control (MAC) Protocols for Wireless Sensor Network: An Energy Aware Survey

Rajoua Anane, Ridha Bouallegue and Kosai Raoof

Abstract Wireless Sensor Network (WSN) is a collection of small electronic nodes capable to sense and collect the information of interest in different application. However, these devices are limited in power supply, that is why energy efficient in a crucial factor of wireless sensor nodes. Many researches were performed in order to minimize the energy consumption in different layers of the network and especially at the Medium Access Control (MAC) layer as it coordinate all the sensor nodes to the share the wireless medium. Therefore, a well-designed MAC protocol can prolong the network life. In our survey, we first outline the sensor network properties that are crucial for the design of MAC layer protocols. Then we have compared a set of MAC techniques in terms of their suitability to be used in WSN emphasizing their strengths and weaknesses. Our goal is to yield a foundation for future MAC protocol design, and to detect important design issues that can allow us to enhance the performance of the wireless sensor network.

Keywords Sensor networks · Medium access control protocols · Energy efficiency

R. Anane (✉) · R. Bouallegue
Innovation of Communication and Cooperative Mobiles,
Innov'COM, University of Carthage, Tunis, Tunisia
e-mail: rajoua.anane.etu@univ-lemans.fr

R. Bouallegue
e-mail: ridha.bouallegue@ieee.org

R. Anane · K. Raoof
Laboratory of Acoustics at University of Maine,
LAUM UMR CNRS n°, 6613 Le Mans, France
e-mail: kosai.raoof@univ-lemans.fr

© Springer International Publishing Switzerland 2016
A. El Oualkadi et al. (eds.), *Proceedings of the Mediterranean Conference on Information & Communication Technologies 2015*, Lecture Notes in Electrical Engineering 380, DOI 10.1007/978-3-319-30301-7_60

1 Introduction

Advances in hardware technology led to the development of low-cost and small sizes wireless sensor nodes. A broad range of applications such as intrusion detection, precision agriculture, medical systems, environment monitoring, climate control, scientific application, use the Wireless Sensor Network [1] consisting of spatially distributed sensors able to collect and process data then transmit information back to specific nodes. These wireless sensors operate autonomously with a low power radio transceiver and limited energy source, usually a small battery. Therefore, energy saving is a fundamental criterion in the design of Wireless sensor network protocols. On the other hand, communication process between sensor nodes is the major power consuming. This process is controlled by the Medium access control (MAC) protocol that enables the successful operation of the network. One important task of the Medium access is to reduce collisions so that two sensors do not transmit at the same time.

Hence, by designing efficient MAC protocols we can reduce energy consumption and of course increase the lifetime of wireless sensor networks. In addition, wireless sensors may die over time, new sensor can join, and more sensors can change position. A good MAC protocol should gracefully accommodate such network changes. A great amount of research on the design and development of wireless sensor networks MAC protocols have been implemented out in the last few years. Typical examples include the time division multiple access (TDMA), code division multiple access (CDMA), and contention-based protocols like IEEE 802.11 [2]. Thus, surveys of MAC techniques are conducted to summarize the varieties of conception and implementations. This article details and compares various medium-access control (MAC) protocols with different objectives for wireless sensor networks.

2 MAC Properties for Wireless Sensor Network

Moreover, to being energy efficient, WSN should be adaptable to change. Indeed, wireless nodes may die over time when their batteries become discharged, new nodes may join, or nodes may move to a different location. Under these circumstances, the MAC technique used should gracefully accommodate such network changes and must be energy efficient by reducing the potential energy wastes presented below [3].

2.1 Factors of Energy Waste

A sensor's major waste of energy in a MAC protocol is due to the following reasons [4, 5]:

- *Collision*: it is caused by contention, when two nearby nodes both try to use the channel at the same time. Data packets that cause the collision have to be dropped. Thus, the retransmissions of these lost packets are required, which increase the energy consumption and latency as well.
- *Overhearing*: this phenomenon is happened when a sensor collects packets that are destined to other sensors. This also leads to unnecessary consume of energy.
- *Control-packet overhead*: Sending and receiving control packets as Ready-to-Send (RTS), Clear-to-Send (CTS), and Ac-knowledge (ACK), etc., can lead to the waste a node's energy reserves.
- *Idle listening*: it is considered as the major source of inefficiency, because sensor must keep its radio in receive mode at all times to listen communication channel in order to receive possible packets. Many measurements indicated that idle mode consumes between 50 and 100 % of the energy required for receiving [6].

To minimize the energy consumption during idle listening mode, several MAC protocols uses wakeup scheduling approach with fixed duty cycle defined as follows [7]:

$$Duty\ Cycle = Listen\ Interval/Frame\ Length$$

To sum up and given the above reasons, a well-designed MAC protocol should prevent these energy wastes.

3 Taxonomy of MAC Protocol

Many research works have been done in the area of energy savings MAC protocols for wireless sensor networks. Current MAC design can be broadly divided into two categories: Schedule based and contention based. On the other hand, authors in [8] classified MAC techniques in three categories namely Asynchronous, Synchronous and Frame-Slotted. In our work, we are going to compare several MAC protocol by considering the two classifications mentioned above. The contention based protocols allows many sensors to use the same radio channel without pre-coordination but using "listen before talk" operating procedure. It relaxes time synchronization requirements and can easily adjust to the topology changes by joining some new nodes.

The schedule based protocols is based on strict time synchronization requirements. They schedule listen periods and transmit, thus avoiding collisions, overhearing and idle listening. Taxonomy of wireless sensor network MAC protocols is presented in Fig. 1.

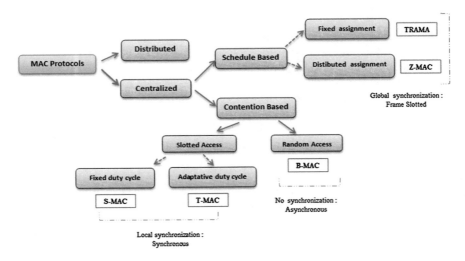

Fig. 1 Wireless sensor network taxonomy

4 MAC Techniques Proposed

In this section, several MAC techniques designed for sensor networks are described briefly by stating their essential behaviors.

4.1 Synchronous Contention-Based Protocol

Sensor-MAC (S-MAC)

To prolong the network lifetime the Sensor-MAC protocol uses periodic listening and sleeping techniques [3, 9]. Wireless sensors in S-MAC create virtual clusters by periodically exchanging sleep schedules with their immediate neighbors. This schedule exchanges are implemented by sending a SYNC packet, which is very short, and contains the address of the sender and the time of its next sleep. The period for each node to send a SYNC packet is called the synchronization period. In addition, RTS and CTS packet exchanges are used for unicast type data packets and to avoid overhearing. In the other hand, SMAC protocol employs a fixed duty cycle. When the traffic load fluctuates, sensor will consume more energy in idle state due to the fixed duty cycle.

S-MACs design includes also the utilization of message passing in which long messages are divided into frames and sent in a burst, this technique may reduce communication overhead and latency (Fig. 2).

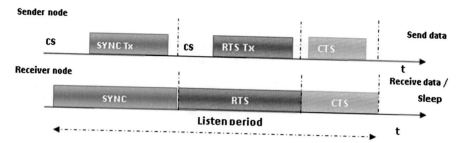

Fig. 2 S-MAC protocol concept

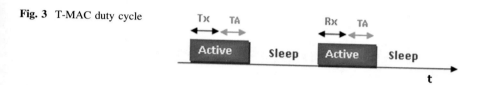

Fig. 3 T-MAC duty cycle

Timeout-MAC (T-MAC)

As stated above, the Sensor MAC does not perform when the traffic load fluctuates. To overcome this weakness, the Timeout MAC adds the timeout value to finish the active period of a wireless sensor [10].

Indeed, Timeout-MAC protocol uses an active/sleep duty cycle such as S-MAC protocol. However, this protocol implements an adaptive duty cycle in which the listen period ends when no activation event has occurred for a time threshold TA (Fig. 3).

Timeout-MAC approach follows the S-MAC in terms of synchronization and virtual clustering schemes. It solves the early sleeping problem using two methods: "Future Request To Send" (FRTS) and "Taking Priority on Full Buffers" [11].

4.2 Asynchronous Contention-Based Protocol

Berkeley-MAC (B-MAC)

The Berkeley-MAC protocol offers a different approach which reduces the overhead generated by control frames (SYN, RTS, CTS, ACK) and does not explicitly synchronize the transmitter and the receiver.

This protocol uses an adaptive preamble to reduce idle listening [12]. Each wireless sensor must verify the channel state periodically. When a sensor has a data to transmit, it waits during a back-off time before checking the channel.

Figure 4 illustrates the B-MAC mechanism. In the scenario presented, sensor 1 sends data to sensor 2 and sensor 3 has data to transmit too.

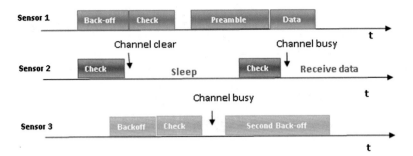

Fig. 4 Communication example based on B-MAC protocol

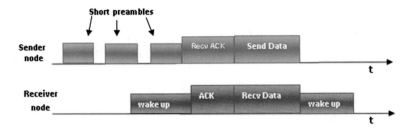

Fig. 5 X-MAC's short preamble technique

Short Preamble MAC (X-MAC)

X-MAC protocol implements a series of short preamble data packets each containing the ID (address) of the target node [7].

The sensor node can transmit data only when, the receiver confirms with an acknowledgment, the receipt of the first preamble. Accordingly, we can decrease the data transmission delay and minimize the energy consumption. More, this approach permits to avoid the idle listening. A visual representation of asynchronous Short Preamble MAC technique is shown in Fig. 5.

4.3 Synchronous Schedule-Based Protocol

Traffic-Adaptive MAC Protocol (TRAMA)

The Traffic-Adaptive MAC Protocol is a TDMA-based algorithm proposed to increase the utilization of classical TDMA in an energy-efficient manner [13].

It switches between random access period and scheduled access period. Random-access period is used to establish two-hop topology information where medium access is contention-based. In this period, nodes that have data to send will claim slots for use. This approach permits to eliminate the hidden-terminal problem

Table 1 Comparison of the several MAC protocols

	Technique	Energy saving	Advantages	Disadvantages
S-MAC	Fixed duty cycle Adaptive listenning (synchronous)	Periodic sleep Message passing	Low energy consumption when traffic is low, adaptive to change in topology	Sleep latency problem with broadcast, need to maintain loos frame control (Syn, RTS, CTS)
T-MAC	Adaptive duty cycle (synchronous)	Uses 20 % of energy used in sensor-MAC	Achieve optimal active period	Early sleeping Overhearing problem
B-MAC	Low power listening Channel assessment (asynchronous)	Better power saving, throughput and latency than S-MAC	RTS, CTS frame not required Delay tolerant, low overhead when network is idle	Long preamble creates large overhead Bad performance at heavy traffic
X-MAC	Reduce preamble length (asynchronous)	Shortened preamble	More energy saving Lower latency	Mistakenly data transmission by neighbor node after seeing gaps in packets
TRAMA	TDMA Adaptive assignment (frame slotted)	Transmission schedule	Better channel utilization More energy efficiency and throughput	Time is divided into random access period Overhearing
Z-MAC	Slot stealing CSMA into TDMA (frame slotted)	Significant energy using DRAND, time slot assignment	High throughput (low contention), high channel utilization (high contention	Additional overhead

and hence ensures that all sensors in the one-hop neighborhood of the transmitter will receive the packet without any collision.

Zebra MAC (Z-MAC)

Zebra MAC protocol is a kind of Frame-Slotted MAC protocols.

It is used to enhance channel utilization by incorporating the CSMA technique into TDMA [14]. This protocol applies DRAND (Distributed randomize) technique to do time slot assignment which ensures that no two nodes within the two-hop communication neighborhood are assigned to the same slot.

Consequently, Z-MAC performs as CSMA under low contention and possesses high channel utilization as TDMA under high contention.

5 Comparative Analysis

Table 1 compares the several MAC protocols discussed in the preceding section. Here, we have shown their comparison based on different parameters: technique used, energy saving, advantages and disadvantages.

6 Conclusion

In this article, we have presented a comparative survey of MAC protocols both synchronous/Asynchronous contention-based and synchronous schedule based categories. The comparative study is carried out on the basis of factors like energy saving, latency, throughput etc. From the various techniques discussed, we can deduce that all the methods with their protocols are application oriented designs.

As there is no standard technique to classify all protocols with the same metrics, the choice of suitable MAC protocol depends on intended application.

The obtained comparative study can be used as a guideline for future MAC protocol design.

References

1. Akyildiz, F., Su, W., Sankarasubramaniam, Y., Cayirci, E.: A survey on sensor networks. IEEE Commun. Mag. 40(8), 102–114 (2002)
2. LAN MAN Standards Committee of the IEEE Computer Society, Wireless LAN medium access control (MAC) and physical layer (PHY) specification, IEEE, New York, USA, IEEE Std 802.11-1997 edition
3. Egea-Lopez, E., Vales-Alonso, J., Martínez-Sala, A.S., García-Haro, J.: A wireless sensor networks MAC protocol for real-time applications. Pers. Ubiquitous Comput. 12(2), 111–122 (2008)
4. Ye, W., Heidemann, J., Estrin, D.: An energy-efficient MAC protocol for wireless sensor network. In: Proceedings of the INFOCOM'02. IEEE Computer Society, San Francisco (2002)
5. Demirkol, I., Ersoy, C., Alagoz, F.: MAC protocols for wireless sensor networks: a survey. IEEE Commun. Mag. 44(4), 115–121 (2002)
6. Stemm, M., Katz, R.H.: Measuring and reducing energy consumption of network interfaces in hand-held devices. IEICE Trans. Commun. E80-B(8), 1125–1131 (1997)
7. Riduan Ahmad, M., Dutkiewicz, E., Huang, X.: A survey of low duty cycle MAC protocols in wireless sensor networks. Emerg. Commun. Wirel. Sensor Netw. (2011). ISBN 978-953-307-082-7
8. Huang, P., Xiao, L., Soltani, S., Mutka, M.W., Xi, N.: The evolution of MAC protocols in wireless sensor networks: a survey. IEEE Commun. Surv. Tutorials 15(1) (2013)
9. Anna, H.: Wireless Sensor Network Design. Wiley, New York (2003)
10. Halkes, G.P., van Dam, T., Langendoen, K.G.: Comparing energy-saving MAC protocols for wireless sensor networks. Mobile Netw. Appl. 10(5), 783–791 (2007)

11. Lin, P., Qiao, C., Wang, X.: Medium access control with a dynamic duty cycle for sensor networks. In: Proceedings of the IEEE Wireless Communications and Networking Conference (WCNC '04), vol. 3, pp. 1534–1539 (2004)
12. Polastre, J., Hill, J., Culler, D.: Versatile low power media access for wireless sensor networks. In: Proceedings of the Second International Conference on Embedded Networked Sensor Systems (SenSys '04), pp. 95–107. ACM Press, New York, Nov 2004
13. Rajendran, V., Obraczka, K., Garcia-Luna-Aceves, J.J.: Energy-efficient, collision-free medium access control for wireless sensor networks. In: Proceedings of ACM SenSys '03, pp. 181–92. Los Angeles, CA, Nov 2003
14. Rhee, I., Warrier, A., Aia, M., Min, J.: Z-MAC: a hybrid MAC for wireless sensor networks. IEEE/ACM Trans. **16**(3), 511–524 (2008)

Strategies to Improve the VoIP over LTE Capacity

Youness Jouihri and Zouhair Guennoun

Abstract In this paper, we analyze the voice communication over Long Term Evolution (LTE) network which is transmitted using a voice over internet protocol (VoIP), as in LTE the core network is purely packet switched, designed for high data throughput. Similarly to legacy circuit-switched network, serving the largest number of users performing simultaneous VoIP calls, and ensuring their satisfaction will be the main objective for LTE operators willing to introduce voice over LTE service. To simulate the impact of scheduling strategies in the VoIP capacity, we use the Open Wireless Network Simulator (OpenWNS). With the obtained results, we show the VoIP capacity bottleneck. We also recommend the best strategies to improve the VoIP capacity.

Keywords Voice over LTE · Fitting strategy · VoIP capacity

1 Introduction

In an LTE network, the core is purely packet-switched [1]. However, with the new core design, the voice service will be transmitted over IP (VoIP), and will be regarded as any data application, with specific requirements for real-time traffic. Consequently, to have an effective voice over LTE service, we should ensure the capability of handling a large number of simultaneous VoIP calls while keeping the user satisfaction above the recommended threshold value. The structure of our paper can be depicted as follows: in Sect. 2 we highlight the simulations' environment. In Sect. 3 we provide a theoretical analysis and we validate our study through the results of simulations showing the impact of fitting strategies on VoIP capacity. Finally, a conclusion is given in Sect. 4.

Y. Jouihri (✉) · Z. Guennoun
Electronics and Communications Laboratory (LEC), Mohammadia School
of Engineering (EMI), Rabat Mohammed V University (UM5R),
BP 765 Ibn Sina Avenue Agdal 10000, Rabat, Morocco
e-mail: Jouihri.Youness@gmail.com

© Springer International Publishing Switzerland 2016   571
A. El Oualkadi et al. (eds.), *Proceedings of the Mediterranean Conference on Information & Communication Technologies 2015*, Lecture Notes in Electrical Engineering 380, DOI 10.1007/978-3-319-30301-7_61

2 Simulation Environment

In this paper, all the simulations' results are obtained using the Open Wireless Network Simulator (OpenWNS) simulator [2].

2.1 Fitting Strategies

For a VoIP call, the technique of Transport Block (TB) selection from the free Resource Blocks (RBs) is very significant in the scheduling process, and can be impacted by interference and resource availability [3]. In the OpenWNS simulator, the available fitting strategies used to choose a TB from the available RB space in a sub- frame are the First Fit strategy, the Best Fit strategy, the Worst Fit or Least Fit strategy, and the Random Fit strategy, described in [4].

2.2 Test Environment

The simulations address the VoIP capacity in the different access nodes' or Base Stations' (BS) scenarios, under a Single-In-Multiple-Out (SIMO) based system. The environment consists of a 120 m by 50 m rectangular area (default spacing used for an Indoor scenario in the OpenWNS). There are two BS sites, each one has an omnidirectional antenna. The center of the graph has the coordinates (1000.0, 1000.0, 0.0). The coordinates of base stations are (0970.0, 1000.0, 0.0) and (1030.0, 1000.0, 0.0). The users performing simultaneous VoIP calls will be added progressively to this environment.

This layout is used for all the types of simulated access nodes (Indoor Hotspot (InH), Urban Micro (UMi), typical Urban Macro (Uma), Rural Macro (RMa), and Suburban Macro, described in [5]). This makes the comparison between the test results of the different BS possible. Consequently, parameters, such as distance of user from BS or user distribution, will be general parameters, and will have similar impact in all performed simulations. Our test consists of a gradual increase in number of users inside the selected test environment; the added users will be randomly and uniformly distributed over the area, using the "IndoorHotspotUE" placer defined within the ituM2135 scenarios of the OpenWNS [2]. The same placer is used in all simulations. For each BS type, we test all the fitting strategies available in the OpenWNS simulator. The simulation will be stopped once the user's satisfaction falls below the threshold of 98 %. This threshold value reflects the user satisfaction according to "the guidelines for evaluation of radio interface technologies for IMT-Advance" [6]. All simulations are carried out while considering the scenario of a Physical Downlink Control Channel (PDCCH) limit of 8, as it is used in a large number of test projects [7].

For the general parameters used in simulations, we used 5 dB for the BS noise figure, 7 dB for the UT noise figure, 0 dBi for BS antenna gain, 0 dBi for UT antenna gain, −174 dBm/Hz for thermal noise level, 5 + 5 MHz (FDD) for the simulation bandwidth, and 20 s for the simulation time span (single drop).

3 Theoretical Analysis and Simulation Results

3.1 Theoretical Analysis: the Impact of the Fitting Strategies on the Voice over LTE Service

The voice over LTE service requires a guaranteed bit rate (GBR) bearer, to establish a mobile originating or terminating call; which adds more constraints on the scheduling of voice service data [8]. However, once the call is scheduled, it will be less vulnerable to outages and drops, as it will benefit from the GBR specifications' advantages. To evaluate the subscriber satisfaction during a voice over LTE (VoLTE) call, we check the possible scenarios that can impact such call. The first impact is the failure during call establishment, either related to coverage or to resource unavailability; all these cases can be referred to as the blocking scenario. The second impact includes drops, noise, silence or any kind of outage that might affect an established call; these cases can be referred to as the outage scenario. The subscriber satisfaction percentage while using VoLTE service can be depicted in the following equation:

$$P_{satisfaction} = (1 - P_{blocking})(1 - P_{outage}). \tag{1}$$

where $P_{satisfaction}$ is the percentage of satisfaction, $P_{blocking}$ is the percentage of calls impacted by the blocking scenario, and the $P_{outage}P_{outage}P_{outage}$ is the percentage of calls affected by the outage scenario. Using a GBR bearer, the VoLTE service is more impacted by the blocking scenario then the outage scenario. The GBR bearer reduces the outage probability, but adversely increases the blocking probability. Based on the description of scheduling strategies given in Sect. 2.1, we realize that the Best Fit strategy results in a minimum number of left over RBs after the scheduling of TBs. But the Worst fit strategy results in a maximum number of left over RBs after the scheduling of TBs. We conclude that the Worst Fit strategy maximizes the number of served and satisfied users, unlike the Best Fit strategy.

3.2 The VoIP Capacity Bottleneck

Using the OpenWNS simulator, we start our simulations with the unrealistic case where we neglect the impact of fading. The simulation is done under an InH BS using the First Fit strategy. Figure 1 shows that the user satisfaction threshold (98 %) is maintained for the Uplink case until we reached about 600 nodes. After that,

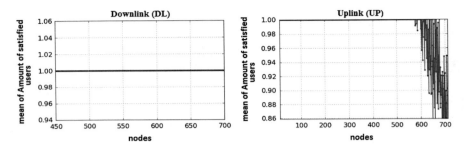

Fig. 1 Average amount of satisfied users under InH BS using First Fit in the uplink and downlink cases without the fading effect

further increase in the number of nodes led to a user satisfaction below the 98 % value. For the Downlink case, the impact was not visible; even after reaching 700 nodes, the user satisfaction remains 1 [Fig. 1 (Downlink)]. We conclude that the Uplink is the bottleneck of VoIP capacity, which is somehow expected. Besides the constraints applied for the Downlink channel, the Uplink channel has the constraint of RB allocation contiguity and a higher interference level. On the remaining simulations, only the Uplink cases will be considered. We will introduce also the fading impact to have a simulation environment near to the real world.

3.3 Simulation Results of the Fitting Strategies Using Different Types of Access Nodes

The simulation results of the First Fit and Random Fit strategies using different types of BS are presented in Fig. 2. The trend of user satisfaction is maintained

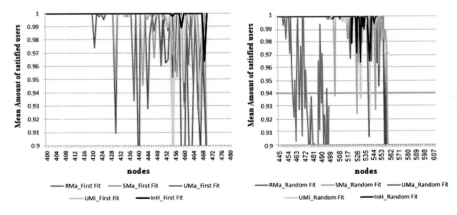

Fig. 2 Average amount of satisfied users using First Fit and Random Fit in the uplink case (with the fading effect) under different types of BS

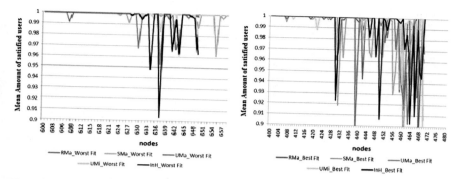

Fig. 3 Average amount of satisfied users using Worst Fit and Best Fit

greater than the 98 % threshold while reaching 421, 430, 440, 455 and 469 nodes for the First Fit case, and 520, 504, 457, 531 and 522 nodes for Random Fit case using RMa, SMa, UMa, UMi, and InH base stations respectively. Any subsequent increase in number of nodes leads to a value of satisfaction below 98 %. In Fig. 3, the Worst Fit and Best Fit cases are simulated, under different types of access nodes. The trend of satisfaction is maintained above 98 % while reaching 643, 655, 630, 643 and 634 nodes for the Worst Fit case, 430, 431, 430, 430 and 430 nodes for the Best Fit case using RMa, SMa, UMa, UMi, and InH base stations respectively.

From the various simulations' results, we conclude that, the Worst Fit strategy provides the highest VoIP capacity. By scheduling the largest number of simultaneous VoIP calls, under all types of simulated access nodes; while keeping the user satisfaction value above the 98 % threshold. Contrarily, the Best Fit strategy provides the lowest VoIP capacity under all types of access nodes. These results are in line with the theoretical analysis done in Sect. 3.1.

4   Conclusion

To evaluate the VoIP over LTE capacity, different types of base stations were considered and different fitting strategies were applied. Using the OpenWNS simulation results, we show the impact of the fading, where a 22 % decrease on VoIP capacity is observed. We also demonstrate that the Uplink traffic is the bottleneck of the VoIP capacity, and that improving it will substantially increase the VoIP capacity. Finally, after comparing the simulations' results, we conclude that assigning the adequate fitting strategy will significantly improve the VoIP capacity. In future work, we will explore other possibilities to improve the VoIP capacity using OpenWNS simulator.

References

1. Puttonen, J., Henttonen, T., Kolehmainen, N., Aschan, K., Moisio, M., Kela, P.: Voice-over-ip performance in UTRA long term evolution downlink, pp. 2502–2506. IEEE, May 2008. http://ieeexplore.ieee.org
2. Bultmann, D., Muhleisen, M., Klagges, K., Schinnenburg, M.: OpenWNS—open wireless network simulator, pp. 205–210. IEEE, May 2009. http://ieeexplore.ieee.org
3. Choi, S., Jun, K., Shin, Y., Kang, S., Choi, B.: MAC scheduling scheme for VoIP traffic service in 3g LTE, pp. 1441–1445. IEEE, Sep 2007. http://ieeexplore.ieee.org
4. Tung, L.C., Lu, Y., Gerla, M.: Priority-based congestion control algorithm for cross-traffic assistance on LTE networks, pp. 1–5. IEEE, Sep 2013. http://ieeexplore.ieee.org
5. Burstrom, P., Furuskar, A., Wanstedt, S., Landstrom, S., Skillermark, P., Anto, A.: System performance challenges of IMT-Advanced test environments, pp. 2080–2084. IEEE, Sep 2009. http://ieeexplore.ieee.org
6. ITU-R: Guidelines for evaluation of radio interface technologies for IMT-Advanced. Technical Report, ITU-R M.2135-1, International Telecommunication Union (ITU), Dec 2009. https://www.itu.int/dms_pub/itu-r/opb/rep/R-REP-M.2135-1-2009-PDF-E.pdf
7. 3GPP: Evolved Universal Terrestrial Radio Access (E-UTRA); Further advancements for E-UTRA physical layer aspects. TR 36.814 V9.0.0, 3rd Generation Partnership Project (3GPP), Mar 2010. http://www.3gpp.org/dynareport/36814.htm
8. Osterbo, O.: Scheduling and capacity estimation in LTE. In: Osterbo, O. (ed.) ITC'11 Proceedings, pp. 63–70. IEEE, San Francisco, CA (2011)

Erratum to: Enhanced Color Image Method for Watermarking in RGB Space Using LSB Substitution

Mouna Bouchane, Mohamed Tarhda and Laamari Hlou

Erratum to:
"Enhanced Color Image Method for Watermarking in RGB Space Using LSB Substitution" in:
A. El Oualkadi et al. (eds.),
Proceedings of the Mediterranean Conference on
Information & Communication Technologies 2015,
Lecture Notes in Electrical Engineering 380,
DOI 10.1007/978-3-319-30301-7_17

The spelling of the author's name was incorrectly given as "Mohamed Taghda" instead of "Mohamed Tarhda" in the book post-publication. The erratum chapter and the book have been updated with the change.

The updated original online version for this chapter can be found at
10.1007/978-3-319-30301-7_17

M. Bouchane (✉) · M. Tarhda · L. Hlou
Electrical Engineering and Energy Systems Laboratory, Ibn Tofail University,
BP. 133, Kenitra, Morocco
e-mail: mouna.bouchane@gmail.com

M. Tarhda
e-mail: tarhdamo@yahoo.fr

L. Hlou
e-mail: hloul@yahoo.com

© Springer International Publishing Switzerland 2016   E1
A. El Oualkadi et al. (eds.), *Proceedings of the Mediterranean Conference
on Information & Communication Technologies 2015*, Lecture Notes
in Electrical Engineering 380, DOI 10.1007/978-3-319-30301-7_62

Printed in the United States
By Bookmasters